Related Titles

Cocks, F. H.

Energy Demand and Climate Change

Issues and Resolutions

2009
ISBN: 978-3-527-32446-0

Coley, D.

Energy and Climate Change

Creating a Sustainable Future

2008
ISBN: 978-0-470-85313-9

Wengenmayr, R., Bührke, T. (eds.)

Renewable Energy

Sustainable Energy Concepts for the Future

2008
ISBN: 978-3-527-40804-7

Paul, B.

Future Energy

How the New Oil Industry Will Change People, Politics and Portfolios

2007
ISBN: 978-0-470-09642-0

Kruger, P.

Alternative Energy Resources

The Quest for Sustainable Energy

2006
ISBN: 978-0-471-77208-8

George A. Olah, Alain Goeppert,
and G. K. Surya Prakash

Beyond Oil and Gas:
The Methanol Economy

Second updated and enlarged edition

WILEY-VCH

WILEY-VCH Verlag GmbH & Co. KGaA

The Authors

Prof. Dr. George Olah
Dr. Alain Goeppert
Prof. Dr. G. K. Surya Prakash
Loker Hydrocarbon Research Institute
University of Southern California
837 W. 37th. Street
Los Angeles, CA 90089-1661
USA

All books published by Wiley-VCH are carefully
produced. Nevertheless, authors, editors, and
publisher do not warrant the information contained
in these books, including this book, to be free of
errors. Readers are advised to keep in mind that
statements, data, illustrations, procedural details or
other items may inadvertently be inaccurate.

Library of Congress Card No.: applied for

British Library Cataloguing-in-Publication Data
A catalogue record for this book is available from the
British Library.

**Bibliographic information published by
the Deutsche Nationalbibliothek**
The Deutsche Nationalbibliothek lists this
publication in the Deutsche Nationalbibliografie;
detailed bibliographic data are available on the
Internet at http://dnb.d-nb.de.

© 2009 WILEY-VCH Verlag GmbH & Co. KGaA,
Weinheim

Cover Design Adam-Design, Weinheim
Typesetting Thomson Digital, Noida, India
Printing and Binding betz-druck GmbH, Darmstadt

Printed in the Federal Republic of Germany
Printed on acid-free paper

ISBN: 978-3-527-32422-4

Contents

Beyond Oil and Gas: The Methanol Economy, Second updated and enlarged edition
George A. Olah, Alain Goeppert, and G. K. Surya Prakash
Copyright © 2009 WILEY-VCH Verlag GmbH & Co. KGaA, Weinheim
ISBN: 978-3-527-32422-4

Preface to the Second Updated Edition

After just three years since the publication of the first edition of our book it is rewarding that favorable reception and interest prompted our publisher to suggest an updated edition. The concept of our proposed "Methanol Economy" in the intervening time has made progress from extended research to practical development in countries around the world. From smaller demonstration plants to full-scale methanol and derived dimethyl ether (DME) plants, practical industrial applications are growing in this field. These include carbon dioxide to methanol (and DME) conversion plants but also large million metric tonnes per year, coal or natural gas based mega-plants using still available large coal and natural gas resources. The full potential of the Methanol Economy will be realized, however, when the chemical recycling of natural and industrial carbon dioxide sources into methanol and its derived products are widely implemented, making their use environmentally carbon neutral and regenerative. This will allow us to mitigate the grave environmental problems linked to global warming. At the same time chemical carbon dioxide recycling, eventually from the air itself, will provide humankind with an inexhaustible carbon source available everywhere on earth. The needed hydrogen for the conversion of CO_2 into methanol can be produced from water using any renewable or atomic energy source. This conversion will allow the continued production of convenient transportation and household fuels, and synthetic hydrocarbons and their products on which we all so much depend on. It should be emphasized that methanol is not an energy source but only a convenient way to store, transport and use any form of energy. We are not suggesting that this approach is necessarily in all aspects the only solution for the future. The Methanol Economy, however, is a new feasible and realistic approach, warranting further development and increasing practical application.

Los Angeles, August 2009

George A. Olah
Alain Goeppert
G.K. Surya Prakash

Beyond Oil and Gas: The Methanol Economy, Second updated and enlarged edition
George A. Olah, Alain Goeppert, and G. K. Surya Prakash
Copyright © 2009 WILEY-VCH Verlag GmbH & Co. KGaA, Weinheim
ISBN: 978-3-527-32422-4

Acronyms and Initialisms

AFC	alkaline fuel cell
BP	British Petroleum
BWR	boiling water reactor
CEA	Commissariat à l'Energie Atomique (France)
CEC	California Energy Commission
CI	compression ignition
CIA	Central Intelligence Agency
DME	dimethyl ether
DMFC	direct methanol fuel cell
DOE	Department of Energy (United States)
EDF	Electricité de France
EIA	Energy Information Administration (DOE)
EPA	Environmental Protection Agency (United States)
EPRI	Electric Power Research Institute
EU	European Union
GDP	gross domestic product
GHG	greenhouse gas
IAEA	International Atomic Energy Agency
ICE	internal combustion engine
IEA	International Energy Agency
IGCC	integrated gasification combined cycle
IPCC	International Panel on Climate Change
ITER	International Thermonuclear Experimental Reactor
JAERI	Japan Atomic Energy Research Institute
LNG	liquefied natural gas
MCFC	molten carbonate fuel cell
MTBE	methyl-*tert*-butyl ether
NRC	National Research Council (United States)
NREL	National Renewable Energy Laboratory (United States)
OECD	Organization for Economic Cooperation and Development
OPEC	Organization of Petroleum Exporting Countries

Beyond Oil and Gas: The Methanol Economy, Second updated and enlarged edition
George A. Olah, Alain Goeppert, and G. K. Surya Prakash
Copyright © 2009 WILEY-VCH Verlag GmbH & Co. KGaA, Weinheim
ISBN: 978-3-527-32422-4

ORNL	Oak Ridge National Laboratory
OTEC	ocean thermal energy conversion
PAFC	phosphoric acid fuel cell
PEMFC	proton exchange membrane fuel cell
PFBC	pressurized fluidized bed combustion
PV	photovoltaics
PWR	pressurized water reactor
R/P	reserve over production ratio
SUV	sport utility vehicle
TPES	total primary energy supply
UNO	United Nations Organization
UNSCEAR	United Nations Scientific Committee on Effects of Atomic Radiation
UNEP	United Nation Environmental Program
URFC	unitized regenerative fuel cell
USCB	United States Census Bureau
USGS	United States Geological Survey
WCD	World Commission on Dams
WCI	World Coal Institute
WEC	World Energy Council
WMO	World Meteorological Organization
ZEV	zero emission vehicle

Units and their Abbreviations

atm	atmosphere
b and bbl	barrel
btu	British thermal unit
°C	degree Celsius
cal	calorie
g	gram
h	hour
ha	hectare
kWh	kilowatt-hour
m	meter
Mb	megabarrel (10^6 barrels)
ppm	parts per million
s	second
Sv	Sievert
t	metric tonne
toe	tonne oil equivalent
W	watt

Prefixes

μ	micro 10^{-6}
m	milli 10^{-3}
k	kilo 10^3
M	mega 10^6
G	giga 10^9
T	tera 10^{12}
P	peta 10^{15}
E	exa 10^{18}

Beyond Oil and Gas: The Methanol Economy, Second updated and enlarged edition
George A. Olah, Alain Goeppert, and G. K. Surya Prakash
Copyright © 2009 WILEY-VCH Verlag GmbH & Co. KGaA, Weinheim
ISBN: 978-3-527-32422-4

Conversion of Units

Volume

1 tonne of crude oil = 7.33 barrels of oil
1 gallon = 3.785 liters
1 barrel of oil = 42 U.S. gallons = 159 liters
1 m^3 = 1000 liters
1 m^3 = 35.3 cubic feet (ft^3)

Energy

1 kcal = 4.1868 kJ = 3.968 Btu
1 kJ = 0.239 kcal = 0.948 Btu
1 kWh = 860 kcal = 3600 kJ
1 toe = 41.87 GJ
Quadrillion Btu, QBtu = 1 × 10^{15} Bt

1
Introduction

Ever since our distant ancestors managed to light fire for providing heat, means for cooking and many essential purposes, humankind's life and survival has been inherently linked with an ever-increasing thirst for energy. From burning wood, vegetation, peat moss and other sources to the use of coal, followed by petroleum oil and natural gas (fossil fuels), we have thrived using Nature's resources [1]. Fossil fuels include coal, oil and gas – all composed of hydrocarbons with varying ratios of carbon and hydrogen.

Hydrocarbons derived from petroleum oil, natural gas or coal are essential in many ways to modern life and its quality. The bulk of the world's hydrocarbons are used as fuels for propulsion, electrical power generation and heating. The chemical, petro-chemical, plastics and rubber industries also depend upon hydrocarbons as raw materials for their products. Indeed, most industrially significant synthetic chemicals are derived from petroleum sources. The overall use of oil in the world is now close to 12 million metric tons per day [2]. An ever-increasing world population (presently nearing 7 billion and projected to increase to 8–11 billion by the middle of the twenty-first century [3]; Table 1.1) and energy consumption, compared with our finite non-renewable fossil fuel resources, which will be increasingly depleted, are clearly on a collision course. New solutions will be needed for the twenty-first century to sustain the standard of living to which the industrialized world has become accustomed and to which the developing world is striving to achieve.

The rapidly growing world population, which stood at 1.6 billion at the beginning of the twentieth century, is now approaching 7 billion. With an increasingly technological society, the world's resources have difficulty keeping up with demands. Satisfying our society's needs while safeguarding the environment and allowing future generations to continue to enjoy planet Earth as a hospitable home is one of the major challenges that we face today. Man needs not only food, water, shelter, clothing and many other prerequisites but also increasingly huge amounts of energy. In 2004 the world used some 1.13×10^{20} calories per year (131 Petawatt-hours), equivalent to a continuous power consumption of about 15 terawatts (TW), which is comparable to the production of 15 000 nuclear power plants each of 1 GW output [4]. With increasing world population, development and higher standards of living, this demand for energy is expected to grow to 21 TW in 2025 (Figure 1.1). In 2050 the demand is expected to reach 30 TW.

Beyond Oil and Gas: The Methanol Economy, Second updated and enlarged edition
George A. Olah, Alain Goeppert, and G. K. Surya Prakash
Copyright © 2009 WILEY-VCH Verlag GmbH & Co. KGaA, Weinheim
ISBN: 978-3-527-32422-4

Table 1.1 World population.

Year	1650	1750	1800	1850	1900	1920	1952	2000	2009	Projection 2050[a]
Population (millions)	545	728	906	1171	1608	1813	2409	6200	6800	8000 to 11 000

[a]Source: United Nations, Department of Economic and Social Affairs, Population Division.

(a)

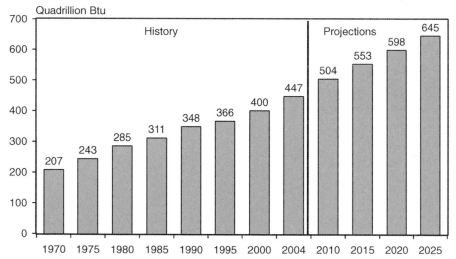

(b)

Figure 1.1 World primary energy consumption 1970–2025 in units of (a) petawatt hours; (b) Btu (British thermal units). (Based on data from: Energy Information Administration (EIA), International Energy Outlook 2007.)

Our early ancestors discovered fire and began to burn wood. The industrial revolution was fueled by coal, and the twentieth century added oil and natural gas and introduced atomic energy.

When fossil fuels such as coal, oil or natural gas (i.e., hydrocarbons) are burnt to generate electricity in power plants, or to heat our houses, propel our cars, airplanes, and so on, they form carbon dioxide and water as the combustion products. They are thus used up, and are non-renewable on the human timescale.

Fossil fuels: petroleum oil, natural gas, tar-sand, shale bitumen, coals

They are mixtures of hydrocarbons (i.e., compounds of the elements carbon and hydrogen). When oxidized (combusted) they form carbon dioxide (CO_2) and water (H_2O) and thus are not renewable on the human timescale.

Nature has given us, in the form of oil and natural gas, a remarkable gift. It has been determined that a single barrel of oil has the energy equivalent of 12 people working all year, or 25 000 man hours [5]. With each American consuming on average about 25 barrels of oil per year, this would amount to each of them having 300 people working all year long to power the industries and man their households to maintain their current standard of living. Considering the present cost of oil, this is truly a bargain. What was created over the ages, however, mankind is consuming rather rapidly. Petroleum and natural gas are used on a massive scale to generate energy, and also as raw materials for diverse man-made materials and products such as the plastics, pharmaceuticals and dyes that have been developed during the twentieth century. The United States energy consumption is heavily based on fossil fuels, with atomic energy and other sources (hydro, geothermal, solar, wind, etc.) representing only a modest 15% of the energy mix (Table 1.2) [6].

With regard to electricity generation, coal still represents about half of the fuel used, with some 19% for natural gas and 19% for nuclear energy (Table 1.3).

Other industrialized countries, in contrast, obtain between 20% and 90% of their electrical energy from non-fossil sources (Table 1.4) [7].

Oil use has grown to the point where the world consumption is around 85 million barrels (1 barrel equals 42 gallons, i.e., some 160 L) a day, or almost 12 million metric

Table 1.2 United States energy consumption by fuel (%).

Energy source	1960	1970	1980	1990	2000	2005
Oil	44.2	43.5	43.7	39.6	38.8	40.4
Natural gas	27.5	32.1	26.1	23.3	24.2	22.6
Coal	21.8	18.1	19.7	22.6	22.8	22.8
Nuclear energy	0.002	0.4	3.5	7.2	7.9	8.1
Hydro-, geothermal, solar, wind, and so on	6.5	6.0	7.0	7.2	6.2	6.1

Source: U.S. Census Bureau, *Statistical Abstract of the United States 2008*, Section 19, Energy and Utilities.

Table 1.3 Electricity generation in the United States by fuel (%).

	1990	2000	2005
Coal	52.5	51.7	49.8
Petroleum	4.2	2.9	3.0
Natural gas	12.6	16.2	19.0
Nuclear	19.0	19.8	19.3
Hydroelectric	9.6	7.2	6.6
Geothermal	0.5	0.4	0.4
Wood	1.1	1.0	0.9
Waste	0.438	0.607	0.594
Wind	0.092	0.147	0.361
Solar	0.013	0.013	0.012

Source: U.S. Census Bureau, *Statistical Abstract of the United States 2008,* Section 19, Energy and Utilities.

tonnes [2]. Fortunately, we still have significant worldwide reserves left, including heavy oils, oil shale and tar-sands and even larger deposits of coal (a mixture of complex carbon compounds more deficient in hydrogen than oil and gas). Our more plentiful coal reserves may last for 200–300 years, but at a higher socio-economical and environmental cost. It is not suggested that our resources will run out in the near future, but it is clear that they will become even scarcer, much more expensive, and will not last for very long. With a world population nearing 7 billion and still growing (as indicated earlier, it may reach 8–11 billion), the demand for oil and gas will only increase. It is also true that, in the past, dire predictions of rapidly disappearing oil and gas reserves have always been incorrect (Table 1.5) [2, 8]. Until fairly recently the reserves have been growing, but lately they seem to have leveled off.

The question is, however, what is meant by "depletion" and what is the real extent of our reserves? Proven oil reserves, instead of being depleted, have in fact almost

Table 1.4 Electricity generated in industrial countries by non-fossil fuels (%, 2004).

Country	Conventional thermal	Hydroelectric	Nuclear	Geothermal, solar, wind, wood and waste	Total non-fossil
France	9.4	10.9	78.6	1.1	90.6
Canada	25.7	58.0	14.7	1.6	74.3
Germany	61.9	3.6	27.5	6.9	38.1
Japan	62.2	9.2	26.4	2.2	37.8
Korea, South	62.8	1.2	35.9	0.1	37.2
United States	71.0	6.7	19.8	2.4	29.0
United Kingdom	75.5	1.3	20.0	3.2	24.5
Italy	81.1	14.1	0.0	4.8	18.9

Source: Energy Information Administration, *International Energy Annual 2007,* World Net Electricity Generation by Type, 2004.

Table 1.5 Proven oil and natural gas reserves (in billion tonnes oil equivalent).

Year	Oil	Natural gas
1960	43	15
1965	50	22
1970	78	33
1975	87	55
1980	91	70
1986	95	87
1987	121	91
1988	124	95
1989	137	96
1990	137	108
1995	140	130
2002	160	160
2003	162	162
2004	162	161
2005	164	162
2006	165	163

Source for 1995–2006: *BP Statistical Review of World Energy* [2].

doubled during the past 30 years and now exceed 150 billion tonnes (more than one trillion barrels) [2]. This seems so impressive that many people assume that there is no real oil shortage in sight. However, increasing consumption due to increasing standards of living, coupled with a growing world population, makes it more realistic to consider per-capita reserves. Based on this consideration, it becomes evident that our known accessible reserves will not last for much more than this century. Even if all other factors are taken into account (new findings, savings, alternate sources, etc.) our overall reserves will inevitably decrease, and thus we will increasingly face a major shortage. Oil and gas will not become exhausted overnight, but market forces of supply and demand will start to drive the prices up to levels that nobody even wants to presently contemplate. Therefore, if we do not find new solutions, we will face a real crisis.

Humankind wants the advantages that an industrial society can give to all of its citizens. We essentially rely on energy, but the level of consumption varies vastly in different parts of the world (industrialized versus developing and underdeveloped countries). At present for example, the annual oil consumption per capita in China is still only two to three barrels, whereas it is about ten-fold this level in the United States [2]. China's oil use is expected to at least double during the next decade, and this alone equals roughly the United States consumption – reminding us of the size of the problem that we will face. Not only the world population growth but also the increasing energy demands from China, India and other developing countries is already putting great pressure on the world's oil reserves, and this in turn contributes to price escalation. Large price fluctuations, with temporary sharp drops, can be expected, but the upward long-term trend in oil prices is inevitable.

Even though the generation of energy by massive burning of non-renewable fossil fuels (including oil, gas and coal) is feasible only for a relatively short period in the future, it is generating serious environmental problems (*vide infra*). The advent of atomic energy opened up a fundamental new possibility, but also created dangers and concerns regarding the safety of radioactive by-products. Regrettably, these considerations brought any further development of atomic energy almost to a standstill, at least in most of the Western world. Whether we like it or not, we clearly have few alternatives and will rely on using nuclear energy, albeit making it safer and cleaner. Problems, including those of the storage and disposal of radioactive waste products, must be solved. Pointing out difficulties and hazards as well as regulating them, within reason, is necessary, but solutions to overcome them are essential and certainly feasible.

As we continue to burn our hydrocarbon reserves to generate energy at an alarming rate, diminishing resources and sharp price increases will inevitably lead to the need to supplement or replace them by feasible alternatives. Alternative energy and fuel sources and synthetic oil products are, however, more costly. Nature's petroleum oil and natural gas are the greatest gifts we will ever have. However, with a barrel of oil presently priced between $30 and $150, within wide market fluctuations, some synthetic manufacturing processes are already becoming economically viable. Regardless, it is clear that we will need to get used to higher prices, not as a matter of any government policy but as a fact of market forces over which free societies have limited control.

Synthetic oil products are feasible. Their production was proven via synthesis-gas (syn-gas), a mixture of carbon monoxide and hydrogen obtained from the incomplete combustion of coal or natural gas, which, however, are themselves non-renewable. Coal conversion was used in Germany during World War II and in South Africa during the boycott years of the Apartheid era [9]. Nevertheless, the size of these operations hardly amounted to 0.3% of the present United States consumption alone. This route – the so-called Fischer–Tropsch synthesis – is also highly energy consuming, giving complex product mixtures and generating large amounts of carbon dioxide, thereby contributing to global warming. It thus can hardly be seen on its own as the technology of the future. To utilize still-existing large natural gas reserves, their conversion into liquid fuels through syn-gas is presently being developed; for example, on a large scale in Qatar, where Shell is spending over $10 billion on the construction of gas-to-liquid (GTL) facilities, to produce about 140 000 barrels per day of liquid hydrocarbon products, mainly sulfur-free diesel fuel. Chevron in partnership with Sasol has already built a GTL unit in Qatar with a capacity of 34 000 barrels per day. However, even when running at full capacity, these plants will provide only a daily total of some 180 000 barrels, compared with present world use of transportation fuels alone in excess of 45 million barrels per day. These figures demonstrate the enormity of the problem that we face. New and more efficient processes are clearly needed. Some of the required basic science and technology is already being developed. As will be discussed below, still abundant natural gas can be, for example, directly converted, without first producing syn-gas, into gasoline or hydrocarbon products. Using our even larger coal resources to produce synthetic oil could extend its availability, but new approaches based on renewable resources are essential for the future. The development of biofuels, primarily by the fermentative conversion of agricultural products (derived from sugar cane, corn, etc.) into ethanol is evolving. Whereas ethanol can be used as a gasoline additive or alternative fuel, the enormous amounts of transportation fuel needed clearly limits the applicability to specific countries and situations. Other plant-based oils are also being developed as renewable equivalents of diesel fuel, although their role in the total energy picture is again limited. Biofuels have also started to affect food prices by competing for the same agricultural resources [10].

When hydrocarbons are burned, as pointed out, they produce carbon dioxide (CO_2) and water (H_2O). It is a great challenge to reverse this process and to chemically produce, efficiently and economically, hydrocarbon fuels from CO_2 and H_2O. Nature, in its process of photosynthesis, recycles CO_2 with water into new plant life using the Sun's energy. While fermentation and other processes can convert plant life into biofuels and products, the natural formation of new fossil fuels takes a very long time, making them non-renewable on the human timescale.

The "Methanol Economy®" [11] – the subject of our book – elaborates a new approach of how humankind can decrease and eventually liberate itself from its dependence on diminishing oil and natural gas (and even coal) reserves while mitigating global warming caused by the carbon dioxide released by their excessive combustion. The "Methanol Economy" is in part based on the more efficient direct conversion of still-existing natural gas resources into methanol or dimethyl ether, and most importantly on their production by chemical recycling of CO_2 from the exhaust gases of fossil fuel-burning power plants as well as other industrial and natural sources. Eventually, even atmospheric CO_2 itself can be captured and recycled using catalytic or electrochemical methods. This represents a chemical regenerative carbon cycle alternative to natural photosynthesis [12]. Methanol and dimethyl ether (DME) are both excellent transportation and industrial fuels on their own for internal combustion engines and household uses, replacing gasoline, diesel fuel and natural gas. Methanol is also a suitable fuel for fuel cells, being capable of producing electric energy by reaction with atmospheric oxygen contained in the air. It should, however, be emphasized that the "Methanol Economy" *per se* is not producing energy. In the form of methanol or DME it only stores energy more conveniently and safely compared to extremely difficult to handle and highly volatile alternative hydrogen gas, which is the basis of the so-called "hydrogen economy" [13, 14]. Besides being most convenient energy storage materials and suitable transportation fuels, methanol and DME can also be catalytically converted into ethylene and/or propylene, the building blocks of synthetic hydrocarbons and their products presently obtained from our diminishing oil and gas resources.

The far-reaching applications of the new "Methanol Economy" approach clearly have great implications and societal benefit for humankind. As mentioned, the world is presently consuming about 85 million barrels of oil each day, and about two-thirds as much natural gas equivalent, both being derived from our declining and non-renewable natural sources. Oil and natural gas (as well as coal) were formed by Nature over the eons in scattered and frequently increasingly difficult-to-access locations such as under desert areas, in the depths of the seas, the inhabitable reaches of the Polar Regions, and so on. In contrast, the recycling of CO_2 from industrial exhausts or natural sources, and eventually from the air itself, which belongs to everybody, opens up an entirely new vista. The energy needs of humankind will, in the foreseeable future, come from any available source, including alternative sources and atomic energy. As we still cannot store energy efficiently on a large scale, new ways of storing energy are also needed. The production of methanol offers a convenient means of energy storage. Even now, our existing power plants, during off-peak periods, could, by the electrolysis of water, generate the hydrogen needed to produce methanol from CO_2. Other means of cleaving water by thermal, biochemical (enzymatic) or photovoltaic (using energy from the Sun, our ultimate clean energy source) pathways are also evolving.

Initially, CO_2 will be recycled from high level industrial emissions to produce methanol and to derive synthetic hydrocarbons and their products. CO_2 accompanying natural gas, geothermal and other natural sources will also be used. The CO_2 content of these emissions is high and can be readily separated and captured. In contrast, the average CO_2 content of air is very low (0.038%) (Table 1.6). Atmospheric CO_2 is therefore presently difficult to utilize on an economic basis. However, these difficulties can be overcome by ongoing developments using selective absorption or other separation technologies. Humankind's ability to technologically recycle CO_2 to useful fuels and products will eventually provide an inexhaustible renewable carbon source.

Carbon dioxide can readily be recovered from industrial sources, such as flue gas emissions of power plants burning carbonaceous fossil fuels (coal, oil and natural gas), fermentation processes, the calcination of limestone in cement production, production of steel and aluminum, and so on, as well as natural CO_2 accompanying natural gas, geothermal sources and others. As these plants and operations emit very large amounts of CO_2 they contribute to the increasing "greenhouse warming effect" of our planet, which is causing grave environmental concern. The relationship between the atmospheric CO_2 content and temperature was first studied scientifically by Arrhenius as early as 1895 [15]. The climate change and warming/cooling trends of our Earth can be evaluated only over longer time periods, but there is clearly a relationship between the CO_2 content in the atmosphere and Earth's surface temperature.

Table 1.6 Composition of air.

Nitrogen	78%
Oxygen	20.90%
Argon	0.90%
Carbon dioxide	0.038%
Water	Few % (variable)
Methane, nitrogen oxides, ozone	Trace amounts of each

Recycling excess CO_2 evolving from human activities into methanol and dimethyl ether, and further developing and transforming them into useful fuels and synthetic hydrocarbons and products, will thus not only help to alleviate the question of our diminishing fossil fuel resources but at the same time help to mitigate global warming caused by human-made greenhouse gases.

One highly efficient method of producing electricity directly from varied fuels is achieved in fuel cells via their catalytic electrochemical oxidation, primarily that of hydrogen (Equation 1.1).

$$\textbf{Fuel} \underset{\text{Fuel Cell}}{\overset{\text{Combustion}}{\longrightarrow}} \begin{array}{l} \textbf{Thermal Energy} \\ \\ \textbf{Electrical Energy} \end{array} \qquad (1.1)$$

The principle of fuel cells was first recognized by William Grove during the early 1800s, but their practical use was only recently developed. Most fuel cell technologies are still based on Grove's approach, that is, hydrogen and oxygen (air) are combined in an electrochemical cell-like device, producing water and electricity. The process is clean, giving only water as a by-product. Hydrogen itself, however, must be first produced in an energy-consuming process, using at present mainly fossil fuels and to a lesser extent the electrolysis of water. The handling of highly volatile hydrogen gas is not only technically difficult, but also dangerous. Nonetheless, the use of hydrogen-based fuel cells is gaining application in static installations or in specific cases, such as space vehicles. Currently, hydrogen gas is produced mainly from still-available fossil fuel sources using reformers, which converts them into a mixture of hydrogen and carbon monoxide from which hydrogen is then separated. Although this process relies mostly on our diminishing fossil fuel sources, electrolysis or other processes to cleave water can also provide hydrogen without any reliance on fossil fuels. Hydrogen-burning fuel cells, by necessity, are still limited in their applicability. In contrast, a new approach (discussed in Chapter 11) uses, directly, convenient liquid methanol, or its derivatives, in fuel cells without first converting it into hydrogen. The direct oxidation liquid-fed methanol fuel cell (DMFC) has been developed in a cooperative effort between our group at the University of Southern California and Caltech-Jet Propulsion Laboratory of NASA, who for a long time developed fuel cells for the U.S. space programs [16, 17]. In such a fuel cell, methanol reacts with oxygen present in the air over a suitable metal catalyst, producing electricity while forming CO_2 and H_2O:

$$CH_3OH + 1.5O_2 \rightarrow CO_2 + 2H_2O + \text{electrical energy} \qquad (1.2)$$

More recently, it was found that the process could also be reversed. Methanol and related oxygenates can be made from CO_2 via aqueous electrocatalytic reduction without prior electrolysis of water to produce hydrogen in what is termed a "regenerative fuel cell." This process can convert CO_2 and H_2O electrocatalytically into oxygenated fuels (i.e., formic acid, formaldehyde and methanol), depending on the electrode material and potential used in the fuel cell in its reverse operation.

The reductive conversion of CO_2 into methanol is primarily carried out by catalytic hydrogenation using hydrogen produced by electrolysis of water (using any available energy sources such as atomic, solar, wind, geothermal, etc.) or other means of cleavage (photolytic, enzymatic, etc.):

$$CO_2 + 3H_2 \rightarrow CH_3OH + H_2O \qquad (1.3)$$

Natural gas, when available, can also be used for the CO_2 to methanol conversion, including improved processes such as our proposed bi-reforming (Chapter 10) [18]:

$$3CH_4 + 2H_2O + CO_2 \rightarrow 4CH_3OH \qquad (1.4)$$

As mentioned, methanol is a convenient energy storage material and an excellent transportation fuel. It is a liquid, with a boiling point of 64.6 °C, allowing it to be transported easily and stored using existing infrastructure. Methanol can also be readily converted into dimethyl ether, which has a higher calorific value and is an excellent diesel fuel and household gas substitute:

$$2CH_3OH \rightarrow CH_3OCH_3 + H_2O \qquad (1.5)$$

Methanol and DME produced directly from methane (natural gas) without going to syn-gas or by recycling of CO_2 can subsequently also be used to produce ethylene as well as propylene (Equation 1.6):

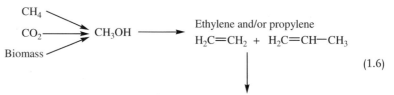

$$(1.6)$$

Synthetic hydrocarbons and their products

These are the building blocks in the petrochemical industry for the ready preparation of synthetic aliphatic and aromatic hydrocarbons, and for the wide variety of derived products and materials, obtained presently from oil and gas, on which we rely so much in our everyday life.

2
History of Coal in the Industrial Revolution and Beyond

Coal was formed during the Carboniferous Period – roughly 360 to 290 million years ago – from the anaerobic decomposition of then-living plants. These plants ended up as coal because, upon their death, they failed to decompose in the usual way, by the action of oxygen to form eventually CO_2 and water. As the carboniferous plants died they often fell into oxygen-poor swamps or mud, or were covered by sediments. Because of the lack of oxygen they only partially decayed. The resulting spongy mass of carbon-rich material first became peat. Then, by action of the heat and pressure of geological forces, peat eventually hardened into coal.

During this process, the plant's carbon content was trapped in coal, together with the sun's energy used in the photosynthesis of plants, and accumulated over millions of years. This energy source was buried until modern man dug it up and made use of it. It is only very recently on the Earth's timescale that humankind has started to use coal. Historically, the use of coal began when the Romans invaded Britain [19]. While it was used occasionally for heating purposes, the main use of this "black stone" was to make jewelry, since it could be easily carved and polished. It was only during the late twelfth century that coal re-emerged as a fuel along the river Tyne in Britain, especially around the rich coal fields of Newcastle. The widespread use of coal, however, would not be significant before the middle of the sixteenth century. At that time, England's population – and that of London especially – was growing rapidly. As the city grew, the nearby land was deforested to a degree where wood had to be hauled from increasingly distant locations. Wood was used not only for home heating and cooking purposes but also in most industries, such as breweries, iron smelters and ship building. As the shortage of wood became increasingly pronounced, its price increased such that the poorest of the population were increasingly unable to afford it. These were particularly hard times because Europe had just entered into a so-called "little ice age," which would last until the eighteenth century. However, a severe energy crisis never materialized thanks to coal, which became increasingly the country's main source of fuel by the beginning of the seventeenth century. This conversion to coal was not without problems; coal's thick smoke upon burning made London's air quality one of the poorest in all of Europe. On some days, the sun was hardly able to penetrate the coal smoke, and travelers could smell the city miles before they actually saw it.

Beyond Oil and Gas: The Methanol Economy, Second updated and enlarged edition
George A. Olah, Alain Goeppert, and G. K. Surya Prakash
Copyright © 2009 WILEY-VCH Verlag GmbH & Co. KGaA, Weinheim
ISBN: 978-3-527-32422-4

What really brought about the power of coal as an energy source came along in the early eighteenth century with the invention of the steam engine. The steam engine was at the heart of the resulting industrial revolution [20], and it was fueled by coal. At the time, one of the main problems facing coal mining was water seepage and flooding from various sources. Rainwater seeping down from the surface accumulated in the tunnels, and once the mines reached below the water table the surrounding groundwater also contributed to the problem. Consequently, the mines became slowly submerged in water. If the mine was located on a hill, simple draining shafts could be used, but as the mines were pushed deeper into the ground, the water had to be removed by other means. The earliest method relied on miners hauling the water up in buckets strapped to their backs. As this was not really convenient, various ways were designed to increase the effectiveness of the human labor. Among these were chains of buckets or primitive forms of pumps powered not only by human muscle but also in some cases by windmills, waterwheels (Figure 2.1) or horse power. However, none of these was very convenient or economical.

One of the most pressing challenges for contemporary England was to find a way to keep its coal mines dry. This led eventually to the introduction of a device invented by Thomas Newcomen, who was not a scholar but a very inventive small-town ironmonger [1]. His device consisted of a piston that moved up by steam generated by heating water with burning coal, and down by reduced pressure resulting from the condensation of steam with cold water. The piston was connected to the rod of a pump used to pump water.

In 1712, one of these Newcomen engines was first used in a coal mine and became an almost immediate hit among mine operators, largely because it was much cheaper to operate than horses and could pump water from a much greater depth than ever before. The drawback was that the engine needed large amounts of coal to generate the steam necessary to keep it operating, and therefore found little use outside of the coal mines.

At about this time James Watt, a carpenter's son from Scotland, improved Newcomen's steam engine dramatically. Watt realized that as steam was injected and then cooled with water, heat was wasted in the constant reheating and cooling of the cylinder. The installation of a separate condenser immersed in cold water connected to the cylinder kept it hot and avoided unnecessary heat losses (Figure 2.2). This improved the efficiency of the steam engine by at least a factor of four, and allowed it to move out from the coal mines and find its place in factories.

To really move the industrial revolution ahead, however, another technological advance was needed: the manufacture of iron using coal-based coke. Until that time, the iron needed to build engines and factories was essentially made using charcoal obtained by burning huge amounts of wood, which was increasingly becoming scarcer in Britain. Charcoal provided both the heat and the carbon needed for the reduction of the iron ore. The use of coal to smelt iron was hindered by the impurities it contained, which made it unsuitable. After more than a century of experimentation, however, the key to making iron using coal was found. In the same way that wood was turned into charcoal, coal had first to be baked to drive off the volatiles and form coke. By the 1770s, the technology had advanced to the point where coke could be used in

A—AXLES.   B—WHEEL WHICH IS TURNED BY TREADING.   C—TOOTHED WHEEL.
D—DRUM MADE OF RUNDLES.   E—DRUM TO WHICH ARE FIXED IRON CLAMPS.
F—SECOND WHEEL.   G—BALLS.

Figure 2.1 Water removal in mines during the middle ages.
From an engraving by Georgius Agricola, *De Re Metallica*: book 6
ill. 36, 1556.

all stages of iron production. With this breakthrough, Britain, rather than being dependent upon iron imports, became in just a few years the most efficient iron producer in the world. This technological advance allowed it to build its powerful industries at home and its vast empire worldwide.

The "coal economy" resulted in a concentration of the ever-larger and mechanized factories, as well as their workforces, into urban areas, making them more efficient. The epicenter of this industrial revolution was Manchester, which became the

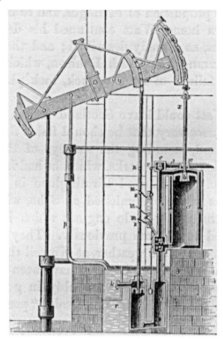

Figure 2.2 Watt's engine, 1774.

premier center of manufacture in England. The city also became home of the first steam locomotive-driven public railway, the Liverpool and Manchester Railway, which opened in 1830 (Figure 2.3). The "father of the railways" was George Stephenson, who first envisioned moving large quantities of coal over land. It was through the steam locomotive that this transport became possible, although this invention would in time have revolutionary consequences far beyond the coal industry.

The Liverpool and Manchester Railway became a huge success, transporting hundreds of thousands of passengers during the first months of operation. This success established a bright future for railway as a transportation system and

Figure 2.3 Stephenson's locomotive, *The Rocket*, 1829.

triggered massive investment in this industry. Although other European nations followed its example, Britain had a good 50 years head start in industrialization, and maintained its lead for most of the nineteenth century. In 1830, Britain produced 80% of the world's coal and, in 1848, more than half of the iron of the world, making the nation the most powerful on Earth until the end of the nineteenth century. Across the Atlantic, however, the United States – having even more coal and other resources than England – also began to undergo an even faster industrial transformation.

Historically, coal has probably been the most important fossil fuel, as it triggered the industrial revolution that led to our present-day modern industrial society. During the twentieth century, coal has been supplemented and displaced progressively by oil and natural gas, as well as nuclear power, for electricity generation. Coal was increasingly considered as a "dirty fuel" of the past, and was deemed to have a limited future. Only with the energy crisis in the 1970s, and the growing concerns about the safety of nuclear energy, did coal again become an attractive energy source, especially for electricity production. Because the reserves of coal are geographically widespread and coal is a heavy and bulky solid which is costly to transport, it is mainly utilized close to its source. The economically recoverable proven coal reserves are enormous, and estimated as being in the order of one trillion tonnes [2, 21, 22] – enough at current rates of consumption to supply our needs for more than 150 years. The reserve over production (R/P) ratio is more than two times as large as that for natural gas, and about four times as high as that for oil. Unlike oil and natural gas, our coal resources should last at least for the next two centuries. Our total coal resources are estimated to be more than 6.2 trillion tonnes [23]. The main reason why the R/P ratio for coal is not even higher is the limited incentive to find new exploitable reserves, given the size of already-known reserves. Production has increased ten-fold over the past 100 years, without any significant increase in coal price. In contrast, the implementation of advanced mining technologies has improved, and will continue to improve productivity and steadily lower the cost of coal extraction and treatment. The efficiency of coal transportation, which can represent as much as 50% of the import cost into Europe or Japan, is also improving [24]. Furthermore, large reserves, coupled with competition between coal-producing countries, make a sustained price increase unlikely and should result in relatively flat coal prices in the foreseeable future. In the case of coal, neither the abundant resources nor the competitive prices are determining factors in the fuel's future. The rate of extraction of coal is presently only a function of its relatively limited demand. In industrialized countries, where coal is used mainly to generate electricity, the demand will be governed by the ability of coal to compete with natural gas, not only from an economic point of view but also increasingly from environmental considerations. One of the reasons why we no longer rely more heavily on coal is that, from an environmental aspect, it is the most polluting fossil fuel compared to oil and gas. It usually emits significant levels of pollutants, especially sulfur dioxide, nitrogen oxides and particulates. Heavy metals such as mercury, lead, arsenic and even uranium are difficult to remove from coal, and are generally released into the air upon combustion [25]. Interestingly, these concerns about pollution are as old as the use of coal itself. An ordinance from 1273 prohibiting the use of coal in London as prejudicial to health is the earliest known attempt to reduce smoke pollution [26]. Present efforts to ban or reduce the use of coal are thus not revolutionary or new.

In a continuous effort to diminish the environmental impact of coal burning, the development and progressive introduction of new separation technologies applied to existing or new power plants can greatly reduce or nearly eliminate the emissions of SO_2, NO_x and particulates [27]. Emission regulations for mercury and other impurities present in coal are under evaluation in several nations, including the United States. Most significantly, the combustion of coal also generates large amounts of CO_2, a harmful greenhouse gas that contributes to a large extent to human-caused global warming. The only presently considered technology to mitigate CO_2 emissions is to capture and subsequently sequester it in underground formations or at the bottom of the seas [27]. At present, however, there are no CO_2 emission capture technologies operating at large-scale power plants. Given the growing concerns about global warming, coal-burning power plants represent a major challenge. Compared to oil and gas, coal is the fuel that produces the most CO_2 and other pollutants per unit of energy released. To tackle this problem, so-called "clean coal technologies" are being developed to improve the thermal efficiency and reduce emissions and, consequently, the environmental impact of coal-fired power plants [23, 28]. Among these technologies, some are already commercially available.

In the atmospheric fluidized bed combustion (AFBC process), coal is burned in a fluidized bed at atmospheric pressure and the heat recovered to power steam turbines [28]. An improved version of that system, the pressurized fluidized bed combustion (PFBC), in which gas produced by the combustion of coal is used to drive directly a gas turbine, is currently under development. Supercritical and ultra-supercritical power plants operate under supercritical conditions at steam pressures above 22.1 MPa and 566 °C, where there is no longer any distinction between the gas and liquid phases of water, as they form a homogeneous fluid. Such plants are well established and operate routinely at pressures up to 30 MPa with efficiencies above 45%. The introduction of special metal alloys that are more resistant to corrosion (but also are more expensive) to increase the operating pressure to 35 MPa and the thermal efficiency of power generation to over 50% are under development.

Integrated gasification combined cycle (IGCC) of coal is another emerging technology that has been demonstrated on a commercial scale, but not yet widely deployed. In this case, the coal is first gasified to produce syn-gas, which is then combusted under high pressure in a gas turbine to generate electricity. The hot exhaust gas from this turbine is used to generate steam that can produce additional electricity using a steam turbine. Using this technology, the goal, for the United States, is to reach 52% efficiency by 2010. Nevertheless, even with this level of efficiency, an IGCC plant would still produce twice as much CO_2 per kWh generated than a combined cycle natural gas turbine [24], the currently favored option for power generation. However, the process also has a longer-term strategic importance because it is the first electric power-generating technology to rely on gasified coal. It could therefore act as a bridge to more advanced coal gas-based power plants that would have zero emissions [29]. Besides electricity, this type of plant could also produce hydrogen from the syn-gas generated during coal gasification, whereas CO_2 would be captured and sequestered underground. Alternatively, the obtained syn-gas can also be used to produce chemicals, including methanol and its derived products. China is investing heavily in this approach with the construction of numerous coal-based methanol plants to generate not only methanol but also dimethyl ether, ethylene, propylene and various other chemicals [30].

From an energy perspective, coal has a major advantage as its resources are still vast and are widely distributed around the world. Furthermore, the outlook for coal supply and prices is subject to less fluctuation than for oil and gas. However, coal could be penalized for its high carbon content, and the key uncertainty affecting the future of coal is the impact of environmental policies. Longer-term prospects for coal may therefore depend on the development and introduction of clean coal technologies and carbon dioxide recycling that would reduce or even eliminate carbon emissions. In any case, coal resources will not last for more than two or three centuries – longer than oil and gas, but still a short period on the timescale of humanity.

As coal (and all other carbon containing fossil fuels) upon combustion forms carbon dioxide, a major greenhouse gas contributing to global warming, serious attacks have recently been directed by environmentalist groups and individuals to abandon altogether coal as a fuel. These attacks are part of an effort to "cure" society from its carbon addiction. We do not consider this a realistic goal, at least in the short term, as with our large coal reserves, lasting centuries, humankind will hardly be able to avoid the use of coal as an energy source, and also as a source for synthetic hydrocarbons and products. A more feasible and practical solution seems to be to capture and chemically recycle carbon dioxide to methanol and derived products as part of our "Methanol Economy" [12]. Capturing and sequestering carbon dioxide emission is already being considered and is starting to be implemented. Sequestering of CO_2 is, however, only a temporary and potentially dangerous solution for the disposal of CO_2 in contrast with its chemical recycling in the context of the "Methanol Economy" (Chapters 10–14).

3
History of Petroleum Oil and Natural Gas

Petroleum oil and natural gas, similarly to coal, are generally the result of the degradation of organic materials, primarily plankton, which settled on the seafloor many millions of years ago. This process occurred in the so-called "source rock," where the biomass was trapped along with other sediments [31]. Depending on the depth at which this source rock was buried in the absence of air during its existence, the biomass will form either oil or gas. If this source rock was buried for a sufficient time between 2500 and 5000 m in depth, where the temperature is around 80 °C, the organic material will break down to form mainly oil. At depths exceeding 5000 m, however, usually no oil is found. At temperatures higher than 145 °C at that depth, over geological time, all carbon–carbon bonds will break to form methane, the dominant component of natural gas. The geological formations from which oil and gas are extracted are usually different from the source rock in which they were originally formed. In fact hydrocarbons, once separated from the source rock, can migrate upward to form shallow oil or gas fields called "reservoirs." They can even appear as surface oil seeps, for example as in the Los Angeles Basin at the La Brea Tar Pits in Southern California.

Notably, an abiological origin of some natural gas and oil sources is also possible and has been suggested [32]. Accordingly, methane could have been formed over the eons deeper in the earth crust by the reaction of a carbon source, such as an asteroid hitting the earth, with metals under elevated pressure and temperature. As in the case of aluminum carbide when reacted with water, methane would have been formed. The fact that volcanoes and fissures in the ocean floor are known to release methane may support this view.

Natural seepage of oil has been used since ancient times in locations in the Middle East and the Americas for various medicinal, lighting and other purposes. Petroleum was referred to as early as the Old Testament. The word petroleum means "rock oil" from the Greek *petros* (rock) and *elaion* (oil). Uses of petroleum oil, however, were very limited and it was not before the mid-nineteenth century that wide use and the real potential of these natural resources began to evolve. America's first commercial oil well was drilled in 1859 by Colonel Edwin Drake to a depth of some 20 meters near Titusville in the State of Pennsylvania (Figure 3.1), yielding about 10 barrels of oil per day [33]. Drake's single well soon surpassed the entire production of Romania,

Beyond Oil and Gas: The Methanol Economy, Second updated and enlarged edition
George A. Olah, Alain Goeppert, and G. K. Surya Prakash
Copyright © 2009 WILEY-VCH Verlag GmbH & Co. KGaA, Weinheim
ISBN: 978-3-527-32422-4

Figure 3.1 Edwin Drake (right) in front of his well in Titusville, Pennsylvania, 1866. (Source: Pennsylvania Historical and Museum Commission, Drake Well Museum, Titusville, Pennsylvania.)

which was at that time a major source of oil for Europe [34]. The area was known previously to contain petroleum, which seeped from the ground and was skimmed from a local creek's surface, which was called therefore "Oil Creek." Drake was a former railway conductor who, because of ill health, was urged by his doctor to move from the east coast to a more rural area. The "Colonel" title was given to him not as a result of any military service but by the Seneca Oil Company, which hired him and believed that such a title would help Drake to get the assistance of the local people. The efforts to find oil grew out of technological evolution and the need for lubrication and illumination products. Without evolving markets and processes for such products Drake would have not been sent to the Pennsylvania hinterland to prospect and develop his oil operation.

In the mid-nineteenth century, the need for illumination in cities such as Boston and New York led to the development of gas lighting. This was not done using natural gas but by employing an illuminating gas that was produced by heating coal at gasworks located at the edge of the towns. Where gas was not available, and before the discovery of oil, whale oil more than any other product was used to fulfill the demand for clean and efficient illumination. Operating out of the north-eastern United States, and later Hawaii and the north-western States, the American whaling fleet mainly searched for sperm whales, which contained large amounts of high-quality oil. However, with a steadily declining whale population, coupled with supply problems during the American Civil War and ever-higher prices, the need for less expensive or more easily available alternatives was growing. The advent of petroleum oil signaled the end of the use of whale oil. These lamp oils were replaced by the more convenient and easily obtained, and seemingly inexhaustible, kerosene.

Kerosene, which was derived initially from bituminous coal and also known as "coal oil," proved to be a competitive illuminant during the 1850s, although its foul odor kept the rate of development relatively slow. The name kerosene was quickly extended to all illuminating oils made from minerals, however. For the production of kerosene, petroleum oil could be used instead of coal in the same distillation process, and the product was distributed over the same existing network, thereby facilitating its development as a fuel. As a further positive aspect, these fossil fuel light sources put an end to sperm whaling for oil, just as the use of coal had helped to save the remaining forests.

Kerosene was the first petroleum product to find a wide market, allowing John D. Rockefeller, a Cleveland (Ohio) entrepreneur, to build his Standard Oil Company into a vast industrial empire that would, during the subsequent rise in the use of gasoline for the internal combustion engine, enjoy a virtual monopoly over the production and distribution of oil in the United States [35, 36]. In 1911, however, Standard Oil, due to antitrust regulations passed by the US Congress, was broken up into separate companies (the "Seven Sisters"), including those which became Chevron, Amoco, Conoco, Sohio and, of course, Mobil and Exxon. The latter two giants recently merged again to form ExxonMobil without causing much public concern. *Tempora mutantur* – such is the effect of changing times.

The discovery and uses for petroleum has paralleled, and to a large extent been responsible for, the growth in oil production. Inspired by the invention of the gasoline-burning engine by Nikolaus Otto in 1876, the combination of Gottlieb Daimler's engine, Carl Benz's electrical ignition and Wilhelm Maybach's floatfeed carburetor resulted in the 1890s in the first successful commercially produced internal combustion engine based automobile (Figure 3.2). Henry Ford's assembly line mass production methods soon made it widely available and changed humankind's life in the twentieth century (Figure 3.3).

The use of oil began to increase dramatically to produce the large quantities of gasoline needed to fuel automobiles. However, the amount of gasoline that could be obtained from crude oil was low, anywhere from 10 to 20%. The original production process was based on simple distillation (fractionation), that is, separating hydrocarbons through differences in their boiling points. Later, owing mainly to growing demand, the refining of crude oil to yield a range of liquid fuels suited to various specific applications, ranging (eventually) from massive diesel locomotives to supersonic airplanes, was transformed by the introduction of cracking and other refining processes. Thermal cracking in combination with high pressure was introduced in 1913. The high temperature and pressure reproduced, on a short time scale, the naturally occurring process in breaking larger molecules into smaller ones.

Figure 3.2 Daimler and Maybach in their first four-wheel automobile.

This process was further improved by the introduction of catalytic cracking in 1936. World War II saw also the introduction of high-octane gasoline produced by alkylation and isomerization. Without these processes it would be impossible to produce, inexpensively, the required large amounts of more valuable lighter fractions from the intermediate and heavy, higher molecular weight compounds of the crude oils. Furthermore, with these processes, the route to petrochemicals was opened up, since cracking provides the ability to produce unsaturated hydrocarbons – molecules that, in contrast to saturated hydrocarbons (paraffins, which are the main components of oil), can be readily used and further transformed in chemical reactions to yield products such as lubricants, detergents, solvents, waxes, pharmaceuticals, insecticides, herbicides, synthetic fibers for clothing, plastics, fertilizers and much more. Today, our daily lives would be unthinkable without all of these products.

Figure 3.3 Ford's first assembly line.

Since the first "black gold" rush, initiated by Drake in Pennsylvania, the search for new oil fields worldwide has never stopped. The most intensive exploration occurred in the United States, where vast quantities of oil were found in several States, including Oklahoma, California, Texas and, more recently, Alaska. In Eurasia, the earliest oil exploitations occurred around the Black Sea and on the Caspian Sea near Baku in the then Russian Empire (now Azerbaijan), where the Swedish Nobel family (remembered mostly for the Prize one of its member's founded at the end of the nineteenth century) had for some years held the early concession for oil production. More recently, in the harsh regions of Siberia, vast reserves of oil have been found and developed. In Sumatra, Java and Borneo, petroleum resources have been exploited since the second half of the nineteenth century, starting with the Royal Dutch Shell Company [36].

In the Middle East, exploration for oil began during the 1930s, and this led to discoveries of the world's largest oilfields, such as al-Burgan in Kuwait in 1938 and a decade later Saudi's al-Ghawar, the world's largest super-giant oil-field that contains almost 7% of the world's oil reserves (according to estimates in 2000). Other major oil production areas include South America, South-East Asia, Africa and, more recently, the North Sea [36].

3.1
Oil Extraction and Exploration

Oil extraction begins with the drilling of a well. Drake's first oil well was only around 20 m deep. With rotary drilling techniques, used successfully for the first time at the Spindeltop well in Beaumont, Texas in 1901, oil wells surpassed 3000 m in the 1930s; drilling production wells deeper than 5000 m is now possible and used in several hydrocarbon reservoirs. Many people played significant roles in the development of modern drilling technologies. H. R. Hughes in particular improved significantly the rotary drill bit, enabling its use in almost any kind of rock formation. In 1933, directional drilling was first used successfully by W.M. Keck in California [37]. Directional and more recently horizontal drilling technologies are now widely used. A single vertical borehole can at certain depth divert to a number of directional and horizontal wells, improving greatly the recovery of oil from diverse strata. For the same reservoir, horizontal wells can produce many times as much oil as traditional vertical wells. The longest horizontal wells now exceed 4000 m in length.

In the early days of oil exploration, drilling took place exclusively on land, but moved to off-shore locations as the land deposits became less abundant and the necessary technologies were developed. The first off-shore well connected to land by a wharf was drilled at Summerland, south east of Santa Barbara, California in 1897. However, it was 40 years later in 1937 that the first off-shore well with no connection to land was successfully drilled in shallow water along the Gulf coast of Louisiana by Pure Oil Co. and the Keck's Superior Oil Co. The first deep-water off-shore oil platform was also built on the Gulf coast in 1947. Today, some of these off-shore platforms are working in waters 2000 m deep or more. Since oil reservoirs are

Figure 3.4 An oil supertanker.

unevenly distributed and often far from major consumption centers, the crude oil must be transported over long distances, sometimes thousands of miles. For the long-distance transport of oil products on land, pipelines and railway tank cars are used. Pipelines are expensive to build and maintain, and breaks along the line can cause severe oil spills. However, they are the most energy-efficient means of transporting oil overland.

When oil must be transported overseas, for example from the Persian Gulf to North America or Europe, it is carried in specialized oil tankers. Since the 1970s the need to carry ever-increasing quantities of oil has resulted in the construction of so-called "supertankers;" these are the largest ships afloat in the world, and larger even than aircraft carriers (Figure 3.4). However, if a tanker is damaged in an accident the spillage can result in severe environmental problems. Some of the oil spills resulting from such accidents have become infamous. For example, the *Amoco Cadiz*, which in 1978 broke up off the coast of France, spilling 1.6 million barrels of crude oil, damaged not only the ecosystem but also the lucrative French tourist industry. In 1989, the *Exxon Valdez* spilled almost 270 000 barrels of crude oil off the coast of Alaska. Despite the increasing oil quantities transported overseas, the amount of oil spilled has decreased over the years, thanks to the development of new technologies and infrastructures such as double-hulled tankers or deep-water "superports."

The twentieth century saw a tremendous increase in the use of oil for various purposes. With the widespread use of automobiles, trucks and diesel locomotives, oil replaced coal as the main fuel for transportation. Oil also replaced coal as the primary home heating fuel because of its greater convenience. By the mid-twentieth century oil had become the world's primary energy source, but after the two oil crises of the 1970s and concerns about the reliability and limitations of our oil sources another fossil fuel, namely natural gas, began gaining wider use.

3.2
Natural Gas

Natural gas used to be regarded as an undesirable by-product of petroleum oil production, and was simply burned, or "flared," at the oil wells. This is still done in some parts of the Middle East or at some off-shore platforms around the world, where no nearby markets exist for natural gas and transportation to more distant markets is not yet economically viable. Most commercially used natural gas, however, comes from wells that are bored solely for natural gas production. Before the nature of natural gas was understood it posed a mystery to mankind. Sometimes lightning strikes would ignite natural gas that was escaping from the Earth's crust; this would create a fire coming from the Earth, burning the natural gas as it seeped out from underground. These fires puzzled most early civilizations, and were the root of much myth and superstition. One of the most famous of these flames was found in ancient Greece, on Mount Parnassus, approximately around 1000 BC. According to the legend, a goat herdsman came across what looked like a "burning spring," a flame rising from a fissure in the rock. The Greeks, believing it to be of divine origin, built a temple over the flame. This temple housed a priestess, who was known as the Oracle of Delphi, giving out prophecies claimed to be inspired by the flame.

Such flames became prominent in the religions of India, Greece and Persia. Since the origin of these fires could not be explained, they were often regarded as divine or supernatural. It was not until about 500 BC that the Chinese discovered the potential to use such fires to their advantage. After first finding places where gas was seeping to the surface, the Chinese built crude pipelines from bamboo stems to transport the gas, which was then used to boil sea water, separating the salt and making the water potable.

In America, naturally occurring gas was identified as early as 1626, when French explorers discovered natives igniting gases that were seeping from the ground around Lake Erie. Indeed, the American natural gas industry began in this area. Actually, the very same first well dug by Drake in 1859 produced not only oil but also natural gas. At the time, a 5 cm (2 inch) diameter pipeline was built, running some 9 km from the well to the village of Titusville, Pennsylvania. The construction of this pipeline proved that natural gas could be brought safely and relatively easily from its underground source to be used for practical purposes. For this reason, most consider this well as the beginning not only of the oil industry but also of the natural gas industry in America.

In fact, the first well specifically intended to obtain natural gas was dug in Fredonia, New York, by William Hart. After noticing gas bubbles rising to the surface of a creek, Hart dug a well about 10 m deep to obtain a larger flow of gas to the surface. Hart is regarded by many as the "father of natural gas" in America. Expanding on Hart's work, the Fredonia Gas Light Company was eventually formed, becoming the first American natural gas company.

In 1885, Robert Bunsen in Germany invented what became known as the Bunsen burner (Figure 3.5). The burner mixed natural gas with air in the correct proportions, creating a blue flame that could be safely used for cooking and heating. The invention of the Bunsen burner opened up new opportunities for the use of natural gas. The subsequent invention of temperature-regulating thermostatic devices made it possible to better use the heating potential of natural gas, allowing the temperature of the flame to be adjusted and controlled.

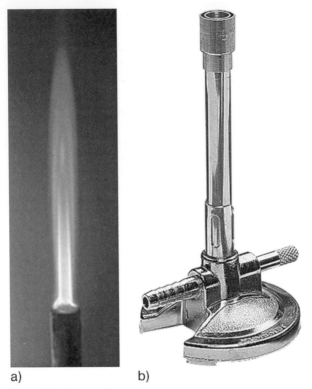

a)   b)

Figure 3.5 A Bunsen burner.

Without any means of transporting it effectively, natural gas discovered prior to World War II was usually just allowed to vent into the atmosphere, or burned when found together with coal and oil, or simply left in the ground. Transportation by pipelines developed only gradually. One of the first natural gas pipelines of considerable length was constructed in 1891; this was 200 km long and carried natural gas from wells in central Indiana to the booming metropolis of Chicago. This early pipeline was very rudimentary and used no artificial compression, relying completely on the natural underground pressure. As might be imagined, the pipeline was not very efficient in transporting natural gas. It was not until the 1920s that any significant effort was put into building suitable pipeline infrastructure. However, it was not until after World War II that newly developed welding techniques, pipe rolling and metallurgical advances permitted the construction of reliable pipelines. The post-war pipeline construction boom lasted well into the 1960s, and allowed many thousands of kilometers of pipeline to be constructed in America and worldwide.

Once its transportation was made feasible and safe, many uses for natural gas were developed. These included its use to heat homes and operate appliances such as water heaters, cooking oven ranges and clothes dryers. These advances were achieved relatively easily because customers were already comfortable with the use of town gas, which was generated from coal and had been widely used around the industrialized world for close to a century.

Industry also began to use natural gas extensively in manufacturing and processing plants. Natural gas was also increasingly used to generate electricity. Its transportation infrastructure had made natural gas easy to obtain, and it increasingly became a most popular and efficient form of energy.

Moving large amounts of natural gas by pipeline is relatively easy, but the process is not adaptable to transporting natural gas across the oceans from remote locations far from consumer markets such as the Middle East to North America or Europe. Because of the gaseous state of natural gas, it occupies enormous volumes compared to liquid or solid fuels. In accordance with Avogadro's law, 16 grams of methane occupies about 22 liters at normal temperature and pressure ($0\,°C$, 1 atm.). For intercontinental transport, across oceans, natural gas is usually liquefied to yield liquefied natural gas (LNG). This process is, however, energy intensive and requires specially designed and highly expensive tankers that keep natural gas at or below its boiling point ($-162\,°C$) in highly insulated, double-walled tanks. LNG is also potentially dangerous. In 1944, the explosion of an LNG storage facility in Cleveland, Ohio, killed 128 people and injured several hundreds. Subsequent accidents were not infrequent, such as in 2003 when an explosion occurred at a natural gas liquefaction plant in Algeria [38]. A study by Sandia National Laboratory indicated that an explosion from a LNG tanker could result in major injuries and significant damage up to 500 m from the leak, while people up to 2 km away could suffer second-degree burns [39]. The effects of an LNG tanker blowing up in a major port city (Boston, Los Angeles, Shanghai, Tokyo, etc.) could be devastating, possibly ending LNG maritime transport. The transformation of natural gas into an easily and safely transportable liquid such as methanol is therefore an attractive alternative. Besides accidents, LNG facilities and tankers are also potential targets for terrorist attacks. For these reasons, LNG terminals are usually not welcomed near major cities or population centers and are increasingly located off-shore.

It should be emphasized that natural gas production wells frequently also release large amounts of carbon dioxide. Natural gas is accompanied, for example, by 10% CO_2 in some North Sea fields off the coast of Norway and up to 40% in some Algerian wells [40]. Whereas this is still mostly vented to the atmosphere, environmental considerations have begun to force its capture and underground storage, that is, sequestration. Sequestration is presently the only technology considered feasible and is already used in some locations. Today, natural gas is extensively used because of its clean-burning properties and convenience for heating purposes and generation of electricity to replace older and more polluting coal-fired power plants. From an environmental point of view, natural gas is also advantageous because it produces the least amount of CO_2 greenhouse gas per unit of energy compared to all other fossil fuels.

As the nineteenth century had been the golden age of coal, the twentieth century saw the rise of petroleum oil and natural gas as the new "kings" of fuels. Advances in the petrochemical industry, which provide us with many of the necessities of modern life, may be credited, more than anything else, for the high standard of living that we enjoy today. As we advance into the twenty-first century, the use of oil and natural gas will continue. The unquestionable fact, however, is that these non-renewable resources are limited. With their increasing use and the growing population of the Earth, together with increasing standards of living, they will become increasingly scarce and expensive. Their combustion also produces very large amounts of carbon dioxide, which contributes to global warming. We have no choice but to develop new sources and technologies to increasingly replace fossil fuels, such as the chemical recycling of CO_2 in the frame of the "Methanol Economy" [12]. The time to do this is now, when we still have extensive sources of fossil fuels available to make the inevitable changes gradually, without major disruptions or crises.

4
Fossil Fuel Resources and Their Use

Today's enormous energy demands are still mainly fulfilled by the extensive use of fossil fuels. In 2000 they contributed to 85% of the energy consumption in the United States, and to 86% worldwide. The fossil fuels used – coal, oil and natural gas – are all hydrocarbons that differ in their hydrogen to carbon ratio. To liberate their energy content and to power our electricity plants, heat our houses and to propel our cars and airplanes the fuels must be burned. In the process, their carbon content forms CO_2 and their hydrogen content forms water. Consequently, they are non-renewable, and as such on the human timescale they cannot be naturally regenerated and are therefore available in a finite amount on Earth. The recurring question is: how much of these reserves are still available? Most estimates put our overall worldwide fossil fuel reserves as lasting not more than 200–300 years, of which readily accessible oil and gas would last for less than a century. These forecasts, however, are based on our present state of knowledge. The dynamic nature of discovery and consumption rates of fossil fuels makes it difficult to provide a definitive estimate. In fact, estimates have changed over the years, as new large oil and gas deposits have been found and more advanced technologies were developed. However, the probability of finding new large deposits is diminishing. At the same time, consumption and future needs are steadily growing – driven by population growth and an increasingly technological society. As our readily recoverable fossil resources are finite, it is certain that we face a major problem.

To better understand how the available supply of a given energy source is determined, the notions of reserve and resource must be defined. In geological terms, and applied to fossil fuels, the *reserve* is the amount that can be recovered economically with known technology. A *resource* is the entire amount known or estimated to exist, regardless of the cost or technological development needed to extract it. The amounts of energy available in these categories change over time. Continued exploration and the discovery of new sources or advances in technology could transfer an energy source from the resource to the reserve category. The criteria of profitability can also change as market conditions vary. This is especially apparent in the case of oil price increases. Much oil that was formerly considered as "unrecoverable" because of high extraction costs became profitable with price increases and thus became categorized as "reserve." Significant price fluctuations are inevitable, but they do not affect the long-range picture.

Beyond Oil and Gas: The Methanol Economy, Second updated and enlarged edition
George A. Olah, Alain Goeppert, and G. K. Surya Prakash
Copyright © 2009 WILEY-VCH Verlag GmbH & Co. KGaA, Weinheim
ISBN: 978-3-527-32422-4

Resource exhaustion is thus rather a question of unacceptable high costs than necessarily actual physical depletion. Notably, reported reserves may vary greatly, depending on the location and the prevailing political and economic conditions. This is apparent for example in the huge and abrupt reported increases in oil reserves for several OPEC nations in the late 1980s [41]. This occurred almost overnight, and was not the result of any major oil discovery at the time. Until then, OPEC assigned a share of the oil market based on each country's annual production capacity. However, during the 1980s OPEC changed the rules to also take into account the oil reserves of each country, leading most of them to promptly increase their reserve estimate to obtain a larger share of the market. On the other hand, the former Soviet Union used to publish wildly optimistic estimates of its reserves, probably to mislead its cold war adversaries. Major oil companies also tended to exaggerate their reserves to increase the value of their stocks. Much re-evaluation is consequently taking place on a regular basis.

Coal and natural gas, due to their different availability and utility, have different markets and their reserves also vary accordingly.

4.1
Coal

Historically, coal was the first fossil fuel to be used on a massive scale for industrializing our societies, and still represents worldwide 26% of our primary energy source (Figure 4.1) [42]. In most industrialized countries, however, coal has been replaced by oil, gas or electricity in household uses. Electricity generation is the one area where coal has retained a major role as an energy source in the industrialized

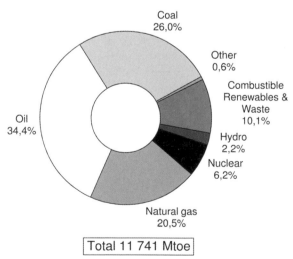

Coal
26,0%

Other
0,6%

Combustible
Renewables &
Waste
10,1%

Oil
34,4%

Hydro
2,2%

Nuclear
6,2%

Natural gas
20,5%

Total 11 741 Mtoe

Figure 4.1 Total world primary energy supply (TPES) in 2006.

world. Its share of the world market has remained almost unchanged over the past few decades. Worldwide, almost 40% of electric power is generated using more than 60% of the global coal production. Many countries are very heavily dependent on coal for electricity production, including Poland (93%), South Africa (93%), Australia (80%), China (78%) and India (69%) [43]. In the coal-rich United States, 92% of the domestic coal production is used to generate 50% of the country's electricity needs in giant coal-burning power plants [44]. The heat created by coal combustion is used to produce steam from water which, under high pressure and temperature, drives turbines connected to generators, the modern steam engines. In addition to steam generation for electricity, coal is used mainly in varied industries such as in steel and cement manufacturing.

In 2006, the world's coal production (hard and brown coal) amounted to 6.2 billion tonnes, with China being the main producer with almost 2.5 billion tonnes, which represented about 38% of global production – more than the United States, the European Union and Japan combined. China is followed in the amount of coal consumed by the United States, and is much ahead of India, Australia, and others (Figure 4.2) [2]. The largest future increase in coal production will come from increasing electricity demand in China, India and other developing countries in Asia. In China alone a new large-scale coal-based power plant opens every week to 10 days, adding significantly to global CO_2 emissions [45].

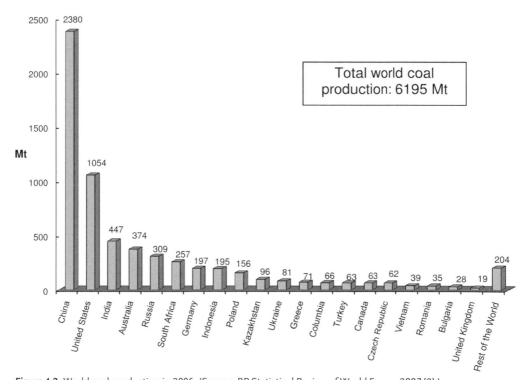

Figure 4.2 World coal production in 2006. (Source: BP Statistical Review of World Energy 2007 [2].)

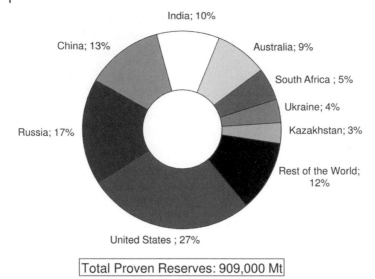

India; 10%

China; 13%

Australia; 9%

South Africa ; 5%

Ukraine; 4%

Kazakhstan; 3%

Russia; 17%

Rest of the World; 12%

United States ; 27%

Total Proven Reserves: 909,000 Mt

Figure 4.3 Distribution of proven coal reserves. (Source: BP Statistical Review of World Energy 2007 [2].)

Importantly, our reserves of coal are still large and geographically widely dispersed. Today's economically recoverable proven reserves are close to a trillion tonnes, representing at least 150 years of supply based on the current rate of consumption. The largest deposits are concentrated in the United States (27% of the proven reserves), Russia (17%), China (13%), India (10%) and Australia (9%) (Figure 4.3) [2]. The amount of coal in the world is so huge that even nations with a small percentage of the proven reserves can be major coal exporters. This includes South Africa, Indonesia, Canada and Poland. The global coal resources are estimated conservatively at more than 6.2 trillion tonnes [25]. Given the size of already-known reserves, there is for now little incentive to engage in extensive exploration to find new exploitable reserves. In any case, coal is our most abundant fossil fuel source.

The quality of coal and the geological characteristics of coal deposits have as much importance as the actual size of the reserves. Quality of coal can vary widely from one region to another. The classification depends mainly on caloric value, carbon and water content, which itself depends on the degree of maturity of a given coal deposit. Initially, peat (the precursor of coal) is converted into lignite or brown coal – both coal-types with low organic maturity. With exposure to high pressures and temperatures over millions of years, lignite is progressively transformed into sub-bituminous coal. Lignite and sub-bituminous coal are lower grades of coals that contain higher concentrations of water. They are softer, friable materials and have a low energy content of 5700–4200 kcal kg^{-1} for sub-bituminous coal, and less for lignite. They burn with smoky flames at a lower temperature. As the process of maturation continues, these coals become harder, forming bituminous or hard coal. Under the right conditions, a further increase in maturity can yield anthracite, the rarest and most desirable form of coal, representing less than 1% of known coal reserves. Anthracite and bituminous coal, the highest grades of coals, have the lowest water content. They burn at high temperatures with no or little smoke and ash formation, have a high caloric value (>5700 kcal kg^{-1}) and are suited for industrial coke production.

Beside the caloric value and water content, the amount of polluting impurities such as sulfur plays an increasingly important role as environmental regulations become more stringent. In the United States, for example, many electricity producers switched to lower grade sub-bituminous coal from the Rocky mountains (which contains up to 85% less sulfur than higher-grade bituminous deposits in the Eastern States) to reduce acid rain-inducing SO_2 emissions and comply with the new environmental laws [46]. Because coal consumption (in tonnes) per kilowatt-hour generated is higher for sub-bituminous than for bituminous coal, the shift to coal mined from the Western States is projected to increase the tonnage consumed per kilowatt-hour of electricity generation. Despite this drawback, and the fact that coal must be transported over longer distances, the switch was also made easier because of the much higher productivity and low cost of coal extraction in western coal mines, which are operated as open-pit surface mines as compared to under-ground mining in the eastern United States. Surface mining, which accounted in 2000 for 65% of the coal extracted in United States and about one-third in the rest of the world, is also safer (in China alone, yearly, thousands of miners perish in coal mine accidents). However, it is only economical when the coal seam is near the surface and the amount of overburden that must be removed is limited. Most of the coal (especially hard coal) reserves are too deep and must be mined underground. Increasingly, highly mechanized processes called "room-and-pillar" and "longwall" mining are used, which have nonetheless a lower productivity than surface mining. Worldwide, about two-thirds of the coal is still extracted from underground mines. Coal mining is a dangerous and physically demanding occupation that also involves many health hazards. The socioeconomics of underground coal mining are thus also a major factor in how much of these coal resources will indeed be recoverable in the future.

Environmental consequences of coal mining are increasingly being taken into account. In West Virginia and other southeastern states 450 Appalachian mountain-tops have already been removed to allow underlying coalbeds to be mined. This mining by "mountaintop removal" has had considerable destructive consequences on the environment. The overburden from the mountaintop is generally scraped into adjacent valleys, changing the topography and polluting water streams. Coal sludge resulting from coal processing is kept behind large earthen impoundment dams [47]. One such dam broke in Buffalo Creek in 1972, resulting in the death of 125 people and the destruction of more than 500 houses. Another sludge spill in Martin County, Kentucky contaminated the water of more than 27 000 residents and killed all aquatic life in the affected rivers.

Once the coal has been extracted from a mine it must be moved to the power plant or other location of use. Because it is a bulky commodity, it is expensive to transport. Consequently, the world coal industry is dominated by production for local use; more than 60% of the coal used for power generation is consumed within 50 km of the mining site. This is made possible by a widespread geographical distribution of coal. Currently, only about 17% of the world's coal supply is traded internationally [48].

During the last decades of the twentieth century coal prices declined steadily as productivity for coal mining and transportation greatly improved. The nominal price for coal imported from OECD countries and used in power plants has fallen from $50 per ton in 1980 to less than $35 in 2000 [22, 49]. Since 2003, however, the price of coal, as other fuels and commodities, has increased to reach levels above $80 per tonne in 2007 [50]. At the same time, environmental consideration for increasingly "clean" coal is making it more expensive for the end-user. Owing to the still extensive reserves and a large number of producing countries the coal market, however, remains very competitive.

Though coal is generally seen as a major fossil-fuel source for the long term, its use presents significantly greater environmental challenges than that of oil or gas. Upon combustion, it produces significant levels of pollutants such as sulfur dioxide, nitrogen oxides and particulates, as well as heavy metals such as mercury, arsenic, lead and even uranium [51]. This is the reason why coal bears the image of being a "dirty" and "polluting" fuel, bringing to mind images of black smoke rising from chimney stacks. Today, however, this view is slowly changing and, with proper processes and treatments, coal can be burned quite cleanly and efficiently. Problems arising from emissions were initially alleviated by building tall chimney stacks to improve dispersion, following the old diction: "The solution to pollution is dilution." This is, though, a view of the past. Technologies to reduce to acceptable levels or practically eliminate emissions, especially particulates responsible for smoke and dust as well as SO_2 and NO_x inducing acid rains, are now available and progressively applied, mainly in developed countries, and this has led to a dramatic decrease in pollution levels [29].

Regardless of eliminating varied pollutants, the inevitable emission that is now at the center of concern regarding coal combustion is the formation of CO_2, because of its potential as a greenhouse gas. As discussed, any carbon-containing material, such as coal, upon combustion produces CO_2. Compared to oil and gas, coal produces the highest amount of CO_2 per unit of energy generated. To reduce CO_2 emissions, extensive efforts are being made to improve the thermal efficiency of coal-fueled power plants (*vide supra*). To dispose of CO_2 emissions, ways to capture and sequester CO_2 from coal-burning power plants and other industrial sources, in subterranean cavities, various geological formations or in the seas, are being developed. As discussed in Chapters 10–14, instead of sequestering, which in our opinion is only a temporary solution, chemical recycling of CO_2 to methanol offers a new feasible, long-range solution to render coal use cleaner. This process also provides economic value for the carbon recycled for fuels and synthetic hydrocarbon products. Complying with environmental and health regulations is necessary, but should not increase substantially the overall cost of power generation from coal to the point that it becomes prohibitive and uncompetitive in the market place.

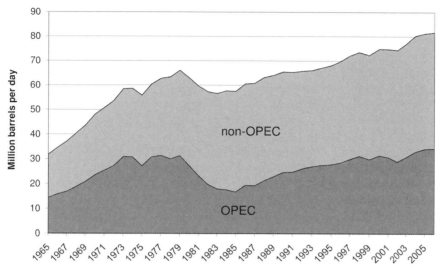

Figure 4.4 Oil production over time. OPEC is the Organization of
Petroleum Exporting Countries. (Based on data from BP Statistical
Review of World Energy 2007 [2].)

4.2
Petroleum Oil

Petroleum oil-derived gasoline and diesel fuels are the most widely used and
familiar fossil fuel products in our daily lives and have become essential to us. Oil
production increased rapidly after World War II, making it the dominant fuel and
energy source over the past half century (Figure 4.4). About 30 billion barrels of oil
were produced in 2006, representing 34% of the world's total primary energy supply
(TPES) [42] – more than coal (26%) and natural gas (21%). Renewable energy
sources accounted for about 13% of the global energy supply, with nuclear energy
providing the remaining 6% (Figure 4.1). Today, the world is using a staggering 85
million barrels of oil (about 12 million tonnes) every day, representing the content of
more than 50 giant supertankers each the length of three football fields. In a span of
two days, the world consumes today as much oil as the yearly oil production was in
1900. The US itself consumes about 25% of that oil. The amount of crude oil
consumed has increased from 58 to 85 million barrels a day between 1973 and 2006.
However, the share of oil in the TPES after the two oil crises of the 1970s has
decreased from 45 to 34%, mainly in favor of natural gas and nuclear power.
Regardless, we are still extremely dependent on oil as the economic prosperity of
our modern society is closely related to the availability of abundant and relatively
cheap oil. Oil – and all products derived from it – is also essential in many other
ways for our daily lives. Worldwide, about 6% of oil is used as the feedstock for the
manufacture of chemicals, dyes, pharmaceuticals, elastomers, paints and a multitude of other products. These petrochemicals provide many of the necessities of

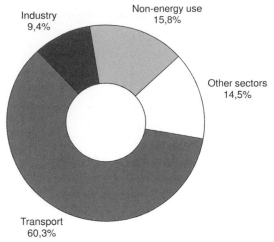

Figure 4.5 World oil consumption by sector (2005). (Source: Key world energy statistics 2007.)

modern life to which we have become so accustomed that we do not even notice our daily dependence on them. The consumption of petrochemicals is still growing at a significant pace. The bulk amount of oil, however, is used as heating fuel, for generating electricity, for industrial processes and predominantly as transportation fuels (Figure 4.5). Transportation, which accounts for about two-thirds of the oil consumed in the United States [52] and more than 60% worldwide [42], is by far the most oil-dependent sector in our modern economy. Automobiles, trucks, buses, locomotives, agricultural machinery and ships all rely on gasoline or diesel fuel. Air transport would be unthinkable without refined jet and aviation fuels. More than 95% of the energy used in transportation comes from oil. Its use in the transportation sector is growing inexorably, with little immediate prospect for short-term alternatives. In the developed countries, the transportation sector accounts for virtually all new growth for oil demand, while power generation and domestic and industrial uses are being switched increasingly to natural gas and to some degree to alternative fuels. Global demand for oil is currently rising by up to 2% per year. From 85 Mb per day in 2006, longer range forecasts predict that the world oil demand in 2015 will be 98 Mb per day, reaching 116 Mb per day in 2030 [48]. Temporary fluctuations or even decreases due to economic downturns are possible, but the long-range trends of increasing oil consumption are fairly predictable. If these estimates are correct, some 800 billion barrels of oil would be necessary from 2006 to 2030 to satisfy the growing demand. Considering that a cumulative total of about 1000 billion barrels have been consumed between the beginning of oil exploitation in the 1850s and today, this is a staggering amount. The question that must be raised is: will we have enough resources to cover the demand, and from where will we obtain all of this oil?

There are still sufficient oil reserves to satisfy demand, at least during the next few decades. However, a significant amount of this oil will come from politically unstable areas of the world such as the Middle East and South America and, increasingly, from inhospitable regions like the Arctic seas. There is also considerable uncertainty about the amount of the de facto oil that exists worldwide and the proportion of this resource that can be economically recovered. Numerous estimates made by various assessors, such as the World Energy Council, IHS Energy, Organization of Petroleum Exporting Countries (OPEC), United States Geological Survey (USGS), *Oil and Gas Journal* and others, have reported current proven oil reserves ranging from 1000 to 1200 billion barrels [22] and overall recoverable resources might be double that but would, however, become increasingly difficult and costly to access. With current global oil production being about 30 billion barrels per year, known reserves in 2006 represent about 30–40 years worth of supply. In practice, there has been little decline in estimated reserves, with most assessments indeed showing increases. Proven reserves, far from being depleted, have almost doubled during the past 40 years. This is because the nature of oil reserves is very dynamic and depends on many factors. New oil fields have been discovered, increasing the amount of proven oil reserves, while rising oil prices have made formerly uneconomical oil sources economical. Technological advances also increase the amount of economically recoverable oil. However, with a growing world population, which is now approaching 7 billion and is still growing (it will reach 8–11 billion by the mid-twenty-first century), as well as increasing standards of living around the world, the demand for oil will continue to expand. Oil will not be exhausted abruptly, but market forces will inevitably drive prices up as demand surpasses supply, creating a true and lasting oil crisis.

In recent history, there have been three distinct periods in oil supply versus demand involving oil-producing and oil-consuming countries. The period from 1960–1973 was one of rapid economic growth and burgeoning oil demand. Wealth in developed countries grew by 90%, while energy demand grew by a similar amount, and oil demand grew by 120%. The transportation sector boomed and oil also cut deeply into coal's markets as a heating fuel. Worldwide, oil demand rose from some 20 million to almost 60 million barrels per day. Many developed countries produced little primary energy or had static or declining production. Thus, they became heavily dependent on imports in great part from OPEC countries, especially from the politically unstable Middle East.

The oil price shock resulting from the 1973 crisis, reinforced by the subsequent Iran/Iraq war of the late 1970s, had profound effects. It abruptly ended a period of extremely rapid growth and prosperity. The world was stricken by high inflation, trade and payment imbalances, high unemployment, a weak business climate and low consumer confidence. The 1973 crisis introduced a new period of oil market development, which lasted until the mid-1980s. It was characterized by vigorous efforts by importing countries to reduce their dependence on oil. During much of this period these efforts were governed by high oil prices. In the first half of the 1980s, however, prices responded to the weakening market as oil supplies increased and demand continued to reflect the oil-saving measures achieved since the mid-1970s. New oil fields also went into production in Alaska and under the North Sea, weakening the monopoly of the OPEC cartel. Nuclear energy, natural gas and coal replaced much of the oil used for electricity generation and energy-saving measures were widely introduced.

The mid-1980s brought an end to the decrease of oil imports. Until the end of the 1990s, with low oil prices for much of the period and steady economic growth, oil consumption rose again and net imports of oil-importing countries are now much higher than they were in the early 1970s. The bulk of additional imports continues to come from the Middle East, Russia and other producers.

Oil prices have increased sharply in recent years, reaching the historical height of more than $140 per barrel in the summer of 2008 before dropping abruptly back, at least temporarily, into the $30–70 range in the first half of 2009. Such price variations, even if excessive, are not unexpected. While experiencing significant, temporary decreases due to economic slowdowns and other socio-economic reasons, with diminishing resources and increasing demand the future trend of ever higher oil prices seems inevitable. Several factors, including increasing oil consumption in Asia and developing nations around the world, especially from the fast-growing economies of China and India, reduced spare capacity in producing countries as well as political uncertainties threatening oil production in the Middle East, Venezuela, Nigeria and others, have contributed and will continue to impact the oil market significantly.

Like in every market, eventually the supply and demand mechanism regulates the price of oil. The supply can be artificially manipulated by decreasing or increasing the oil production to affect prices (as OPEC does frequently). This has serious consequences on the economies of oil-importing countries, but the process is reversible. It also accelerates the importing countries' efforts to find alternative solutions to the use of oil, which is contrary to the interest of many oil-producing countries. Regardless, the inevitably permanent and irreversible decline in world oil reserves and production due to depletion of oil resources will have a lasting and significant effect on our modern societies for which we must plan and prepare ourselves by finding solutions.

Presently, the Middle East is the epicenter of many of the "oil battles," because it has the largest oil reserves in the world. The top five countries in terms of oil reserves are in this region: Saudi Arabia, Iraq, Iran, United Arab Emirates and Kuwait. Russia, Venezuela and African countries such as Nigeria and Libya also have significant reserves. Although the Middle East contains about 62% of the world's oil reserves [2], it produces at present only about one-third of the oil. Oil is cheap to extract in the Middle East, and production costs are among the lowest in the world (as low as $2 per barrel). This compares with a cost of at least $10 to produce a barrel of oil in the North Sea [49]. Saudi Arabia has the world's largest oil reserves, and is the largest oil producer. Given its proven reserves of some 260 billion barrels, Saudi Arabia could produce 8 million barrels per day for 75 years without the discovery of any additional reserves. The largest oil field ever found is the Ghawar field in Saudi Arabia, which contains about 7% of the known world oil reserves. There are some 80 oil fields in Saudi Arabia, but oil is being produced from only a few. Ghawar and five other giant oil fields, all concentrated in a relatively small eastern corner of the Kingdom, produce some 90% of the Saudi oil [53]. Together, all of the Saudi oil fields contain about one-quarter of the world's oil reserves. Not only are these oil fields huge they are also highly productive. Saudi state-controlled Aramco, the world's largest integrated oil company, accounted for 12.8% of the world's oil production in 2007. In contrast,

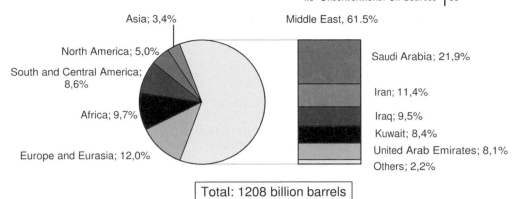

Figure 4.6 Regional distribution of world oil reserves in 2006.
(Based on data from BP Statistical Review of World Energy
2007 [2].)

the largest Western oil company, ExxonMobil, accounted at that time for only 3.2% of
world oil production [54]. Having more than half of the world's remaining oil reserves
concentrated in a relatively small geographical area in the Middle East has significant
economic and geopolitical consequences (Figure 4.6). The market share enjoyed by
the Middle Eastern countries is even increasing because fewer significant new oil
fields are being discovered elsewhere.

The rate of new oil discoveries has dropped mainly because most areas of the world
have already been explored. From the hot Arabian deserts to the freezing Arctic Circle
and the depth of the seas, exploration has gone to the most remote and inhospitable
parts of the world in search of oil. The last major new oil discovery areas have been the
North Sea and Alaska's North Slope. Today, it seems unlikely that there will be many
similar new large oil discoveries. Some of the remaining promising regions are off
the shores of Brazil, in Arctic regions and in the South China Sea, some of which are
starting to be developed. However, it seems highly unlikely that these can replicate the
scope of the Middle East. The Caspian Sea region also has large reserves that are being
developed, although the amount of reserves in place may have been overestimated.

4.3
Unconventional Oil Sources

The way to increase or at least maintain the level of oil reserves for the future is to use
unconventional oil sources, which are starting to play an increasingly important role.
Oil is considered unconventional if it is not produced from underground hydrocarbon reservoirs by means of usual production wells, or if it needs additional processing in producing crude oil-like compounds (sometimes called synthetic crude).
Unconventional oil sources include tar sands, oil shales, heavy oils, coal-based
liquids, biomass-based oils, processed gas liquid (GTL) and oil obtained from the
chemical processing of natural gas.

Figure 4.7 Oil sand mining in Athabaska, Alberta. (Courtesy of Suncor Energy Inc.)

4.3.1
Tar Sands

Among unconventional oil sources, recovery of oil from tar sands is presently being expanded the most rapidly. Tar sands, as the name suggests, are sand deposits containing a high-sulfur tar known as bitumen or very viscous, heavy oils. They are formed when, by erosion, an oil field is brought to the surface and the lighter components evaporate, leaving an almost solid tar behind. Therefore, in a way it can be considered as a "dead" oil field. There are tar sand deposits all around the world, but the largest are located in Canada and Venezuela. Canada's Alberta province contains two enormous deposits: Athabaska (Figures 4.7 and 4.8) and Cold Lake. Of the 2.5 trillion barrels of crude bitumen estimated at these locations, about 12% (or 300 billion barrels) is thought to be recoverable. This is an amount comparable to the present proven reserves of Saudi Arabia. In Venezuela, over 1.2 trillion barrels of bitumen is thought to exist in the Orinoco belt and Maracaibo sedimentary basin, of which 270

Figure 4.8 Suncor oil sand plant in Athabaska, Alberta. (Courtesy of Suncor Energy Inc.)

billion barrels are presumed to be economically recoverable with current technology [49]. The potential for unconventional tar sand oil depends largely on production costs. Thermally extracting oil from these heavy tars presently involves the burning of large amounts of natural gas. In Canada, due to massive investments and technological innovations, the operating costs (using *in situ* recovery) fell from $22 to less than $10 per barrel between 1980 and 2003, making this unconventional oil supply competitive with conventional oil [49]. In 2004, one million barrels of oil were processed daily in the Athabaska region. With ongoing investments, the production should reach 2 million barrels per day within a decade. Decreasing cheap natural gas resources used to provide the heat for oil sand processing represents, however, an increasing concern. The use of other heat sources, including atomic energy, may be needed in the future [55].

4.3.2
Oil Shale

If a tar sand deposit can be considered a "dead" oil field, then oil shale beds can be considered as "unborn" oil fields. Oil shale is a source rock that never sank deep enough for the organic matter to produce petroleum oil. Shale oil indeed contains neither oil nor shale, and the name was probably given to attract investors. It contains kerogen, a solid bituminous material that can be used as a substitute for crude oil. When kerogen is heated to high temperatures, thermal cracking occurs, producing oil that can be further refined to petroleum distillates.

In small quantities, oil shale has long been extracted. By the seventeenth century, oil shales were being exploited in several countries in Europe. In Sweden, oil shales were roasted over wood fires, not to extract the oil but to obtain potassium aluminum sulfate, a salt contained in these deposits and used in tanning leather and for fixing colors in fabrics. During the late 1800s, the same deposits were used to produce hydrocarbons. An oil shale deposit in Autun, France was exploited as early as 1839. The industrial-scale extraction of oil shale, however, began in Scotland around 1859. As many as 20 oil shale beds were mined at different times and by 1881 oil shale production had reached 1 million metric tonnes per year. With the exception of the World War II years, between 1 and 4 million metric tonnes of oil were mined yearly from 1881 to 1955, when production started to decline and finally ceased in 1962. Other commercial operations exist in Brazil, Estonia, Russia and China. From the early 1930s, production of oil shale increased until peaking at 46 million metric tons in the early 1980s, with Estonia producing 70% of that oil shale. In 2006, Estonia still obtained 90% of its electricity from oil shale. Production of oil shale subsequently declined steadily to a low of 15 million metric tons in 2000. It was not because of a diminishing supply but rather because oil shale could not compete economically with petroleum oil. The world's potential shale oil resources are enormous; estimated at 3.2 trillion barrels, they have barely been touched [56]. By far the largest known deposit is the Green River oil shale area in the Rocky Mountains of the western United States, containing a total estimated resource of 1.5 trillion barrels of oil. In Colorado alone, the resource reaches 1 trillion barrels of oil.

These values represent only those resources that are rich enough in oil and close enough to the surface to be open-pit mined and therefore thought to be economically recoverable. Numerous pilot plants using different extraction technologies have been operated over the past decades in this region by oil companies such as Shell, Exxon, Amoco, Unocal and Occidental. At the time these projects, however, were found to be uneconomical and were terminated. Unocal (now part of Chevron) operated the last large-scale experimental mining facility from 1980 until its closure in 1991, producing 4.5 million barrels of oil from oil shale, and averaging 34 gallons of oil per ton of rock extracted. The future development of the oil shale industry is clearly dependent on the availability and price of petroleum crude oil as well as cost and improvements of processing technologies. When the price of extraction of oil from oil shale stays competitive with the persistent price of oil from conventional sources either by technological improvements or higher oil prices, then shale oil will find a place in the fossil fuel energy market. The recent level of oil prices is already prompting renewed interest in oil shale [57–59]. Shell for example is testing a new method to extract oil shale from the ground. Historically, oil shale was mined, hauled aboveground and submitted to heat processing to extract the oil out of the rock. During extraction the shale expands and mountains of expanded shale rock were left behind, causing significant harmful environmental consequences. Instead of mining, Shell is using an *in situ* heating process in which the oil shale is heated at above 350 °C while still in the ground. The heated oil shale slowly releases liquids (mostly light oil and natural gas liquids) that are gathered in wells, pumped to the surface and refined to valuable transportation fuels much like petroleum. Significant problems must still be overcome, such as the large amounts and source of heat needed for the process, which could be generated, as proposed in Canada for the exploitation of tar sands, by nuclear energy. The environmental impact, including greenhouse gas emissions, must be carefully assessed. The availability of large amounts of water needed for the extraction of oil shale in the relatively arid western United States could also be problematic, as will be the protection of groundwater close to processing sites.

The highly successful development of the Athabaska tar sand resources of Canada, with an oil production in excess of 1 Mb per day, could serve as a model for the initiation and growth of the shale oil industry, especially in the western United States, as the basic technologies for the exploitation of these two resources are very similar.

4.4
Natural Gas

Global demand for natural gas has grown much faster than that for petroleum oil and coal over the past several decades. In 2006, world natural gas consumption reached 2.8 trillion cubic meters (Figure 4.9), representing 21% of the world's total primary energy source (TPES) compared to 26% for coal and 34% for oil. The United States alone accounts for about one-quarter of the global natural gas demand, and is the

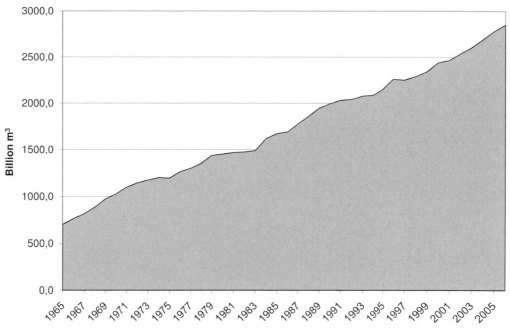

Figure 4.9 World natural gas consumption. (Based on data from
BP Statistical Review of World Energy 2007 [2].)

second largest gas producer after Russia. However, because of its huge demand, the
USA is also the largest natural gas importer. The global market for natural gas is
expected to continue to expand rapidly due to its still ample availability, convenience,
cost-competitiveness and environmental advantages over other fossil fuels. Natural
gas is considered as a premium fuel. It has a high caloric value and is clean-burning
and relatively easy to handle. For these reasons it is an excellent fuel for domestic use
and heating. The bulk increment of natural gas demand will, however, come from
new power plants. In recent years most of the new electricity generating capacity
constructed in the United States uses natural gas as a fuel although, as mentioned,
coal is making a strong comeback. Natural gas has especially become the preferred
fuel for the so-called "peak-load" power plants, which use combustion turbines to
operate the generators and can be rapidly switched on or off following the variation of
demand. Owing to the significant improvement in efficiency of gas-fired electricity
generation, the de facto natural gas consumption in electric plants has increased less
rapidly than the amount of electricity produced. Besides its flexibility and relative
cleanness, natural gas also has the advantage of adding the smallest amount of CO_2 to
the atmosphere per unit of released energy. Typical emissions are about 105 kg-
carbon Gcal^{-1} for bituminous coal, 80 kg-carbon Gcal^{-1} for refined fuel and less than
59 kg-carbon Gcal^{-1} for methane (the major component of natural gas). The reason is
that methane has a H : C ratio of 4, the highest of all hydrocarbons, and contains only
75% carbon by weight compared to some 85% for crude oil and more than 90% for
coal. Considering the growing concerns about global warming, a reduction of CO_2

emissions by using natural gas is certainly a step in the right direction. However, one should bear in mind that methane itself is a very potent greenhouse gas, which has a global warming potential 23 times higher than CO_2 over a 100-year period. Methane accounted for about 18% of the cumulative greenhouse effect from 1750 to 2005, the largest sources of anthropogenic methane being derived from livestock, rice fields and landfills [60, 61]. Oil and gas production, even before their combustion, were responsible for some 15% of the methane emissions, due mainly to leaks during natural gas transport. Pipelines in developed countries do not lose more than 1% of the carried gas. In less-developed countries, however, losses of 5% or more are quite common. With rising natural gas extraction and long-distance transport, these leakage problems should be resolved, not only to reduce the contribution of methane to global warming but also from an economic point of view, to make the best use of this valuable resource.

Natural gas is an abundant source of energy, with its proven reserves estimated at some 180 trillion cubic meters in 2006 (Figure 4.10) [2], exceeding the energy content of the world's proven reserves of oil. Despite ever-increasing production, the reserves of natural gas tripled in the past 30 years because important new fields have been discovered in many parts of the world and technological developments such as deeper and directional drilling have allowed existing reserves to be upgraded and new ones to be realized. At the actual rate of consumption in 2006, proven reserves would last more than 60 years, compared to 40 years for oil. Our overall gas resources are estimated at 450–530 trillion cubic meters. Since the beginning of

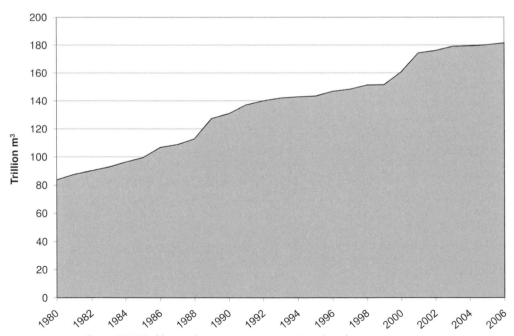

Figure 4.10 World natural gas proven reserves. (Based on data from BP Statistical Review of World Energy 2007 [2].)

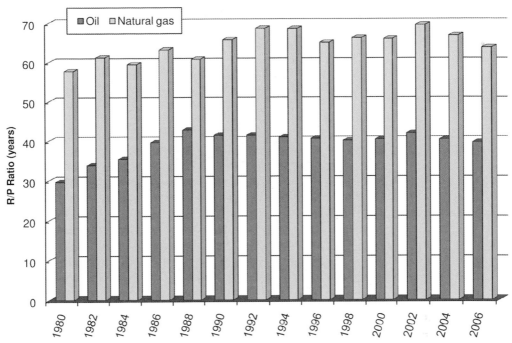

Figure 4.11 Oil and natural gas reserve to production (R/P) ratio.
(Based on data from BP Statistical Review of World Energy
2007 [2].)

fossil fuel exploration, only slightly more than 10% of the world's gas resources have
been consumed, compared to 25% of the estimated world's oil resources
(Figure 4.11).

As with oil, a few countries dominate the natural gas reserves. The largest known
reservoirs of gas are in western Siberia and the Persian Gulf. In 2006, more than
half of the global gas reserves were found in only three countries: Russia with 26%,
and Iran and Qatar, accounting for 15% and 14%, respectively (Figure 4.12). North
America and Europe represent only about 4% each. Like oil, known gas reserves
are concentrated in a relatively small number of giant fields. About 190 giant
reservoirs contain 57% of the global gas reserves, with some 28 000 other reservoirs
holding 28%. The remaining 15% represent marginal fields [25]. Although proven
reserves are abundant, new resources tend to be discovered in more remote areas,
far from consuming centers, in difficult terrain, or as small fields. Even today, the
geographical distribution of natural gas-producing regions is thus far from even.
Most of the demand is concentrated in North America and Europe, but these
regions have relatively limited sources of natural gas, in part because they have
already exploited significant amounts of their original reserves. In Europe, countries such as Norway, The Netherlands and the United Kingdom still have
significant Northern Sea reserves, which are, however, decreasing. A newly
discovered reserve of natural gas found in central Europe (Makó Trough in

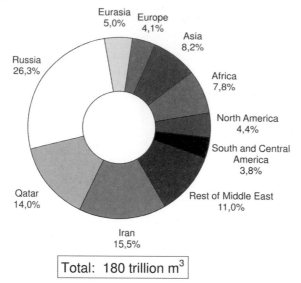

Figure 4.12 Distribution of world natural gas proven reserves in 2006. (Based on data from BP Statistical Review of World Energy 2007 [2].)

Southern Hungary) is promising but due to its depth (>5000 m) and consequent high temperature its exploitation is challenging. A growing part of Europe's natural gas consumption, due to decreasing North Sea production, has to be imported from the rich but remote western Siberian fields in Russia, the Caucasian region as well as from Algeria. North America is still largely self-sufficient in natural gas. The larger part of gas imports in the United States is supplied by pipelines from Canada. However, due to expected declining rates of production as well as a steadily increasing consumption, North America is also starting to rely more on imports, mainly in the form of liquefied natural gas (LNG) from countries such as Algeria, Trinidad and Tobago and Nigeria as well as the Middle East. LNG, however, has its own limitations and hazards.

To be shipped overseas as LNG, natural gas must first be liquefied at a port-side liquefaction plant in the producing country. Liquefaction removes almost all impurities (CO_2, sulfur compounds, water, etc.) and the gas becomes almost pure methane. Once liquefied at low temperature, LNG is piped onto giant tankers, the size of three football fields, with a carrying capacity of 150 000–200 000 m^3, where it is stored in highly insulated, double-walled large containers (Figure 4.13). After a trip of 10–15 days, the LNG is transferred from the ship to an on- or offshore regasification terminal where the liquid is reheated to gas ready for distribution. The construction of a LNG infrastructure is very capital intensive, with liquefaction and regasification plant costing between $1 and $2 billion each [41], particularly if located offshore. Nevertheless, ExxonMobil and Qatar Petroleum, for example, have signed two $12 billion agreements to build a LNG infrastructure to deliver 100 000 m^3 of natural gas per day to both the United States and the United Kingdom. In the United States, the

Figure 4.13 Liquid natural gas (LNG) tanker for transportation of natural gas across the seas.

construction of some ten new LNG regasification terminals is planned, compared to only about a half-dozen operating today. Because of the safety problems associated with the transport and handling of LNG, most future LNG terminals are expected to be built offshore. In the US and other major countries, much discussion and planning is taking place about new LNG terminals. Because of the potential danger involved, the trend is to locate away from major population centers. Regardless, LNG tankers and facilities are also potentially vulnerable to sabotage and terrorist attacks, which could have devastating effects.

In 2007, Japan, South Korea and Taiwan, all countries with no gas reserves, imported together 60% of the world's LNG production, mainly from Indonesia and Malaysia but also increasingly from Persian Gulf countries such as Qatar and Oman [62]. For them, LNG is the only economical solution because they are far from gas-producing areas and the cost of constructing long-distance offshore pipelines would be prohibitive. Undersea links are usually only practical when the distances are relatively short and the sea is not too deep, as in the case of the North Sea gas network and the projected Nord Stream pipeline under the shallow waters of the Baltic Sea that will connect Russia directly to Germany to supply Western Europe. During the past decade, however, development of offshore-pipeline technology has contributed to lower the costs and make possible deep-water projects that were previously impossible, such as the Bluestream pipeline under the Black Sea, which reached a maximum depth of 2150 m. Depths of 3000–3500 m will probably be feasible in years to come, with costs competitive with LNG. The natural gas liquefaction network, which is expensive and highly energy consuming, includes liquefaction, shipping, regasification and storage facilities. The liquefaction step, during which the gas must be cooled to $-162\,^\circ$C, accounts for about half of the total cost, with shipping and regasification accounting for about 25% each. Liquefaction is usually chosen when the gas has to be transported overseas and/or distances overland longer than 4000–5000 km, in which case the construction cost of a pipeline would be higher than for LNG. This is especially true for the Middle East, which accounts for 40% of the world's natural gas reserves, but is too far from the major consuming centers of Europe, the United States and the Far East. In 2001 this region represented only 21% of the LNG export market and 4.5% of the overall natural gas export market. As mentioned, however, there are also serious safety problems associated with LNG transportation and distribution.

Another way of transporting and using remote reserves is to convert natural gas into a liquid product near the gas wells, such that it can be more easily shipped in regular tankers. This is referred to as gas-to-liquid (GTL) technology. Advances in technology and increasing oil prices are stimulating new interest in these GTL processes. Within this broad strategy, two approaches seem to be the most feasible: (i) the conversion of natural gas into a liquid product resembling the light hydrocarbons present in petroleum or (ii) its conversion into methanol. These processes have so far been based on the Fischer–Tropsch technology using synthesis gas (syngas), which was originally developed using coal as a feedstock in Germany during the 1920s. Coal is first converted into syn-gas (a mixture of hydrogen and carbon monoxide), which subsequently will yield, depending on the catalyst and reaction conditions, methanol or synthetic liquid hydrocarbons. Because of the high energy needs and relatively high production costs, the syn-gas-based production of fuels, derived originally from coal, but later from natural gas, was in the past limited to special situations, such as in Germany during World War II and South Africa during the Apartheid era, where there were restrictions on oil imports. Recent technical advances, including reactor design and improved catalysts, have significantly increased yields and thermal efficiency and reduced construction and operating costs. South Africa, due to its decades of experience in this field, opened in the 1990s the Mossgas GTL plants, yielding 30 000–60 000 barrels per day of mainly diesel fuel and gasoline. GTL plants are, however, complex, expensive to build and very energy intensive, consuming up to 45% of the gas feedstock, thereby raising concerns about high CO_2 emissions. In most of the world, such as in New Zealand and Malaysia, syngas-based plants use natural gas (methane). Methane conversion into products in the diesel fuel or gasoline range is currently not used on any large scale. With the anticipated depletion of crude oil reserves, however, several oil companies are presently investing massively in GTL technologies. Sasol, in a joint venture with Qatar Petroleum, has recently completed the construction of a 34 000 barrel per day unit in Qatar, near the giant North Field gas reservoir (containing some 15% of the world's gas reserves) [63]. Shell has also started the construction of an $18 billion (triple the originally estimated cost) GTL facility in Qatar, with a production capacity of 140 000 barrels per day. Owing to increasing costs ExxonMobil, however, decided to abandon a similar 154 000 barrels per day GTL project [64, 65]. The GTL facilities in Qatar will produce mainly diesel fuel and small amounts of other products, including lubricants and motor oil. Owing to the economy of scale of such large installations, the cost to produce a barrel of GTL diesel is expected to be in the $20 range, which is highly profitable considering recent oil prices. Furthermore, diesel fuel obtained from GTL processes has the advantage of being almost free of most pollutants (especially sulfur) contained in regular diesel fuel, and is thus cleaner burning. The use of syn-gas based technologies produces, however, large amounts of carbon dioxide, thereby contributing to global warming and necessitating mitigation.

The proposed "Methanol Economy" also offers other feasible new vistas to natural gas. To be more easily handled and shipped over long distances, natural gas (i.e., methane) can be converted into methanol, a convenient liquid. Today the production of most of the methanol in the world comes from the conversion of natural gas (methane) into syn-gas and consequently to methanol. The direct oxidative conversion of methane into methanol, without going through syn-gas, is, however, the highly desirable technology of the future and is being developed as part of the "Methanol Economy" (Chapter 12).

The amount of natural gas reserves generally pertains to conventional natural sources – that is, gas that accumulates under non-permeable rocks alone or in association with oil. There are, however, also considerable resources of so-called unconventional natural gas sources; some of these are already being recovered. They involve gas extracted from *coalbeds* (coalbed methane), from low permeability sandstone called "*tight sands*" and shale formations. Gas hydrates, primarily *methane hydrate*, present at the continental shelves of the seas and underlying the tundra of arctic areas such as Siberia in large amounts are also significant sources for the future. Although no suitable technology to utilize these resources effectively exists today, they represent a very large untapped unconventional natural gas source. In the United States, during the past 20 years, the use of unconventional natural gas has grown rapidly, helped significantly by tax incentives. Mainly coalbed methane and tight-sand gas currently account for almost half of the total US domestic production, helped by the decreasing production of conventional gas and higher natural gas prices. Just ten years earlier they represented only a quarter of the US domestic production. Despite this increase they remain more expensive to produce than still-available, conventional natural gas and are presently only of modest interest in the rest of the world [44].

4.5
Coalbed Methane

Coalbed methane, as its name indicates, is derived from coal, which acts in both the role of source rock and reservoir for methane. During the geological formation of coal under high temperature and pressure, volatile substances such as water and methane are liberated. Some of these gases may escape from the coal. However, if the pressure of the formation is sufficient, significant quantities of methane are retained in the pressurized coal matrix in an adsorbed form. Because of its large surface area, coal can store up to six or seven times more natural gas than the equivalent rock volume of a conventional reservoir. Gas content generally increases with the depth of burial of the coalbed and reservoir pressure. Fractures (or cleats as they are called in this case) that permeate the coalbeds are usually filled with water. The deeper the coalbed, the less water is present, but the more saline it becomes. To produce methane, wells are drilled through the coal and the pressure is reduced by pumping out water. This allows the methane to desorb and pass into the gaseous phase, so that it can be produced in a conventional manner and transported by pipeline. While economic recovery of methane is possible, water disposal from the exploitation of coalbeds raises environmental concerns. Coalbed accumulations are widespread and characterized by their large sizes. Because of the heterogeneous nature of coalbeds, the production rates are, however, highly variable even within a small area. The USGS estimates worldwide resources at up to 210 trillion cubic meters, but this number is uncertain because of the scarcity of basic data on coalbed resources and their gas content. The largest resources are located in regions rich in coal such as the United States, China, Russia, Australia, Germany and Canada. Numerous pilot projects are exploiting these resources, and many countries have active coalbed methane wells, but the United States is currently the only country where commercial production on a large scale is taking place. In 2006, production from this source represented about 10% of the methane production in the United States. Most of this coalbed methane originated from the Rocky Mountains region. Australia is presently planning a large-scale liquefaction facility to produce LNG from coalbed methane for export.

Coalbed methane is also of interest for providing the needed hydrogen for the chemical recycling of carbon dioxide formed by coal burning power plants to allow CO_2-emission mitigation.

4.6
Tight Sands and Shales

Tight sands and shales are low-permeability geological formations that sometimes contain large accumulations of natural gas. This means that the underground rock layers holding the gas are very dense, and thus the gas does not flow easily towards wells drilled to collect it, resulting in low recovery rates. This recovery can be enhanced by horizontal drilling and by fracturing the rocks with explosives, or with hydraulic pressure to provide pathways for the trapped gas to flow more easily to the well bores to be pumped to the surface. In the United States, exploitation of these sources accounted for more than 30% of domestic methane production in 2006, with the development of tight gas sands representing most of the production. Despite the fact that gas resources from tight sands and shales are immense there are many uncertainties about the cost of their production. In 2006, the US Energy Information Administration estimated the total technically recoverable natural gas from these sources in the United States to be around 12 trillion cubic meters, or about 20 years of natural gas needs.

4.7
Methane Hydrates

Another unconventional gas source that has attracted increasing attention in recent years is methane hydrates (Figure 4.14). Gas hydrates are naturally occurring crystalline, ice-like solids in which the gas molecules are trapped by water in cage-like structures called clathrates. Although many gases can form hydrates in nature, methane hydrate is by far the most common. Various clathrate compounds have been discovered and studied in laboratories, beginning in the 1800s. However, as no natural occurrences were known, the subject remained a purely academic

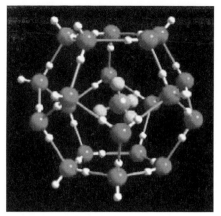

Figure 4.14 Methane hydrate. A molecule of methane is trapped
in a cage made out of water molecules (Source: NETL.)

curiosity until the 1930s, when solid hydrates were found to be plugging natural gas
pipelines. In the 1960s, "solid natural gas" or methane hydrate was observed as a
naturally occurring constituent in Siberian gas fields. Since then, gas hydrates have
been found worldwide in oceanic sediments of continental and insular slopes,
deepwater sediments of inland seas and lakes, and in polar sediments on both
continents and continental shelves of the seas (Figure 4.15). In permafrost Arctic
regions, gas hydrates can be present at depths ranging from 150 to 2000 m, but most

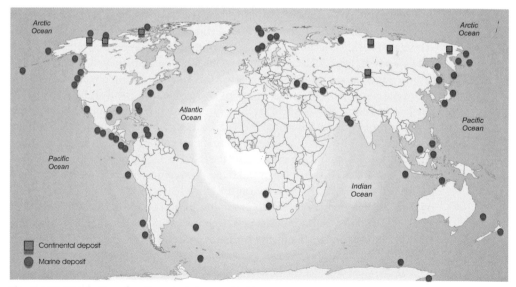

Figure 4.15 Distribution of methane hydrates, worldwide.
(Source: Lawrence Livermore National Laboratory.)

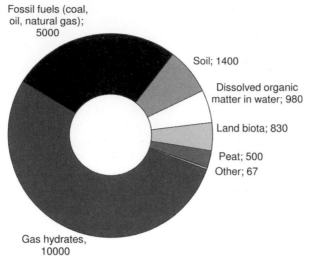

Figure 4.16 Estimated distribution of organic carbon in Earth reservoirs (Gigatons). (Source: USGS.)

occur in oceanic sediments hundreds of meters below the sea floor at water depths greater than 500 m.

Whilst the exact size of the world's methane hydrate deposits remains the subject of debate, largely due to the lack of field data, there is a consensus that the overall amount is huge. The estimated content of methane in methane hydrates, at the present time, is of the order of 21 000 trillion cubic meters (Figure 4.16). This represents more than 100 times our conventional proven gas reserves! However, in consideration of methane hydrates as a future energy source, the overall amount of methane hydrates is only a part of the question. As is the case for the development and production of oil from tar sands, large investments in research and development will be necessary to resolve significant technical issues before methane hydrates can be considered as an affordable methane source. Furthermore, the recoverable gas hydrate reserves are likely to represent only a small fraction of the gas hydrate resources because the deposits are too dispersed or the gas concentration is too low to be economically recovered. Regardless, the amounts of natural gas hydrates seems to be so enormous that even the exploitation of a small part of the resources could provide all of our energy needs for centuries to come. Japan and India – two countries with large energy needs but limited energy resources, as well as the United States and Canada – are pursuing methane hydrate recovery research programs. Japan in particular has already spent $300 million since 2001 in developing technologies to tap methane hydrates. By lowering the pressure in a methane-hydrate deposit about 1100 m below the Canadian permafrost, a Japanese–Canadian team managed to extract methane in a steady and constant flow for days for the first time in 2007. Using heat to release the gas is also effective but requires too much energy. For the time being, methane hydrates remain only a potential and still distant energy source.

4.8
Outlook

At current production rates, there are sufficient proven reserves to meet oil demand for a conservative low estimate of some 40 years, and gas demand for more than 60 years. With expected discoveries and technical improvements their availability will likely be extended by several decades. Reserves indeed have been steadily growing over time and are now higher than ever before in history. However, worldwide consumption is also increasing due to growing population and per-capita consumption. The rate of new oil and gas discoveries is now nearing the point where consumption exceeds new discoveries, that is, replacement – the so-called Hubbert's peak, named after the geologist M. King Hubbert. This means that we will start to use more fossil fuels (except coal) than the amount we can add to our reserves (Chapter 5). Unconventional sources of petroleum oil and natural gas, such as the tar sands, heavy oil, oil shales, coalbed methane or the future development of gas hydrates, will play an increasingly important role in extending our resources. Concerning the high abundance of coal, its future is not only a matter of resource availability or production costs but also one of environmental acceptability. New combustion and emission control technologies have to be introduced or improved to make cleaner use of coal. The greatest challenge, however, not only for coal but also for all carbon-based fossil fuels is the need for mitigation or elimination of CO_2 and other emissions responsible for a significant part of human activity-caused greenhouse gases and resulting climate changes. The ways to manage and prepare ourselves for the time beyond readily available oil and gas is also a challenge that is discussed in the subsequent chapters of this book, with particular emphasis on the suggested "Methanol Economy." We think that hydrocarbons have and should continue to fulfill a major and essential role in our lives. They should, however, be made environmentally CO_2 neutral and regenerative through their chemical recycling in a technical carbon cycle.

5
Diminishing Oil and Natural Gas Reserves

The world is greatly dependent on oil. The transportation sector in particular, so vital in our society for moving people, goods, food and materials, relies for more than 95% of its needs on gasoline, diesel oil and kerosene being derived from petroleum, and consumes about 60% of the oil produced. We also depend on oil for the large variety of petrochemicals and derived products, including plastics, synthetic fabrics and materials and varied chemicals, that are so ubiquitous in our daily lives that we hardly think about where they come from.

Oil is particularly important because it is the most versatile among our three primary fossil fuels. It has a high-energy content, is easy to transport and is relatively compact. Coal is heavier, more bulky and more polluting. Natural gas on the other hand is cleaner burning but bulky in volume and requires pipelines or expensive liquefaction for transportation, which raises safety concerns.

Unlike the case of coal, daily reports of oil as well as gas prices, reserves and production rates are highly publicized and have an immediate effect on our economies and lives. The outcome of every OPEC meeting is covered by the world media like a national or presidential election.

Chapter 4 discussed that proven oil reserves are estimated at between 1000 and 1200 billion barrels [22]. Dividing that figure by an annual production rate of 30 billion barrels suggests an availability of some 30–40 years. This number is called a reserve/production or R/P ratio. It is commonly used, especially in industrial reports issued by major oil companies, and can give a useful warning sign of resource exhaustion when the number starts to decline. Today, the R/P ratio is still actually higher than it has stood at any time in the past 50 years. This fact frequently leads to forecasts that anticipate a still long and comfortable era of acceptably priced and abundant oil, allowing a smooth transition to other energy sources. This optimistic vision of our oil future, supported by some economists, has been championed by those who believe in the crucial role of market prices and in continued improvements in exploration, drilling and production technologies to provide adequate oil production supplies for many years to come. Accordingly, it is assumed that although there is a finite amount of accessible petroleum oil on planet Earth this alone is not decisive because commercial oil extraction could cease long before its actual physical exhaustion. The point that matters is the cost to find and exploit new reserves.

Beyond Oil and Gas: The Methanol Economy, Second updated and enlarged edition
George A. Olah, Alain Goeppert, and G. K. Surya Prakash
Copyright © 2009 WILEY-VCH Verlag GmbH & Co. KGaA, Weinheim
ISBN: 978-3-527-32422-4

When this cost of exploration and exploitation becomes too high, then oil will be replaced by some other source of energy, leaving part of the oil left in the Earth's crust. The challenge is to find acceptable substitutes before oil becomes so expensive to produce that it would disrupt the economic and social fabric of our society. It is argued that similar transitions have taken place in the past when wood was replaced by coal, or coal by oil. Based on this argument, running low on oil *per se* will not have major direct relevance. This view, however, is not realistic. A more appropriate evaluation is the inevitable depletion and therefore ending of the era of relatively cheap, accessible oil. We are probably already entering into an irreversible decline in oil production following its peak. This assumption is based on consumption data compared with new oil discoveries, reserves and extraction data. In fact, the R/P ratio gives little information about the long-term fate of our resources. Furthermore, it assumes that production will remain constant over the years, which is improbable; so is the assumption that, up to the last barrels, oil can be pumped from the ground as easily and quickly as the oil coming out of the wells today. Globally, the demand for oil is expected to grow, with fluctuations, by some 1.3% per year over the next decades [66]. Occasional economic downturns or political events can periodically decrease demand but this should not affect the overall long-range picture. Three important parameters must be considered to project the future of oil production: (i) the cumulative production based on how much oil has been produced to date, (ii) the amount of recoverable reserves present in the known oil fields and (iii) a reasonable estimate of the oil that can still be discovered and extracted. The sum of these represents the ultimate recovery, which is the amount of oil that will have been extracted by the time that economically extractable oil production comes to an end. The current mean estimate by the United States Geological Survey (USGS) for ultimate oil recovery is 3 trillion barrels (3000 Gbbl) [67]. This estimate has, however, been questioned as unrealistically high by many geologists, who have set the ultimate recovery at about 2000 Gbbl (Figure 5.1). Naturally, the yet to be discovered – and therefore presently unknown – oil fields are the most speculative and controversial part of these estimates. However, the amount of oil is clearly finite and the question is not whether we will run out of readily available oil, but rather how soon.

In reality, the rate at which an oil well, field or even a country can produce oil rises to a maximum and then, after more than half of the oil reserve has been produced, declines gradually to unproductive levels. This trend was realized by M. King Hubbert, a noted American geophysicist who worked as a research scientist for USGS and Shell. He found that in any large region the extraction of a finite resource rises along a bell-shaped curve that peaks when about half of the resource has been exploited. This assumes that no outside regulatory, legislative or other major restraints have been placed on extraction and exploitation. Following this concept, Hubbert predicted in 1956 that oil production in the lower 48 states of the United States would reach a maximum sometime between 1965 and 1972 (Figure 5.2) [68]. US oil production indeed peaked in 1970, and has declined ever since. Larger oil fields are generally found and exploited first, giving a discovery peak that, in the lower 48 states of the US, occurred during the 1930s. The production peak then follows after a time lag (40 years in the case of the United States) depending on the amount

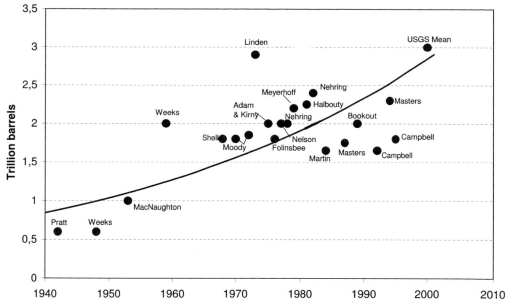

Figure 5.1 Predicted estimates of world ultimate oil recovery over time. (Source: EIA, USGS and Colin Campbell.)

of reserves in place and the rate of exploitation. A similar pattern of peak and decline has been observed in many other countries, including the former Soviet Union, the United Kingdom (North Sea) and more recently Norway. Oil production in the UK peaked in 1999 at 2.7 Mb per day. Since then it has been in sharp decline to only 1.9 Mb per day in 2005 and an expected 1.4 Mb per day in 2009 [69]. From a net

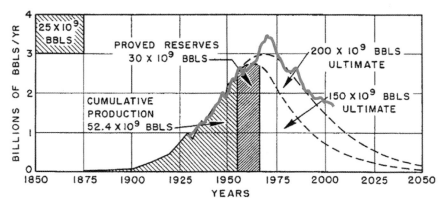

Figure 5.2 Hubbert's original 1956 graph showing crude oil production in the 48 lower States of the US based on assumed initial reserves of 150 and 200 billion barrels. The bold, superimposed line indicates actual oil production until 2004 following Hubbert's 200 billion barrels forecast.

exporter, the UK has now become a net oil importer in less than a decade. Other countries such as Mexico are close to the midpoint, whereas a few states, particularly in the Middle East – namely Saudi Arabia, the United Emirates, Iraq and Kuwait – are still at earlier stages towards depletion [34].

Hubbert, as well as several other analysts, including Campbell and Laherrère [45] and Deffeyes [70], also applied the method to global world oil production, to determine when the peak for world oil production, the so-called "Hubbert's peak," will occur. For their calculations they estimated ultimate world oil recovery at between 1900 and 2100 Gbbl. To date, about 1000 billion barrels of oil have been produced. Accordingly, we are close to the midpoint of the oil era, corresponding to half of the ultimate recovery. The estimated peak for world oil production will occur between 2005 and 2015 [71, 72]. Remarkably, these predictions do not shift significantly even if reserve and production estimates are off by some hundred billions of barrels [70]. After the peak, world oil production will start to decline. What clearly matters is not so much when our oil will be significantly depleted (not necessarily "gone") but when the demand will begin to surpass production. Beyond that point, prices will inevitably rise sharply unless demand permanently declines commensurably. This arguments leads to the conclusion that a permanent oil crisis is rather close and inevitable.

In the forecasts it was assumed that around 90% of the oil that can be recovered has already been discovered, putting the reserves at 900 billion barrels and the yet-to-find oil at only 150 billion barrels. This conclusion arises from the fact that, today, about three-quarters of the world's oil reserves are located in about 370 giant fields (each containing more than 500 million barrels of oil) that are relatively well studied [25]. As discoveries of these giant fields peaked in the 1960s, large additions to the known oil reserves are unlikely, unless new major oil fields in as-yet unexplored regions of the world are discovered [70]. However, as mentioned, mankind has already increasingly gone to the "end of the world," even under the most hostile climatic conditions in search of oil, and only a few areas, such as Antarctica or the South China Sea, remain to be explored. This reality leaves little room for the discovery of a new Middle East. Today, 80% of the produced oil flows from fields discovered before the early 1970s, and a large portion of those fields are already in declining production.

These rather pessimistic – though not unreasonable – forecasts, however, usually do not take into account several elements, in particular economics and better recovery factors. The oil recovery factor (which is the percentage of oil recoverable in a field) has increased roughly from 10 to 35% over the past few decades due to the introduction of new technological advances such as three-dimensional seismic surveys and directional drilling that are now applied in most oil-producing fields. With continuing progress, including so-called enhanced oil recovery (EOR) techniques, recovery factors from 40 to 50% are expected in the future, extending significantly the global oil reserves. Oil prices can also play an important role. Higher prices trigger, besides more economical use and savings, exploration to find new oil resources, the exploitation of fields considered previously non-economical at lower prices and the development of new extraction technologies. There is much more economically recoverable oil within the $30–$150 or higher per barrel price range than there was at $20, thereby adding to our reserves.

As discussed, besides conventional oil there are also many non-conventional oil sources, including heavy oils, tar sands and oil shales. These add significantly to petroleum oil sources as their exploitation is becoming profitable with increasing oil prices. These reserves, as discussed, are many. The Orinoco belt in Venezuela has been assessed to contain a whopping 1.2 trillion barrels of heavy oil, of which 270 billion barrels are thought to be economically recoverable [73]. The Athabaska and Cold Lake tar sand deposits in Canada may contain the equivalent of 300 billion barrels of economically recoverable oil [73]. Colorado oil shale resources are potentially very significant, as are other heavy oil sources. Non-conventional oil, because of its nature, is more difficult to extract than conventional oil. However, with technological innovations and massive investments, sources that were considered before as too expensive to exploit are becoming economically viable. In Canada, the operating costs to produce a barrel of oil from tar sands (by *in situ* recovery) fell from $22 to less than $10 between 1980 and 2003, making this non-conventional oil supply presently competitive with conventional oil [22]. It necessitates, however, very large quantities of natural gas or other energy sources such as atomic for the thermal recovery process, which limits this favorable picture. In the short term, the amount of oil extracted from the Athabaska region is expected to grow from 1 million barrels a day in 2004 to 2 million barrels a day within a decade. While proponents of "Hubbert's peak" theory acknowledge the very large amounts of non-conventional oil resources, they also think that industry would be hard-pressed for the energy, capital and time needed to extract non-conventional oil at a level to make up for the declining conventional oil production. In their view, the exploitation of non-conventional oil would have only a limited effect on the timing of the world oil production peak. Production of these substitutes is, as also mentioned earlier, highly energy intensive. For example, tar sands must presently be treated thermally, whether *in situ* or in treatment plants, to extract oil, using non-renewable, limited and valuable natural gas, and generating overall more CO_2 than the production of conventional oil. These factors must be seriously considered. Eventually, atomic energy and all other sources of alternative, non-fossil energies could be used to allow the exploitation of these heavy hydrocarbon sources [63, 87, 89].

Numerous predictions have been made in the past concerning the peak point of global oil production. Mankind has been said to be running out of oil repeatedly since the beginning of its use on an industrial scale. As early as 1874, in Pennsylvania, which was the largest oil producer at the time, the state geologist estimated that there was only enough oil to keep the kerosene lamps of the nation burning for four more years. In 1919, the US Geological Survey, using data based on R/P ratios, predicted that the "end of oil" would come within 10 years. In fact, the 1920s and 1930s saw the discoveries of the largest oil fields known at that time. Since then, many have prophesized the soon to come (generally within a few years or decades) end of oil. Among others, BP predicted in 1979 that the world production peak would occur in 1985 [76], while others forecasted that the peak would occur between 1996 and 2000.

A long history of too-often wrong predictions has led to general skepticism for new and generally pessimistic forecasts for oil's future [77].

While doomsday scenarios are usually given preferential publicity by the media (always ready to report bad news), there are also more optimistic views forwarded for future of oil production [78].

Using an estimated "ultimate" recovery for conventional oil of 3 trillion barrels, similar to the USGS estimate, Odell and Rosing, for example, predicted in 1984 that global oil production from conventional oil would peak only around 2025 [76]. The steadily increasing exploitation of non-conventional oil sources could push the peak further, as far as 2060, but at a cost clearly much beyond the period of relatively "cheap oil" that we currently still enjoy.

As we can see, predicting the future is not easy, as quoted by many, including the writer Mark Twain. What is certain, however, is that there is only a finite amount of accessible petroleum oil (and natural gas) on planet Earth. Combined oil production from conventional and unconventional sources will soon – and certainly not later then the next few decades – reach a maximum and then decline (Figure 5.3). It is thus imperative to start switching progressively away from oil-based fuels to alternatives.

This switch cannot be delayed for too long, so as not to get caught in a real crisis with very high oil and gas prices and no reasonable substitutes, resulting in serious economic and geopolitical consequences. In that sense, higher oil prices are not necessarily detrimental, as they flatten out or decrease excessive demand, encouraging at the same time savings and the development and transition to alternative

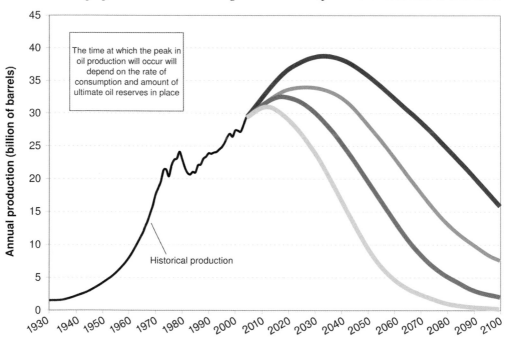

Figure 5.3 Hubbert's peak for world oil production. (Source: BP for historical values.)

fuel sources. Oil and gas are needed for essential services, fuels and the materials they provide us with, including heating, transportation and mobility. If other sources providing the same services are found to be competitive or even better, we will switch to them. In the past, in England and in most industrialized nations, coal replaced wood when the latter became increasingly scarce and therefore expensive, as well as being inferior to coal in its caloric value and convenience of use. During the twentieth century, oil took the place of coal in many uses, not only due to its lower cost but also because it was easier to transport, cleaner, more flexible and had a higher energy density. Oil was convenient as a transportation fuel, in household use, industrial production and in electricity generation.

Following the oil crises of the 1970s, the utilization of oil started to decrease in favor of natural gas, together with nuclear power for electricity generation as well as a renewed interest in coal. Currently, the largest oil-consuming sector in most industrialized countries is that of transportation, with a 95% reliance on oil. Therefore, a major reduction in oil consumption will have to come from this sector through more efficient internal combustion engines, the introduction of new technologies such as hybrid propulsion and fuel cells, or the use of alternative fuels. The production of synthetic liquid fuels (called syn-fuels) from coal was shown to be technically feasible during the 1930s. It was used on a significant industrial scale during World War II by Germany and later in South Africa during the Apartheid boycott era. Considering its large available coal reserves, similar operations have been studied by the United States as a response to the decreasing domestic oil production and increasing oil prices, particularly following the oil crises of the 1970s. These plans, however, were rapidly given up when oil prices stabilized in the mid-1980s. It was generally thought that only when the price of a barrel of oil rose above \$35–40, and remained at that level, would it become economically feasible to consider producing synthetic fuels.

Nevertheless, even if the production of liquid fuels from coal or natural gas is becoming economically viable on a large scale, it will necessitate vastly increased production of these non-renewable fossil fuels and still be very wasteful from an energy point of view. It would also generate large amounts of CO_2, a major greenhouse gas, SO_2 and other polluting gases, as well as solid waste. Natural gas liquefaction, as discussed earlier, is gaining increasing significance as a means of easier transportation from remote areas, and also as a source for the production of liquid hydrocarbon fuels and products. Synthetic Fischer–Tropsch chemistry first converts natural gas or coal into syn-gas and then into hydrocarbon fuels with large amounts of CO_2 produced as a by-product. It can also produce methanol and its derived products. The direct conversion of natural gas (methane) into liquid fuels, primarily through methanol, without first producing syn-gas, is therefore of great interest and promise (Chapter 12). Such processes, which are still under development, could result in the direct commercial production of methanol from still abundant methane. Methanol, as a liquid, can also be transported much more easily than methane. It is also a convenient fuel for internal combustion engines or fuel cells, and can be converted catalytically into ethylene and propylene, and through these into synthetic hydrocarbons and their products.

Presently, natural gas is considered in many ways as a successor to oil because it burns more cleanly, releases less CO_2 per energy unit than any other fossil fuel and there are still large reserves available. It is, however, more expensive to transport and store than oil, necessitating huge pipeline networks on land and liquefaction to LNG for overseas shipping. Furthermore, owing to its lower energy density its use as a transportation fuel in the form of compressed natural gas (CNG) has been generally limited to vehicles able to accommodate large pressurized tanks, such as buses and trucks. Natural gas is consequently employed mainly in stationary applications such as heating, cooking and electricity generation. Despite the dramatic increase in consumption, having more than doubled since 1970, our proven natural gas reserves are now three times larger than 30 years ago, with a R/P ratio above 60 years. Regarding the ultimate recoverable amount and the future of natural gas as a fuel, similar to petroleum oil, there are two major opposing points of view. One side claims that the amounts of natural gas to be ultimately recovered are only equal to or even less than those of petroleum oil. The other side asserts that there is still enough natural gas left, using new advanced recovery techniques and unconventional sources, to fill our growing needs for a long time. By adapting Hubbert's concept to predict the future of world gas production, Campbell and Laherrère – the most prominent proponents of this method – forecast that gas field discoveries and production would follow the same pattern as that of oil, and that global supplies of natural gas will decline not long after that of oil. Laherrère estimates the world ultimate natural gas reserves at $340\,Tm^3$, and conventional sources as representing $280\,Tm^3$. Unconventional sources such as coalbed methane, tight sand or shale gas are estimated to be around $70\,Tm^3$. Based on these estimates, if the consumption rate continues to increase at the current pace, world gas production would peak around the 2030 [79]. Hubbertians therefore dismiss gas as a viable long-term alternative to oil to fulfill our future energy needs. However, new discoveries and production methods as well as economic conditions may extend the availability of natural gas. As in the case of oil, there are other more optimistic estimates. The USGS assessed the global conventional natural gas resources as over $430\,Tm^3$, the energy equivalent of almost 2600 billion barrels or 345 Gt of oil [67]. Other estimates for ultimately recoverable conventional natural gas range between 380 and 490 Gt oil equivalent [80]. Consequently, Odell forecasts a conventional natural gas production peak at around 2050 [81, 82]. Taking also into account unconventional natural gas sources, he predicts the extraction peak for combined conventional and unconventional natural gas by 2090. As the recoverable resources for unconventional natural gas are known to an even lesser extent than those for conventional natural gas, this prediction is at best speculative. Some unconventional gas sources such as coalbed methane and tight gas are already exploited on a large scale, principally in the United States. There is also a very large potential for the utilization of methane hydrates, which are considered to be present in staggering amounts under the sea floor and arctic permafrost. First, however, these must be assessed more accurately. Further, their practical exploitation must be demonstrated by the development of new and effective technologies. Until then, methane hydrates remain a promising but as-yet uncertain energy source.

While considering unconventional sources of natural gas, a further interesting but as yet unproven theory that was suggested and vigorously defended by Thomas Gold (a noted late astrophysicist) should also be discussed. This involves the possible availability of large natural gas resources of abiological or abiotic origin at greater depths in the Earth's crust (abiological deep methane) [35]. According to this suggestion, extraterrestrial carbon of asteroids could have combined, under high temperature and pressure deep under the Earth's surface, with hydrogen to form hydrocarbons. The most stable of them, methane, would then migrate upwards and accumulate in suitable geological formations. Following this suggestion, at least part of natural gas sources could be of non-biological origin. For the time being, however, abiological natural gas remains controversial. Observations that some vents deep on the bottom of the oceans discharge methane were later recognized to be due to the discharge of hydrogen sulfide which then converts the CO_2 of sea water into methane under the influence of microorganisms that are effective even at these depths and despite the absence of light as an energy source. Regardless, abiological methane is a real possibility. It may also explain the recently observed presence of methane in the atmosphere of Mars and on Saturn's moon Titan, which in all likelihood may not have a biological origin [83, 84].

Based on our present knowledge, natural gas must be considered as a finite resource, the production of which will, like oil, reach a peak, most probably during the latter part of the twenty-first century. As in the case of petroleum oil, new solutions must therefore be found to replace progressively declining natural gas reserves used for generating electricity, as a transportation and household fuel and as a hydrocarbon material source. CO_2-based methanol and DME are feasible and represent possible substitutes.

6
The Continuing Need for Carbon Fuels, Hydrocarbons and their Products

Besides still providing the bulk of our energy needs, fossil fuels are also the sources for our hydrocarbon fuels and derived products. Hydrocarbons, the compounds of carbon and hydrogen, contain these elements in various ratios. In methane (CH_4), the simplest saturated hydrocarbon (alkane) and the main component of natural gas, a single carbon atom is bonded to four hydrogen atoms. The higher homologues of methane – ethane, propane, butane and so on – have the general formula C_nH_{2n+2}, displaying the tendency of carbon to form chains involving C–C bonds. These can be either straight-chained or branched. Carbon can also form multiple bonds with other carbon atoms, resulting in unsaturated hydrocarbons with double (C=C) or triple (C≡C) bonds. Carbon atoms can also form rings. Cyclic ring compounds of carbons involving both saturated and unsaturated systems are abundant and involve aromatic hydrocarbons, a class of hydrocarbons of which benzene is the parent (Figure 6.1).

All fossil fuels, that is, natural gas, petroleum oil and coal, are basically hydrocarbons, but they deviate significantly in their hydrogen-to-carbon ratio and composition. Natural gas, depending on its origin, contains besides methane (usually in concentrations above 80–90%) some of the higher homologous alkanes (ethane, propane, butane). In "wet" natural gases, the amount of C_2–C_6 alkanes is more significant. These so-called natural gas liquids, which were generally only considered for their thermal value, are increasingly perceived as feedstocks for more valuable products such as gasoline. Methane itself, though used mainly as a fuel, is also today's primary source of hydrogen and can be transformed (albeit at a considerable energy cost, via syn-gas) into products otherwise obtained from petroleum. Petroleum or crude oil is a remarkably complex substance, both in its composition and varied uses. Depending on the source, its color can range from almost transparent clear to amber, brown, black or even green. It may flow like water or be a semisolid viscous liquid. Crude oil contains hundreds, if not thousands, of individual kinds of hydrocarbons but is predominantly constituted of saturated straight-chain compounds (alkanes) and smaller amounts of branched alkanes, cycloalkanes and aromatics. Petroleum is the most versatile of our three primary fossil fuels. It can be transformed economically and easily into a vast palette of useful products. Coal, in contrast, is a more hydrogen-deficient solid material containing large, complex hydrocarbon systems

Beyond Oil and Gas: The Methanol Economy, Second updated and enlarged edition
George A. Olah, Alain Goeppert, and G. K. Surya Prakash
Copyright © 2009 WILEY-VCH Verlag GmbH & Co. KGaA, Weinheim
ISBN: 978-3-527-32422-4

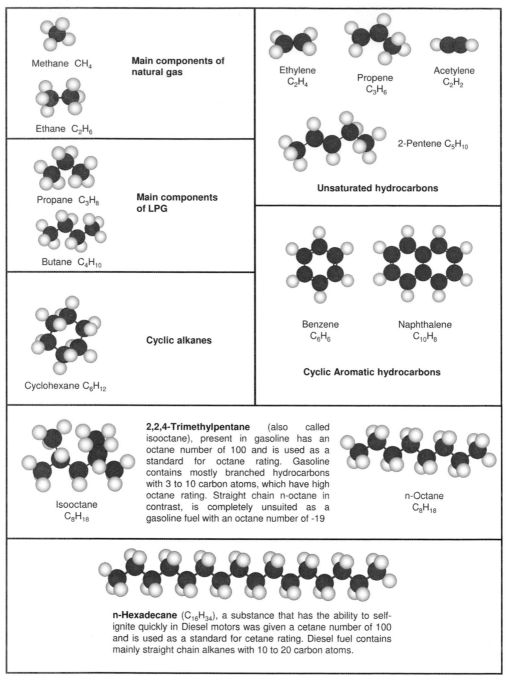

Methane CH$_4$

Main components of natural gas

Ethane C$_2$H$_6$

Propane C$_3$H$_8$

Main components of LPG

Butane C$_4$H$_{10}$

Cyclohexane C$_6$H$_{12}$

Cyclic alkanes

Ethylene C$_2$H$_4$

Propene C$_3$H$_6$

Acetylene C$_2$H$_2$

2-Pentene C$_5$H$_{10}$

Unsaturated hydrocarbons

Benzene C$_6$H$_6$

Naphthalene C$_{10}$H$_8$

Cyclic Aromatic hydrocarbons

2,2,4-Trimethylpentane (also called isooctane), present in gasoline has an octane number of 100 and is used as a standard for octane rating. Gasoline contains mostly branched hydrocarbons with 3 to 10 carbon atoms, which have high octane rating. Straight chain n-octane in contrast, is completely unsuited as a gasoline fuel with an octane number of -19

Isooctane C$_8$H$_{18}$

n-Octane C$_8$H$_{18}$

n-Hexadecane (C$_{16}$H$_{34}$), a substance that has the ability to self-ignite quickly in Diesel motors was given a cetane number of 100 and is used as a standard for cetane rating. Diesel fuel contains mainly straight chain alkanes with 10 to 20 carbon atoms.

Figure 6.1 Examples of hydrocarbons.

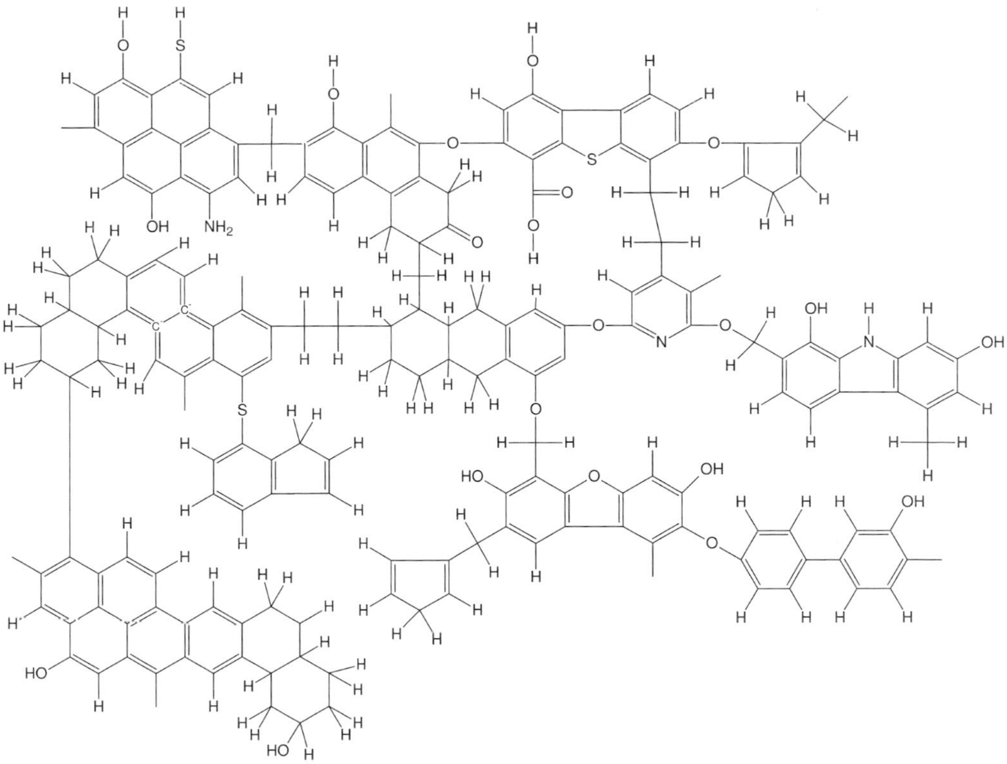

Figure 6.2 Schematic representation of structural groups and connecting bridges in bituminous coal.

composed mainly from aromatic cycles (Figure 6.2). It would probably be more appropriate to use the plural naming "coals" instead of coal as it is not a single material with a well-defined structure. Coal, besides its combustion to provide energy, can be transformed into liquid fuels as well as other petrochemicals (coal liquefaction). This route is feasible and has been used on a large scale. It involves chemical oxidation, cleavage, hydrogenation and so on – processes that are usually energy intensive. Coal is therefore mainly used as a fuel for energy production and heating.

In all hydrocarbons, chemical energy is stored in the C–H and C–C bonds. When hydrocarbons react with oxygen in the air in a combustion reaction they form CO_2 and water. At the same time, energy in the form of heat is released because there is more chemical energy stored in the hydrocarbons and oxygen than in the resulting CO_2 and water. This energy difference is the basis for the utilization of fossil fuels as a source for energy. The burning of hydrocarbons for energy generation is their main use today. A great variety of petrochemicals and chemical products derived from fossil fuels are also produced for numerous other applications.

Our increasingly technological society is today especially dependent on oil because of its versatility. Oil and its derived products are so embedded in our daily lives that we hardly recognize their essential role. As eluded to in earlier chapters, though, humankind's use of petroleum is as old as recorded history. Ancient cultures such as the Sumerians and Mesopotamians used asphalt and bitumen from natural pools to seal joints in wooden boats, line water canals and inlay mosaics in walls and floors. At that time, liquid oil was also the fuel of choice for oil lamps. The Egyptians embalmed mummies with asphalt, while the Romans used ignited containers filled with oil as weapons. Native Americans used crude oil for medicinal ointments. The modern petroleum industry, however, was only born in the middle of the nineteenth century. In America, it was prompted by the use of the kerosene lamp, leading to the formation of the first oil companies in Pennsylvania. Commercial production was first aimed at fulfilling the growing demand for kerosene. At that time, the lighter gasoline was mainly a wasted by-product of the distillation of kerosene from crude oil, until the early 1900s when automobiles with gasoline and diesel engines became commonplace. Similarly, farm equipment soon also became popular, dramatically increasing agricultural productivity. During the 1930s and 1940s, a substantial market for heating oil developed. Today, we are utterly dependent on oil products that we can find in every area of our lives. The most common are the gasoline used to fuel our automobiles, and the heating oil to warm our homes and offices. Gasoline, diesel and jet fuel provide more than 95% of all the energy consumed in the transportation field by automobiles, trucks, farm and industrial machinery, trains, ships and aircraft. Transportation fuels alone account for more than 60% of the petroleum consumed worldwide [42]. Oil is also an essential raw material (together with natural gas) for the synthesis of fertilizers on which agriculture depends, and for all the chemicals, dyes, cosmetics, fibers, pharmaceuticals, plastics and an innumerable host of other products that are essential for everyday life. As an indicator of our dependence, and probably overdependence, on petroleum products we can take the example of the United States, which uses on average more than 20 million barrels of oil per day, representing about 11 L, close to 3 gallons, per day for each person in the country.

Crude oil in its raw state has limited uses, and its processing via refining is first necessary to unlock its full potential. The earliest refineries simply used distillation to produce mainly kerosene. Distillation remains the initial processing step for oil refining to this day, though many more complex processes such as cracking, reforming, alkylation steps and others have been added to convert crude oil into a wide array of desired products.

6.1
Fractional Distillation

Atmospheric fractional distillation, the first and core step in the refining process, uses heat to separate the numerous different hydrocarbon components present in oil into several fractions, depending on their boiling range. The first volatile products of the crude oil are gaseous hydrocarbons containing one to four carbon atoms that are dissolved in the oil. These liquid petroleum gases (LPG) can be used as fuel or converted into useful petrochemicals. With increasingly higher boiling range and number of carbon atoms contained, the next fractions are gasoline, naphtha, jet fuel, kerosene, gas oil and heating oil. Finally, the heaviest products, with boiling temperatures above 600 °C, known as residual oil, can be separated into such individual constituents as coke, asphalt, tars and waxes. The heavier products can be further processed by cracking to produce lighter fractions and maximize the output of the most desirable products such as gasoline and diesel fuel. After atmospheric distillation, the relative amounts of fractions obtained from crude oil, however, do not coincide with the commercial needs. Therefore additional "downstream" processing is needed. In general, these processes are designed to increase the yield of lighter, higher-value products such as gasoline. Downstream operations include vacuum distillation, cracking, reforming, alkylation, isomerization and oligomerization.

6.2
Thermal Cracking

Thermal cracking, the first downstream process that changed the petroleum industry, permitted, by the use of higher temperature and pressure, the heavy, low-value part of the feedstock to be broken into lighter, higher-value heating oil, diesel and gasoline. Thermal cracking was followed by other developments in the 1920s and 1930s. Polymerization (oligomerization) yields high-octane gasoline from unsaturated olefins produced as by-products in thermal cracking units. Vacuum distillation takes the residual oil, which is left at the bottom of the column after atmospheric distillation, and allows its further separation. By heating, the visbreaking unit can reduce the viscosity of the residues from the distillation columns, allowing much easier flow and processing. Coking further uses the products left from atmospheric and vacuum distillation and produces, via thermal treatment at high temperature, gasoline, heavy oil, fuel gas and petroleum coke, an almost pure carbon residue.

During World War II the petroleum industry shifted to products that were essential for the war effort, and especially advanced high-octane aviation fuels. This resulted in the development of the alkylation process in which a catalyst (usually sulfuric acid or hydrofluoric acid) is used to combine a branched saturated alkane with an unsaturated olefin (alkene) to produce high-octane compounds in the gasoline range. Nowadays, this process is one of the most important steps in the production of high-octane gasoline for motor cars. Other advances during that period included catalytic cracking, in which catalysts are used to accelerate the cracking process, and isomerization, converting straight-chain alkanes into branched ones having a much higher octane number. Catalytic reforming produces higher-octane components for gasoline from lower-octane naphtha feedstock recovered in the distillation process.

Over time, these processes have been continuously improved and new ones, such as dehydrogenation (to produce useful alkenes from alkanes) and hydrocracking, have been added. Without these processes it would be impossible to produce economically the large amounts of valuable lighter fractions from the intermediate and heavy compounds that constitute most crude oils.

Figure 6.3 BP Grangemouth refinery, UK. ((© BP p.l.c.).)

Crude oil, as mentioned, is a complex mixture of hydrocarbons. Depending on its source, it varies in (among other properties) color, viscosity and content of sulfur, nitrogen and other impurities. Most commonly, crude oils are classified by their density and sulfur content. Less dense, or lighter, crude oils have a higher share of the more valuable lighter hydrocarbons that can be recovered by simple distillation. Denser or heavier crude oils contain more heavy hydrocarbons of lower value and require additional processing steps to produce the desired range of products. Some crude oils also contain significant amounts of sulfur and heavy metals, which are detrimental as they act as contaminants for most refining processes and finished products. They are also pollutants, necessitating additional purification steps. Because the quality of oil varies so widely, refineries differ in their complexity according to the type of crude oil to be processed as well as the range of products desired. To allow the most flexibility, modern refineries (Figure 6.3) are usually designed to process various blends of different crude oils. Many refineries, however, are not suited to handle heavy oils, limiting for example the processing of Venezuelan or Iranian oil.

The processing of crude oil opened up the route to petrochemicals, because cracking also produces, beside fuels, unsaturated hydrocarbons containing one or more C=C double bonds, in particular ethylene, propylene, butylene and butadiene. These are called olefins or alkenes and, unlike paraffins (alkanes, the main saturated components of petroleum oil), can be readily used and further transformed by chemical reactions. They constitute the basic building blocks for numerous products and are produced in very large quantities. Yearly, some 100 million tons of ethylene and 60 million tons of propylene are manufactured worldwide. They are used mainly for the production of synthetic polymers, and also for many other products. Ethylene, for example, is the starting material for polyethylene, propylene for polypropylene, and synthetic rubber can be produced from butadiene. Besides olefins, aromatic compounds – principally benzene, toluene and xylenes – are also obtained during crude oil refining. These aromatic compounds are important starting materials for synthetic products such as polystyrene, nylon, polyurethane and polyesters. In total, about 6% of crude oil is used today to produce petrochemicals. Crude oil together with natural gas are the sources for some 95% of organic chemicals, yielding products such as lubricants, detergents, solvents, waxes, rubbers, insulation materials, insecticides, herbicides, synthetic fibers for clothing, plastics, fertilizers and many others. The advance of chemistry in the twentieth century has depended – and still depends to a large extent today – on the ready availability of petrochemical building blocks.

The history of petrochemistry started around the 1900s, at which time the demand for natural rubber collected from hevea trees began to surpass the supply when new applications such as motor car tires were introduced. Replacement materials were needed, and this led to the invention of synthetic rubbers; the process began with the polymerization of butadiene, which turned out to be superior to the natural products.

In 1907, the first fully synthetic plastic, named "Bakelite," was created by the reaction of phenol and formaldehyde. This new liquid resin, when hardened, took the shape of the vessel in which it was formed. Unlike earlier plastics such as celluloid, it could not be remelted, it retained its shape under any circumstance and it would not readily burn, melt or decompose in common acids or solvents. Bakelite is still used today as an electric insulator. During the next decade, cellophane, the first clear, flexible and waterproof packaging material, was developed. The 1920s and 1930s witnessed the introduction of petrochemical solvents and the discoveries of numerous new plastics and polymers, including nylon, poly(vinyl chloride) (PVC), Teflon, polyesters and polyethylene. The petrochemical industry grew especially rapidly during the 1940s when, during World War II, the demand for synthetic materials to replace costly and often difficult to obtain, less-efficient natural products led the industry to develop into what would become a major factor in today's technological society. During that time, many other synthetic materials such as acrylics, neoprene, styrene-butadiene rubber (SBR) and others came into use, taking the place of dwindling natural material supplies. Among other applications, nylon was used to make parachutes and to reinforce tires, besides its use for synthetic fibers, especially for nylon stockings. Plexiglas was initially introduced during World War II for airplane windows. Lightweight polyethylene insulation made it possible to mount otherwise too-heavy radar units on airplanes. From then on, petrochemical products, and especially polymers, moved into an astonishing variety of areas. Together with oil- and natural gas-based fuels they touch our daily lives in countless ways. In fact, we are so used to them that we no longer notice their unique nature! Notably, simple olefins, including ethylene, propylene and butylenes, can be efficiently and readily made from methanol and used to produce practically any hydrocarbon and their products, including polymers.

Today, our households are full of products derived from hydrocarbons. In the bathroom, shampoo and shower gel are composed of synthetic soap formulations and their bottles are made out of polyethylene, polypropylene or PVC, which have the advantage of being unbreakable. The toothbrushes, hair dryers, combs, shower curtains and toilet brushes are all made of plastics. In the kitchen, the refrigerators, coffee machines, toasters, microwaves and other appliances are all composed in part

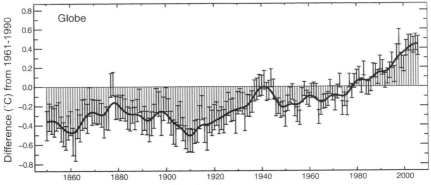

Figure 7.1 Variations of Earth's annual land-surface air temperature for 1850 to 2006 relative to the 1961 to 1990 mean. The smooth curve shows decadal variations.(Source: IPCC, Fourth Assessment Report, *Climate Change 2007: The Physical Scientific Basis.*)

temperature that are still present and measurable today, such as tree rings, ice cores or corals. The collected data, even if less accurate than direct measurements, indicates that the increase in temperature in the twentieth and beginning of the twenty-first century was probably the largest in the past 1000 years. Owing to its shape, this temperature plot is generally known as the "hockey stick" (Figure 7.2). Recently, this widely accepted hockey stick plot has been challenged by several

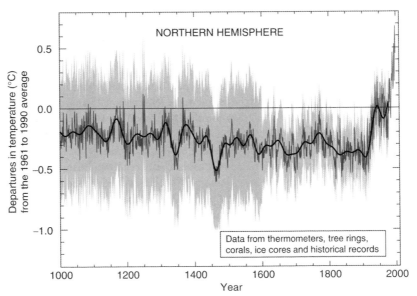

Figure 7.2 Variations of Earth's surface temperature for the last 1000 years.(Source: IPCC, Third Assessment Report, *Climate Change 2001: The Scientific Basis.*)

scientists citing flawed data and the analysis of data using sophisticated statistical processing methods [86]. The interpretation of data collected from proxy to determine the temperature before the 1700s has been especially criticized. Nevertheless, it is undeniable that recent years have been indicative of a warming trend [87]. As a consequence of this trend, there has been a widespread retreat of mountain glaciers in the non-polar regions as well as a 20–25% decrease in the sea-ice extent during spring and summer in the Arctic since the 1950s. The global average sea level rose between 0.1 and 0.2 m during the twentieth century, and the global ocean heat content has increased since the 1950s. However, some parts of the globe, mainly in the Southern hemisphere, have not warmed in recent decades and no warming trends in the sea-ice extent of Antarctica are apparent since the end of the 1970s [85]. Furthermore, no significant trends or changes in storm activity, frequency of tornadoes, thunder or hail were noticed over the twentieth century. It should also be recognized that long before human activity on Earth there were many ice-age periods followed by warming. Thus, human activity-caused climate change, must be considered as superimposed on that caused by natural cycles. We can not affect nature's cycles, such as the degree of alignment of the Earth's axis relative to the Sun or variations in the intensity of the Sun's activity (Sun spots), but should do everything possible to mitigate human activity that causes adverse environmental effects. This mitigation should be carried out whenever possible using reasonable energy conservation, new technologies, reducing CO_2 emissions and so on and not on unrealistic regulations purely based on politico-economic approaches.

Considering the variations from the past, the question that must be raised is, can we predict climate changes for the future? First, the reasons for past warming periods must be explained. Global warming is now recognized as being based significantly on the greenhouse effect caused by heat-absorbing gases that are present in the atmosphere; these trap some of Earth's reflected infrared radiation of the Sun and act like a giant blanket around our planet. These so-called "greenhouse gases" include water vapor, CO_2, methane, nitrous oxide, ozone and some others. More recently, it was established that man-made chlorofluorocarbons (CFCs) contributed to the depletion of the ozone layer, which protects the Earth from excessive, damaging ultraviolet (UV) radiations from the sun. The depletion was most pronounced in polar regions, where holes were detected in the Earth's protective ozone layer. Without naturally occurring greenhouse gases such as CO_2, water vapor and methane in the atmosphere, the Earth's average temperature would be much cooler, comparable to the atmosphere on Mars. At the expected $-18\,°C$ on average, most of the water would be frozen all year long and the emergence and evolution of life as we know it would have been much more difficult, if possible at all. However, too much of the greenhouse effect is also detrimental. Such a situation is encountered on Venus, which has an atmosphere rich in CO_2, where temperatures reaching above the melting point of metallic lead make it as hostile to life as the cold Mars. The humankind-caused increased greenhouse effect is real, and of concern. It is important to be concerned about the greenhouse gas concentrations in our atmosphere in order to safeguard the temperature of the Earth from adverse human effects and to maintain life as we know it. If the concentration of CO_2 or other greenhouse

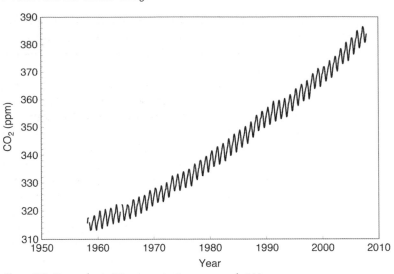

Figure 7.3 Atmospheric CO_2 concentration measured at Mauna
Loa, Hawaii from 1958 to 2008. (Source: CDIAC, Carbon
Dioxide Information Analysis Center.)

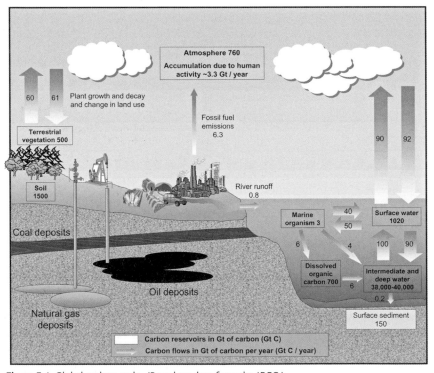

Figure 7.4 Global carbon cycle. (Based on data from the IPCC.)

gases in the atmosphere continues to increase substantially, the effect would most certainly lead to further increases in average global temperature. As early as 1895, the Swedish chemist Svante Arrhenius calculated that a doubling in CO_2 concentration, through human action, would result in an increase of 5–6 °C in the Earth's average surface temperature [15]. He even estimated the amount of coal that would have to be burned, and how long such a process would take. Of course since then we have also burned large amounts of other carbon-based fossil fuels, such as oil and gas. In fact, CO_2 concentrations in the atmosphere have been increasing steadily for more than a century. During several thousand years before the industrial era, the CO_2 concentration was relatively stable around 270 ppm, but since 1750 the atmospheric concentration of CO_2 has increased by 36% to reach some 380 ppm today (Figures 7.3 and 7.4). The present CO_2 concentration has not been exceeded in the past 420 000 years, and the current rate of increase is unprecedented in at least the past 20 000 years. It is now generally accepted that the observed increase in CO_2 concentration is significantly due to anthropogenic effects that is, due to human activities and is an effect superimposed on nature's own cyclic climate changes. The combustion of fossil fuels is by far the largest contributor to man-made anthropogenic CO_2 emissions, with the remainder being mainly a result of land-use change, especially deforestation (Figures 7.5 and 7.6). Whereas the US was for a long time the largest CO_2 emitting country, China was reported to have exceeded it in 2008. About half of

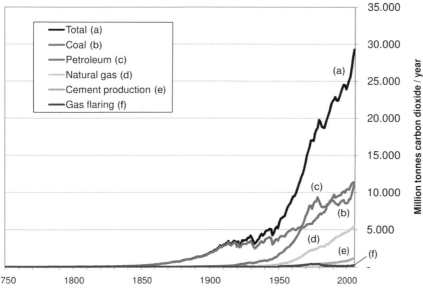

Figure 7.5 Global CO_2 emissions from fossil fuel burning, cement production and gas flaring for the period 1750 until 2005. (Source: G. Marland, T.A. Boden and R.J. Andres. 2008. Global, regional, and national CO_2 emissions, in *Trends: A Compendium of Data on Global Change*, Carbon Dioxide Information Analysis Center, Oak Ridge National Laboratory, US Department of Energy, Oak Ridge, Tenn., USA.)

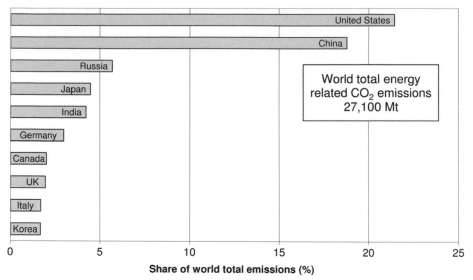

United States

China

Russia

Japan

India

Germany

Canada

UK

Italy

Korea

World total energy
related CO_2 emissions
27,100 Mt

0   5   10   15   20   25

Share of world total emissions (%)

Figure 7.6 World's largest energy related CO_2 emitters in 2005. (Source: IEA.)

the anthropogenic CO_2 emissions is absorbed again by the oceans and vegetation of land areas, whereas the remainder is added to the atmosphere, increasing the CO_2 concentration. It should be reiterated that the overall concentration of CO_2 in the air is only about 0.038% (380 ppm), but this plays an essential role in maintaining life on Earth. Carbon dioxide emissions from the harmful burning of fossil fuels by humans is also a relatively short-term problem as they will last at most a few centuries, representing only a fleeting instant on the geological timescale, and forcing us to find other energy and carbon sources.

So far, most of the concerns about greenhouse gases have focused on CO_2, which represents about 63% of the human-caused greenhouse gases present in the atmosphere (Figures 7.7) [61]. Less attention has been given to other greenhouse gases. Atmospheric water vapor is the most abundant of the greenhouse gases; the concentrations of others have also been increasing, especially methane, nitrous oxide and, more recently, synthetic chlorofluorocarbons and other halocarbons. Atmospheric methane concentrations have risen by 150% since 1750, and continue to increase, being now at the highest in the past 650 000 years. Methane has both natural and human-related origins. The decay of plant matter in wetlands and rice paddies accounts for some three-quarters of natural methane emissions. Other sources include oceans, gas hydrates releasing methane due to changes in temperature or pressure, and termites, which produce methane as a part of their normal digestive process. Similarly, in the digestive system of domesticated animals, such as cattle, buffalo, sheep, goats and camels, cellulose from plant material is broken down by microbial fermentation in their digestion process, producing significant amounts of methane as a by-product. This constitutes a major source of human-related methane emissions besides wet rice agriculture, oil and gas production, and landfills. Methane

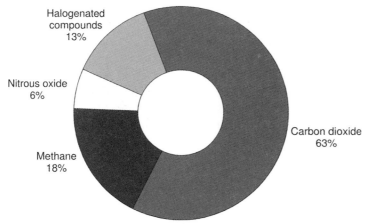

Halogenated
compounds
13%

Nitrous oxide
6%

Carbon dioxide
63%

Methane
18%

Figure 7.7 Relative contributions of greenhouse gases to the
increased greenhouse effect induced by human activity. (source:
IPCC Fourth Assessment Report. *Climate Change 2007: The
Physical Science Basis*, Table 2.1. Total forcing change between
1750 and 2005 is 2.63 W m^{-2}.)

generated in landfills, as waste decomposes under anaerobic (oxygen-lacking)
conditions, is increasingly used as an energy source and is not allowed to escape
into the atmosphere. In total, about half of the current methane emissions are
anthropogenic. Even with a concentration increase of only 1060 ppb in the atmosphere (much less than that of CO_2), methane contributes to 18% of the increased
greenhouse effect [61]. Methane is a more effective greenhouse gas because it has a
higher global warming potential (GWP) (Table 7.1). The GWP is a measure of the

Table 7.1 Global warming potentials (GWPs) of some greenhouse gases.

Gas	Formula	Global warming potential[a]	Atmospheric lifetime (years)
Carbon dioxide	CO_2	1	
Methane	CH_4	23	12
Nitrous oxide	N_2O	296	114
Hydrofluorocarbons (HFCs)		12–12 000	0.3–260
Examples:			
HFC-23	CHF_3	12 000	260
HFC-32	CH_2F_2	550	5
HFC-134a	CH_2FCF_3	1300	14
Fully fluorinated species		5700–22 200	2600–50 000
Examples:			
Perfluoromethane	CF_4	5700	50 000
Perfluoroethane	C_2F_6	11 900	10 000
Sulfur hexafluoride	SF_6	22 200	3200

[a]Over a 100 year time horizon.
[b]Based on data from IPCC, Third Assessment Report, 2001 [60].

relative heat absorption capability of a given substance compared to CO_2 and integrated over a chosen time. Over a 100-year period, methane has a 23-fold higher GWP than CO_2.

The atmospheric concentration of nitrous oxide (so-called "laughing gas," N_2O) has also increased by about 16% during the industrial era to levels not seen in the past thousand years. Because of its high GWP factor of 296, N_2O contributes to 6% of the increased greenhouse effect [61].

Chlorofluorocarbons (CFCs) destroy the Earth's ozone layer and also absorb infrared radiation. The latter contributes to global warming. Owing to their ozone-depleting properties, CFCs have been phased out by the Montreal Protocol, and their concentrations in the atmosphere are now diminishing, or at least increasing more slowly. Substitutes for CFCs may be less harmful for the ozone layer, but some compounds such as hydrochlorofluorocarbons (HCFC) and hydro-fluorocarbons (HFC) still have GWP factors reaching up to 12 000. The concentration of HCFC and HFC in the atmosphere is increasing. All halocarbons combined represent today some 13% of the increased greenhouse effect [61].

Atmospheric aerosols are small airborne particles and droplets produced by various processes that can be natural (such as volcanic eruptions or sand dusts) or anthropogenic (mainly through fossil fuel and biomass burning). Sulfate aerosols from fossil-fuel burning and other aerosols from volcanoes and biomass burning reflect and scatter solar energy before it can reach the Earth's surface, and thus have a cooling effect on the atmosphere (global cooling). Aerosols can also enhance the condensation of water droplets and consequently favor cloud formation, increasing the reflection of incoming sunlight back into space. Whereas aerosols are cooling the atmosphere, the magnitude of this effect does not compensate for the heating induced by greenhouse gases. Furthermore, aerosols have a much shorter lifetime than greenhouse gases in the atmosphere.

When considering the question of global temperature change, it must be remembered that the greenhouse gas of overwhelming concentration is the moisture in the air, and that we have only extremely limited control over it. Nature, therefore, has its own major controlling effect on the climate, including the cycle of celestial alignment of the Earth relative to the Sun [86, 88]. Although temperature and the CO_2 content of the air are clearly related, some point out that nature's own warming cycles themselves would cause increasing CO_2 levels, and not necessarily in the reverse sequence [89].

By using the vast amount of data collected about greenhouse gases and other elements able to influence the Earth's atmosphere, computer simulations have been developed to investigate the causes of climate changes and, most importantly, the presently observed warming trend of our planet. These models have become more sophisticated over the years, but some aspects can still not be simulated effectively due to the extreme complexity of climate systems. Nevertheless, they now provide a reasonably accurate model of climate variations that have occurred during the past 150 years. Over the past 50 years, the rate and increase of warming seems to be directly correlated to the increase in greenhouse gas concentration in the atmosphere. This led the IPCC to state in its fourth assessment report that:

> Most of the global average warming over the past 50 years is *very likely* due to anthropogenic GHG increases . . . . Anthropogenic warming over the last three decades has *likely* had a discernible influence at the global scale on observed changes in many physical and biological systems. [90].

Whereas there is no question that the human activities of a growing population on Earth will continue to affect our climate, the degree to which the human effect is superimposed on Nature's own cycles remains to be questioned. By continuing to emit greenhouse gases, humanity will continue to influence the atmospheric composition into the twenty-first century and eventually even further. Carbon dioxide emissions from fossil fuel combustion especially are expected to remain a major factor for coming trends in global warming. However, as our fossil-fuel reserves are being steadily depleted, this trend will not go on indefinitely as fossil fuel, even in the case of coal, will not last for much longer than a century or two.

Indeed, although peaks in oil and natural gas production have received limited attention in the climate change debate they will have important implications on future CO_2 emissions [91]. Based on mathematical models similar to the ones used to describe the past, projections for future climate changes have been made. Of course they are highly dependent upon assumptions such as future greenhouse gas emissions, and therefore numerous potential scenarios can be envisioned. Nevertheless, they project a global average temperature increase, from 1.1–6.4 °C, over the remainder of this century, with atmospheric CO_2 concentrations ranging from 450 to more than 1000 ppm. Realistically, however, a rise of 3 °C or less is more likely than a 6 °C increase. Nonetheless, even such a temperature change will have widespread repercussions on the Earth. Initially, it would affect the sea level, expected to rise up to 0.6 m during the next century, primarily due to thermal expansion and water added from melting glaciers and ice caps [85]. Even if the overall greenhouse gas concentrations were to be stabilized, the local warming of Greenland, for example, if sustained, could lead to a further meltdown of the ice sheet with a resulting increase of sea level. This would certainly be very bad news for low-lying islands and countries such as Bangladesh, The Maldives and The Netherlands. Globally, due to warming of the oceans and increased evaporation more intense and heavy precipitations can be expected with increased risks of flooding. At the same time, more cloud formation and precipitation could mitigate atmospheric warming.

As a result of rising temperatures, not only sensitive ecosystems but also agricultural productivity in many regions will be affected. The consequences will differ greatly between the industrialized world and developing countries. Globally, industrialized countries in temperate climates will gain longer growing seasons, while the higher CO_2 concentrations will act as a fertilizer, increasing crop growth and yield [92]. Higher temperatures could also extend the growing range of some crops such as wheat further north in Canada or parts of Siberia. At the same time, crop yields in other areas may be reduced by more frequent droughts. Improved irrigation, changes in farming methods or selection of more adapted crops would be necessary.

Overall, it is likely that the adverse impact of global warming will be the greatest for lower-income populations, particularly in tropical countries where fewer resources are available to adapt to the negative effects of climate change. But with the advantages that can also be expected from a moderate temperature increase, the overall balance of the global change is difficult to predict. Clearly, man's contribution to climate change should be of great concern to all of us, and it is important that steps be taken as soon as possible to mitigate greenhouse gas emissions and their consequences. It is also necessary to point out that, on a longer timescale, the human-made causes of climate change, primarily considered presently as global warming, will in many instances be only of a temporary nature and be reversed in the more distant future to give way to a new cooling period. As the world's fossil fuel reserves are finite and not renewable, and when there will be a leveling of the world's population, the excessive CO_2 release will by necessity decrease. It is not only Nature's own photosynthetic recycling but also effective chemical CO_2 recycling technologies (Chapters 10–14 on the Methanol Economy) that will tend to keep CO_2 levels balanced.

7.2
Mitigation

Several options are available in the short term to limit or reduce human-caused greenhouse gas emissions. The efficiency of current processes that consume fossil fuels can be improved to increase the amount of useful energy per unit of CO_2 emitted. This improved efficiency can be achieved by the construction of more efficient electric power plants using improved technologies, by driving cars with higher fuel efficiencies and by employing appliances that consume less energy. Conservation, for example through better insulation of commercial buildings as well as private homes, can also result in significant cuts in heating or air-conditioning energy consumption. These measures will not only reduce CO_2 emissions and other environmental impacts but they are also economically advantageous as they reduce the amounts of fuel necessary, and therefore the costs. At the same time they can reduce the dependency of energy-importing countries, such as Europe, the United States and Japan, on foreign oil and gas suppliers. Such savings alone, however, can only extend the availability of needed and accessible fossil fuels in the relatively short term.

A switch to fuels that emit less or no CO_2 per unit of energy produced will be necessary. Natural gas will continue to play, as long as it is readily available, an important role in emission reduction by partially replacing coal in electricity generation. Non-fossil fuel energy sources will, however, need to play an increasingly important role to provide for our future energy needs. Among them, hydropower is already widely used and well developed for the generation of electricity, but suitable hydropower resources (rivers, waterfalls, etc.) are limited by their nature. Wind, solar and geothermal energy and energy from the combustion of biomass represent a rapidly increasing yet still small fraction of our energy needs. One of the main obstacles to a wider application of these renewable energy sources remains their cost, as well as technological limitations, although wind energy is now competitive in many locations with fossil-fuel based energy generation. Besides renewables, the use of nuclear fission power, which is a well-established and reliable source of energy that does not emit CO_2, is expected to be extended in the future. Of course, nuclear power should be made even safer, and problems of storage and disposal of radioactive waste must be solved. There is also a need to develop new generations of nuclear reactors, including breeder reactors and, eventually, controlled fusion.

As long as we use fossil fuels, the CO_2 produced during their combustion should be captured from industrial exhausts, particularly flue gases from coal or gas-fired power plants, cement and other factories, to avoid large excess emissions to the atmosphere. CO_2 can be captured and removed from exhausts using various techniques, including chemical absorption, adsorption onto solids and permeation through membranes. Until now, none of these has been applied on the scale of large commercial power plants, but the system is feasible and has been tested on a limited scale at a few locations. Billions of dollars recently pledged for carbon capture and sequestration (CCS) by governments around the world are directed at making CO_2 capture practical and economical on a large scale [93]. The removed CO_2 can then be compressed and, according to present plans, sequestered underground in geological formations, depleted oil or gas reservoirs, or even at the bottom of the sea (Figure 7.8) [94]. For example, since 1996, Statoil in Norway has been re-injecting CO_2 contained in the natural gas produced at its Sleipner platform back into a deep saline aquifer beneath the North Sea [95]. Carbon dioxide can also be used in existing oil fields for

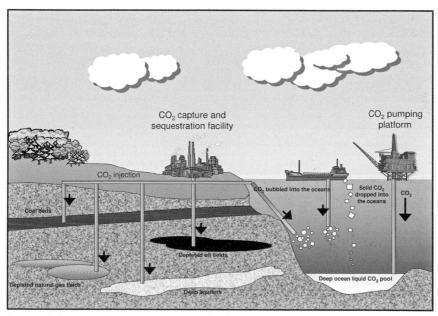

Figure 7.8 Overview of CO_2 sequestration technologies.

enhanced oil recovery. The environmental impacts of large-scale sequestration must be carefully assessed before we start pumping billions of tons of CO_2 into subterranean cavities and under the seas. Catastrophic natural releases of large quantities of CO_2 with lethal consequences have occurred in the recent past, notably in Cameroon. In 1986, CO_2 of volcanic origin, which was present in high concentration at the bottom of Lake Nyos, was suddenly released. The cloud of CO_2 rapidly spread to adjacent valleys killing by asphyxiation 1700 people and thousands of animals [77]. The exact reason for this release is unclear but could have been due to volcanic activity, an earthquake or rapid accumulation of rainwater in the lake. A similar CO_2 "eruption" that occurred in 1984 in the same region in Lake Monoun resulted in the death of 34 people.

In many aspects, CO_2 sequestering provides only a temporary solution, and in the long run the suggested "Methanol Economy" (Chapters 10–14) will allow recovered CO_2 to be chemically recycled to methanol for efficient energy storage, as fuel, or to be converted into synthetic hydrocarbons and their products. This solution will also provide economic value for the re-use of CO_2, thus lowering the cost of CO_2 removal and mitigating the harmful environmental effect. Most importantly, it will also provide a renewable carbon source for future generations.

Among the greenhouse gases other than CO_2, methane and nitrous oxide produced from agricultural origins (livestock, animal waste, rice culture, nitrogen fertilizer) could be reduced by making adequate changes in farming procedures. Methane also can eventually be separated and recycled. Methane from landfills is already used as a source of energy and for other uses. N_2O emissions from industrial processes (e.g., adipic acid production) are being progressively eliminated.

Fluorinated gases are only man-made and produced industrially. Their presence in the atmosphere is purely of human origin. Owing to their very high global warming potential, these gases should be monitored and controlled rigorously. Emissions can be minimized through process changes, improved recovery, recycling or containment, or avoided by the use of alternative compounds and processes.

Two decades ago, with the discovery of a hole in the Earth's ozone layer caused by the emission of CFCs, it was clear that we already faced an environmental challenge of global dimension. This problem threatened to have long-term consequences and could only be resolved by a concerted worldwide response. For the first time, international action was taken through the ratification in 1987 of the Montreal Protocol, progressively phasing out the manufacture and use of CFCs, which had been linked to ozone depletion.

Today, scientific evidence for humanity's responsibility for at least part of the observed climate change is becoming clear. However, this problem is infinitely more complex than that of CFCs, which could be relatively easily banned and replaced. This is not the case with CO_2 emissions. Controlling CO_2 emissions was the intention of the Kyoto Protocol, which was ratified by some 140 nations, but not by the United States, and some others. Australia only ratified the Kyoto treaty after a change of government at the beginning of 2008. It also exempted developing countries such as India, China and Brazil, which are industrializing rapidly. The Kyoto Protocol rightly stated the problem, and suggested regulatory limits for greenhouse gas emissions but, besides the trading of carbon quotas, it did not offer any technical solution. The Stern Review, published in 2006, analyzed the effects of climate change on the world economy. It discussed possible pathways and measures to minimize the economic and social disruption caused by global warming and stabilize greenhouse gas concentration in the atmosphere [96, 97]. The Copenhagen United nations Climate Change Conference, where a deal is expected to be reached to replace the Kyoto Protocol (running out in 2012), will hopefully be more comprehensive in its approach to fight global warming [98, 99]. Today, new technological answers to the problem are needed, and one of these is offered by the suggested "Methanol Economy," which is essentially based on the chemical recycling of CO_2 to produce, via methanol, new fuels and materials. Moreover, such an approach would provide a renewable, inexhaustible carbon source, whilst mitigating human-caused global climate change and liberating mankind from its dependence on diminishing fossil fuel reserves.

Considering that CO_2 is a major greenhouse gas, contributing significantly to human activity-caused global warming, many people, including Nobel Peace Prize winner Al Gore and other environmentalists, are increasingly calling for abandoning the use of carbon fuels and their industrial use to get off our "carbon addiction." Others, including US president Barack Obama, have more recently suggested a concerted effort to find solutions encompassing all reasonable approaches. We believe that the "Methanol Economy" fits into that approach and is starting to gain consideration.

Considering the essential role that carbon plays in our existence and life on planet Earth, it is clearly unrealistic to consider a "ban on carbon," probably prompted in part by the success of banning man-made CFCs with the Montreal Protocol. Carbon is a fundamental building block of all aspects of our life and an essential material for diverse products.

To mitigate the CO_2 emission problem, the only presently offered remedy besides carbon trading quotas assigned to nations and major users is capturing CO_2 emissions and sequestering them underground or at the bottom of the sea. As emphasized, this is, however, only a temporary and potentially unsafe storage technology.

In our view, instead of trying to "cure" us from dreaded carbon dioxide emissions we must find a reasonable solution for the problem. Carbon fuels and derived products can be rendered environmentally carbon neutral and renewable on the human timescale. We suggest that chemical recycling is a feasible, economical and permanent solution for excess CO_2 emissions that also offers humankind a regenerative inexhaustible carbon source for the future beyond fossil fuels.

8
Renewable Energy Sources and Atomic Energy

8.1
Introduction

As discussed earlier, we rely today for a significant part of our energy needs and related hydrocarbon fuels and products primarily on non-renewable fossil fuel sources. In pre-industrial times, as is still the case today in some developing countries, the energy supply was based primarily on renewable sources. The power of watermills and windmills was used to grind grain, press oil or to pump water, while wind energy at sea moved ships, and biomass energy sources such as wood and dung warmed us and cooked our food. With industrialization, however, the role of renewable energy in the global energy supply was gradually taken over by fossil fuels – first coal, and later by oil and natural gas. During the past two centuries – which is a relatively short time in human history – our energy needs have relied predominantly on fossil fuels. While the reserves of these fossil fuels are still significant, they are nevertheless limited and diminishing, and they cannot sustain our life-style and development in a permanent manner over the centuries to come. Therefore, to satisfy our future energy requirements, the use of energy alternatives to fossil fuels, including renewable sources and nuclear energy, must be increasingly relied upon and developed. There is also increasing need to produce synthetic hydrocarbons for transportation fuels and varied materials and products, necessitating large amounts of energy that must be obtained from non-fossil fuel sources. Therefore, it is necessary not only to discuss the availability and feasibility of alternative energy sources and atomic energy upon which we will need to rely in the future, but also to identify new, efficient ways in which to store, transport, dispense and use energy, specifically highlighting the advantages of our proposed "Methanol Economy."

Unlike fossil fuels, renewable energy is derived from sources that are not subject to depletion, and can replenish themselves. These include heat and light (radiation) from the sun and wind, organic matter, hydropower, tides, waves and geothermal heat from the earth's crust. "Renewable," however, does not necessarily mean non-polluting or "green." Renewable energy gained increasing significance, especially in the United States, after the oil crises of the 1970s. The sudden oil shortages, which led to sharp increases in gasoline and electricity prices, shocked those nations that were

Beyond Oil and Gas: The Methanol Economy, Second updated and enlarged edition
George A. Olah, Alain Goeppert, and G. K. Surya Prakash
Copyright © 2009 WILEY-VCH Verlag GmbH & Co. KGaA, Weinheim
ISBN: 978-3-527-32422-4

used to cheap energy and prompted governments to start finding solutions to become more energy independent. Steps were taken to restructure the energy industry and to reduce dependence on imported oil, in part by encouraging the development of renewable and alternative energy sources through government incentives and tax credits. In the early 1980s, many of these measures were in place, but by the mid-1980s fossil fuel prices had fallen substantially and the need for renewable energy sources had faded as the core issue of energy policies because of their higher costs. More recently, however, many countries with few or no fossil-fuel resources, driven by the desire for energy independence and adopting the Kyoto Protocol, which mandates greenhouse gas emission reductions, have begun to invest again intensively in renewable energy sources. The increased use of renewable energy technologies can contribute to mitigating and eventually meeting both environmental and energy security goals. However, as the overall energy needs of humankind are enormous, these technologies cannot on their own be considered as the only solution, at least not for the foreseeable future. Since few of the alternative energy sources depend on combustion to generate heat or electricity, they also offer substantial environmental benefits compared to fossil-fuel technologies (*vide infra*). Renewable energies are typically based on indigenous sources whose supply is not easily disrupted, therefore their development and use also enhances energy security.

There is high hope that renewable sources of energy could become a major factor in developing a secure energy supply for the future. In principle, renewables having the advantage of not being based on limited natural resources have indeed an enormous potential and promise. About ten thousand times more energy from the sun continuously reaches the surface of the earth than is generated by all of the fossil fuels consumed by humans. Although a large part of exploitable renewable primary energy is often scattered, it can be converted in many ways into usable heat and power. However, widespread usage is faced with numerous challenges, and many forms of renewable energy are not currently economically viable. Moreover, there are also serious technical problems associated with the integration of renewables into existing systems. These energy sources are also often less environmentally friendly than usually believed. Indeed, despite popular belief, no form of energy, whether renewable or subject to depletion, is pollution free. Harmful effects occur either during the construction, operation or disposal of generating facilities and fuels:

- Geothermal energy, for example, is not strictly a renewable resource, as underground reservoirs become depleted over longer periods, and solid waste and harmful gases can also be generated.

- The construction of large hydroelectric plants requires the daming of rivers and the consequent submerging of vast areas of land.

- The use of solar energy necessitates the production of photovoltaic solar cells. The energy needed for their manufacture may itself still be provided by fossil fuels. Their production also often involves the use of hazardous materials such as cadmium and arsenic. At the same time extensive use of solar energy would involve large land areas to be covered by light-absorbing panels and facilities.

At present, the main obstacle against the massive and widespread use of renewable energy sources, such as wind, solar or geothermal energy, is their cost when compared to still available conventional fossil fuels. These renewable processes are relatively capital intensive, and require major investment to capture diffuse energy sources, making most of them unattractive for the short term. However, in the long term, after the initial investment has been made, the economics of renewables improves, since operation and maintenance costs are relatively low in comparison with conventional energy sources using fossil fuels, particularly if the latter are subject to significant price increases. In recent years, the progressive deregulation of electricity markets in the United States and Europe, resulting in rising competition between energy suppliers, was directed more towards the short term, as a cost-minimizing strategy. These policies made renewable energies appear at a disadvantage. Government support, in the form of research funds, incentives, subsidies and tax credits, is therefore essential to develop sustainable and meaningful energy policies. Governments must play an essential role in promoting the development and use of renewable energies, which still account today for only a relatively small part of the overall global energy use.

In 2005, according to the International Energy Agency, renewable sources accounted for only 12.8% of the total primary energy supply (TPES) of the world (Figure 8.1). More traditional sources, such as combustible biomass and waste, made up the bulk at 10.1%, with another 2.2% contributed by hydropower. New renewable energy sources for electricity or heat generation, including geothermal, wind and solar, accounted for less than 0.5% of the TPES.

Today, renewable energies are at various stages of development. Some are well established, such as the use of hydropower, steam from geothermal wells and burning of biomass and waste. Others are newer but actively developed, such as wind power, photovoltaic cells (which convert sunlight directly into electricity) and the conversion of biomass into gaseous or liquid fuels. Still others are only emerging concepts in the research and development stage, as in the case of power production from ocean tides, waves, currents and temperature gradients. Some of the advantages, limitations and possible future of these energy sources are discussed below.

8.2
Hydropower

Hydropower is the world's most used renewable energy source for the production of electricity. The use of water power dates back to antiquity, with the use of water wheels turned either by the flow of a river or the weight of water falling from a dam or waterfall. Following the invention of the electric generator and hydraulic turbine, hydropower has been an important part of electricity generation since the nineteenth century. Indeed, today it supplies 17% of the world's electricity needs. For several countries, hydropower is the major electricity source (Figure 8.2); for example, it generates 50% of electricity in Canada and Sweden, over 80% in Brazil and nearly

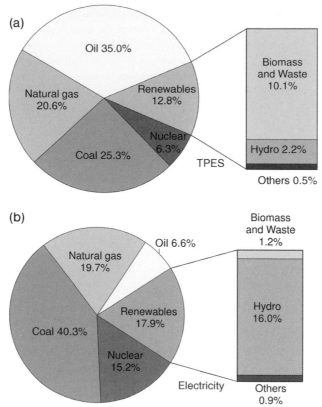

Figure 8.1 (a) Share of renewables in the world total primary energy supply (TPES) in 2005; (b) share of renewables in world electricity production in 2005. (Source: IEA Renewables Information 2007.)

100% in Norway. Hydroelectric power, although still capable of being improved, is a mature technology. New large-scale projects in developed countries, where the most suitable sites have already been developed, are, however, limited by necessity. Current world hydroelectricity production exploits only an estimated 18% of the overall potential [100], and in developing countries and transition economies the unexploited resources are still vast. Latin America has tapped only 20% of its potential, Asia 11% and Africa about 4%. Therefore, 90% of future hydroelectric plants are expected to be constructed on these continents. A rational utilization of hydroelectric resources could help these developing regions to modernize their economies and raise their standard of living. Even if hydropower has lower operating costs and longer life expectancy than most other modes of electricity generation, resulting in generally lower energy prices for the consumer, it is also highly capital intensive and frequently environmentally intrusive. The large initial investment necessary for the construction of large-scale hydropower plants is an important issue, and many developing nations may find it difficult to finance such projects.

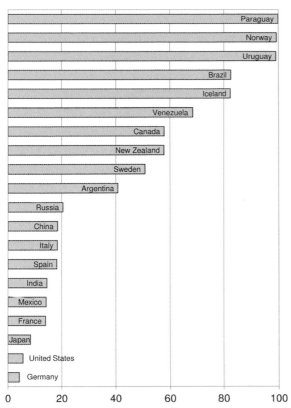

Figure 8.2 Percentage of electricity produced from hydropower in different countries. (Source: *CIA World Factbook*, December 2003.)

Between the 1930s and the 1970s, the construction of large dams was, in the eyes of many, synonymous with development and economic progress. These dams not only produced electric power but also provided water for irrigation and helped in flood control. In the United States, the Hoover Dam on the Colorado River (Figure 8.3), constructed during the height of the Great Depression and completed in 1936, was the largest dam of its time and viewed as a symbol of modernization and man's ability to harness nature. It opened the way to the development of the western United States by powering cities as far away as Los Angeles and nearby Las Vegas, which was at that time hardly more than a water refilling stop for steam engine locomotives on the Union Pacific railroad. In the former Soviet Union, major hydroelectric projects were essential to the industrialization of selected areas. The construction of dams accelerated dramatically after World War II, peaking during the 1970s when numerous large dams were commissioned around the world, not only for hydro-electric power generation but also for flood control and irrigation purposes. Today, there are an estimated 45 000 large dams on our planet, of which 22 000 are located in China alone [101]! Large dams are some of the biggest structures built by humankind.

Figure 8.3 Hoover Dam on the Colorado River in the United States.

The Itaipú hydroelectric plant on the Paraná River in South America, until recently the largest in the world, has an output of 12 600 MW, which is equivalent to about 12 large commercial nuclear reactors, and provides 25% of the energy supply in Brazil and 78% in Paraguay. In recent times, however, large hydro projects have raised many criticisms on environmental and social grounds. They inevitably require large dams and reservoirs that inundate the affected extensive land areas with water and can disturb local ecosystems, reduce biodiversity, modify water quality and cause major socio-economic damage through the displacement and relocation of local populations. Land inundations are large, some running into thousands of square kilometers. The world's two largest dam reservoirs, Ghana's Akosombo on the Volta River, with an area of $8730\,km^2$, and Russia's Kuybyshev on the Volga River, with an area of $6500\,km^2$ areas, approach the size of small countries such as Lebanon or Cyprus. In tropical regions, these waters can also create feeding grounds for malaria-bearing mosquitoes and other water-borne diseases. Until the 1960s, most hydro-electric megaprojects did not involve the massive relocation of populations, but as large dams began to be constructed in more densely populated areas, especially in Asia, the population to be resettled grew in size. It is estimated that during the twentieth century between 40 and 80 million people have been displaced because of large dam construction projects. The Three Gorge Dam (Figure 8.4) on the Yangtze River in China caused the relocation of over one million people. This 2-km long, 200-m high dam is the world's largest hydroelectric plant, with a production capacity of 18 000 MW, which is much needed for China's growing economy, with a price tag close to $30 billion. To make way for this enormous project, the over 500-km long

Figure 8.4 Three Gorge Dam in China.

artificial lake behind the dam submerges more than 4500 towns and villages, ancient temples, burial grounds and spectacular scenic canyons that have inspired poets and painters for centuries and attracted tourists from around the world. Environmentalists argue that the dam will doom migratory fish and wipe out several rare species. The Yangtze River dolphin is now already considered as extinct. There are also concerns that the lake will trap millions of tonnes of pollutants spewing from Chongqing, one of China's largest industrial cities. In Egypt, the High Aswan Dam was completed in 1970, and captures the Nile River in the world's third largest reservoir, Lake Nasser. Before the dam was built, the Nile overflowed its banks once a year, depositing millions of tonnes of nutrient-rich silt on the valley floor, and making the otherwise dry land fertile and productive. In some years, however, when the river did not rise, it caused drought and famine. By constructing the dam the floods could be controlled, as could the drought by water release. At the same time, huge amounts of electricity were generated. Unfortunately, the rich silt that normally fertilized the dry desert land during annual floods is now stuck at the bottom of Lake Nasser, forcing Egyptians to use about one million tonnes of artificial fertilizer to substitute for natural nutrients that once fertilized the arid floodplains [102]. This exemplifies one of the major problems associated with the construction of dams: the progressive silting of reservoirs over time, especially in erosion-prone regions such as the high mountain ranges of the Himalayas and China's Loess plateau. Possible technical solutions are to de-silt reservoirs by flushing or adapted dredging or periodic heightening of the dam walls.

A significant potential for expanding hydropower lies in small systems that have relatively modest and localized effects on the environment and populations. They have generating capacities in the range 1–30 MW, or even less for mini- and micro-hydro installations, and are especially attractive to supply electricity to remote areas far from any electrical grid. These units are typically "run-of-river" plants that transform the kinetic energy of rivers and streams into electricity, and have little or no storage capacity. Consequently, their environmental impact is lower when compared to large hydroelectric installations, but it also means that they are vulnerable to seasonal fluctuations and will not be able to produce electricity in dry seasons. Today, "small-hydro" represents only about 5% of the worldwide hydroelectricity generation.

Most of the concerns associated with the development of hydroelectric power have been addressed, and many were successfully mitigated. The key to future projects will be proper prior planning not only of the economical feasibility but also of environmental and social factors. Potential benefits as well as the downsides of each project should be carefully weighed. Adverse publicity and negative media comments on dams should not overshadow the many benefits associated with their construction. Hydropower installations, besides providing electricity at the lowest cost compared to any other energy source, have also provided other important benefits such as flood control, irrigation, drinking water and improved navigation. The human displacement of thousands of people associated with the construction of a new dam should be compared against the benefit to millions provided with electricity. The further expansion of hydropower will continue to be a key step in the modernization of many developing countries as a reliable and affordable source of energy. The construction of hydroelectric instead of fossil fuel-burning plants also reduces the emission of significant air pollutants, mainly SO_2 and NO_x as well as CO_2, the main greenhouse gas.

8.3
Geothermal Energy

The temperature of the earth gradually increases with depth, reaching $4500\,°C$ at its core. Some of this heat is a relic of the formation of our planet about 4.5 billion years ago, but most is a result of the decay of naturally occurring radioactive elements. As heat flows from warmer to cooler regions, the earth's heat gradually flows from the core to the surface, where an estimated 42 million thermal MW are continually radiated away, averaging merely $0.087\,\text{Wm}^{-2}$. Most of this immense heat cannot be practically captured. However, in some locations heat reaches the surface more rapidly, generally on the margins of the tectonic plates. There, the concentrated energy can be released by natural vents: volcanoes, as hot steam or springs. It is in these locations that geothermal plants can preferably be built and be the most efficient. While hot springs and pools for bathing were used long ago by the Romans, Turks and others, the utilization of geothermal resources for other uses is relatively recent. In 1812, steam from geothermal fields in Larderello, Italy, was used for the manufacture of boric acid. The same geothermal sources were first exploited to produce electricity in 1904, but significant development in other parts of the world did not start before the second half of the twentieth century. Despite its short history, geothermal energy is now relatively common in many parts of the globe (Figures 8.5 and 8.6). In New Zealand, two power plants were built in the early 1960s to produce 170 MW, while in the United States, California's Geysers (Figure 8.7), the world's largest geothermal power development, came online in 1960. Today, the United States is the leading producer of geothermal electricity, followed by the Philippines, Mexico and Indonesia. The total generating capacity is, however, only close to 10 000 MW, and production stands around 57 TWh, which represents 0.3% of the global electricity production. Nevertheless, in some developing countries geothermal

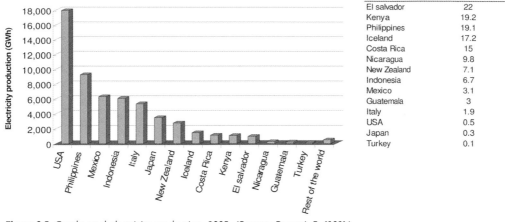

Percentage of geothermy in the
country's total electricity
generation

El salvador	22
Kenya	19.2
Philippines	19.1
Iceland	17.2
Costa Rica	15
Nicaragua	9.8
New Zealand	7.1
Indonesia	6.7
Mexico	3.1
Guatemala	3
Italy	1.9
USA	0.5
Japan	0.3
Turkey	0.1

Figure 8.5 Geothermal electricity production, 2005. (Source: Bertani, R. [103].)

power plays a key role, accounting in 2005 for 22% of the electricity generated in El Salvador, 19% in both the Philippines and Kenya and 17% in Iceland [103].

The conventional type of natural geothermal reservoir used for electricity generation is hydrothermal, consisting of the accumulation of hot water or steam trapped in fractured porous rock. The most profitable and valuable, but also the rarest, are vapor reservoirs, which yield mainly high-temperature, super-heated steam above 220 °C (also called dry steam). This steam is produced from wells up to 4 km deep and

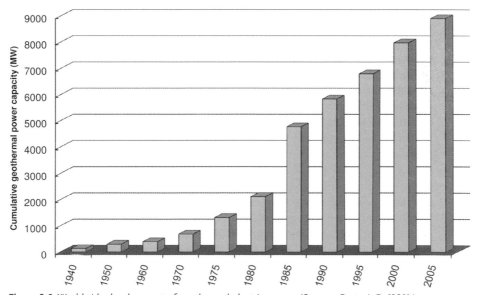

Figure 8.6 Worldwide development of geothermal electric power. (Source: Bertani, R. [103].)

Figure 8.7 California's Geysers. (Source: NREL.)

is able to directly power gas turbines to generate electricity. Notable examples are Larderello in Italy, Geysers in northern California and Matsukawa in Japan. More common are systems based on hot water at temperatures in the range 150–300 °C. In this case the hot water, when brought to the surface, depressurizes and boils explosively, forming large quantities of steam that is fed to turbines to produce electricity in so-called flash-steam power plants. In both cases, residual water and the condensed steam after utilization are re-injected into the reservoir to maintain pressure and prolong productivity. Lower temperature geothermal systems, between 100 and 150 °C, do not allow efficient flash-steam power production. There, the extracted hot water can be used to vaporize another fluid with a lower boiling point (e.g., isopentane) that will drive the power-producing turbine/generator units in so-called binary-cycle plants.

Even if some geothermally active areas have extensive hydrothermal systems, most of them consist only of hot dry rocks with no water or steam. Collecting the heat present in these formations is much more challenging. At least a pair of wells must be drilled at depths typically more than 4 km, and the rock artificially fractured to allow water to circulate into the injection well. Steam or hot water is then returned to the surface via the production well. The potential for this resource is enormous, and technical feasibility studies are under investigation in Japan, the United States and Europe. One of the most advanced research projects on this topic at present is being conducted on a hot dry rock unit in Alsace, France [104]. Injecting water under high pressure in underground rock formations is, however, not without risks. In December 2006 the injection of water to fracture the rock under a geothermal plant, under construction in Basel, Switzerland, triggered a series of earthquakes, one of which was 3.4 in magnitude. The project was immediately halted and it is still unclear if the plant will ever be completed.

Low- to moderate-temperature (35–150 °C) geothermal resources can also make direct use of thermal energy to heat buildings, offices and greenhouses, and even for farming fish in colder areas. For some time now, most of Iceland, which sits right on the geologically very active Mid-Atlantic Ridge, has been heated by geothermal energy. Geothermal heat can also be produced by heat pumps that use the relatively constant soil temperature to provide heating for buildings in winter and cooling in summer. Today, geothermal heat is used by more than 55 countries with an installed total capacity over 17 000 MW of thermal power. The largest installed capacity in the world is presently in the United States, followed by China, Iceland and Japan [100].

Geothermal energy is, in the strictest sense, not a renewable resource. The energy that is taken out of the earth in the form of heat will not be replaced in the future. At a given location, a resource can be depleted relatively quickly if too much hot water or steam is extracted and not reinjected after heat extraction, as happened in the Geysers of California. However, geothermal resources around the globe are so widespread and immense that there is little possibility of their exhaustion on a human timescale. Today, for electricity generation, only the highest grade geothermal resources can be utilized economically, averaging a production cost of $0.03–$0.10 per kWh [100], the lower end of this scale being quite competitive with fossil fuels. Geothermal resources have the advantage of being available all the time, with little or no fluctuation in power output, allowing reliable and predictable electricity production. Their environmental impact is low because the water used is generally re-injected in the reservoir. The emission of non-condensable gases (NCGs) such as H_2S, CH_4, NH_3 accompanying the hot water or steam are limited and generally addressed by regulations. Carbon dioxide emissions, representing 90% of these NCG, are, however, for the most part not regulated [105]. Significant quantities of CO_2, which can amount to 10 wt% of the extracted hot water or steam are therefore in most cases simply vented to the atmosphere in geothermal operations [105, 106]. CO_2 can be relatively easily captured from such a high concentration source (90% of the NCG) with existing technologies, and sequestered or recycled to produce methanol and other chemicals. Compared to emissions from fossil fuel-burning power plants, CO_2 emissions from geothermal plants are much lower. It was estimated that an average of 122 g-CO_2 per kWh is emitted from geothermal power plants while three to ten times this amount is generated in fossil-fuel power plants [105, 106].

Geothermal facilities also produce small amounts of solid by-product materials such as salts or heavy metals that require disposal, while others such as silica, sulfur or zinc can be extracted for sale. The principal barriers to faster worldwide geothermal development are mostly technological, with high costs of exploration, drilling and plant construction, but there are also a limited numbers of sites that are economically exploitable with present technologies. Most future development is expected in already known fields, principally along the tectonically active margins of the Pacific region, notably in East Asia. As mentioned, geothermal energy represents today only 0.3% of the global electricity generation. Despite forecasted regular growth, this share is only expected to increase slightly during the next decades. Therefore, from a global aspect, geothermal energy will remain a limited source of electricity, despite playing an important role on a local scale in some countries. The successful development of hot dry rock technology might alter this picture, however, by allowing the installation of geothermal units in a larger number of formerly unsuitable locations. A 2006 study conducted by MIT concluded that 10% or more of the electricity in the US could be produced economically from geothermal energy with further development of these enhanced geothermal systems [107]. Europe and other areas could also substantially benefit from these new technologies.

8.4
Wind Energy

Wind, which actually is also a form of solar energy, is generated by the uneven heating of the earth and its atmosphere by the sun. As air is heated, it expands and rises because its density decreases. To replace the rising hot air, cooler air from elsewhere flows into the region, producing wind. Topographic features such as hills, mountains, valleys or large open spaces can also influence the speed, density and direction of the wind.

The utilization of wind power, not unlike water power, dates back thousands of years. The Egyptians built and used sailing boats in 2000 BC, while later sailing ships made world exploration and travel possible, allowing Columbus to discover America and Ferdinand Magellan to complete the first voyage around the world. It was not until the end of the nineteenth century that wind, as the prime energy source for maritime transportation, declined rapidly with the development of steam-powered vessels. On land, windmills provided energy for milling and pumping of water in early Mediterranean and Eastern civilizations, as well as in medieval Europe, for example in Holland and Spain. Windmills continued to play an important role well into the twentieth century in the rural United States and Australia, providing energy for isolated farms, mostly to pump water for irrigation and livestock. In the late nineteenth century, windmills also began to be used to produce electricity, making it possible for farmers and ranchers far from electricity distribution lines to generate their own electricity and to operate small electrical devices such as a radio. Their use was, however, progressively eliminated before World War II by the availability of cheaper, more reliable, fossil fuel or water-based electricity from central generating plants delivered through steadily extending electrical transmission grids. Until the early 1970s, the era of cheap oil and gas induced very little interest in electricity generated by wind, and it was only with the oil price shocks of the 1970s and early 1980s that efforts were made to develop and design new wind machines capable of producing electrical power efficiently at a competitive price. In the early 1980s, state subsidies, incentives and tax credits began to revitalize the wind energy industry. Almost all growth was concentrated in the United States, as well as Denmark. In 1985, the United States alone accounted for more than 90% of the installed generating capacity, with 1 GW, and had the world's largest windmill facility at the Altamont Pass, near San Francisco. With the expiration of tax incentives in the United States in 1985 and declining oil prices, however, wind power development came to a standstill. Subsequently, thanks to improvements in turbine designs as well as the active promotion of wind energy, development was restarted in the 1990s and was this time led by European countries such as Germany, Denmark and Spain. The installed capacity has increased from a modest 2500 MW in 1992 to over 90 000 MW at the end of 2007, at an annual growth rate of almost 30% [108]. Today, wind power is the fastest-growing renewable energy resource, and indeed the fastest growing of any energy source. In Denmark, wind produces 20% of the country's electricity needs. The spectacular increase in wind power capacity in recent years has placed Germany, with an installed capacity of more than 22 000 MW meeting 7% of the national electricity needs, as the world leader in power production from wind [108]. Spain, with an installed capacity of over 15 000 MW, obtains 12% of its electricity from wind. More than 60% of the world's wind-power capacity is presently concentrated in the European Union, where wind met 3.8% of the electricity demand in 2007 compared to close to zero only 15 years earlier [109]. Nine of the ten largest wind turbine manufacturers are based in Europe, and European companies control some 90% of the global wind energy market. Outside of Europe, the United States, India and China are the leading markets. In 2007, installation of new wind-power producing capacity was the highest in the United States with over 5000 MW of new capacity added. This addition was mainly due to the availability of a tax credit of 1.9 US cents per kWh for electricity generated from wind turbines over the first 10 years of a project's operation [110]. Substantial development of wind energy is also expected in other countries, including Brazil, Canada and Australia (Figure 8.8).

When considering the installation of a wind-power generating facility, the single most important factors are the wind speed and its consistency. As the power increases in relation to the cube of the speed, a doubling of the average wind speed induces a wind power increase by a factor of eight. Consequently, even small changes in wind

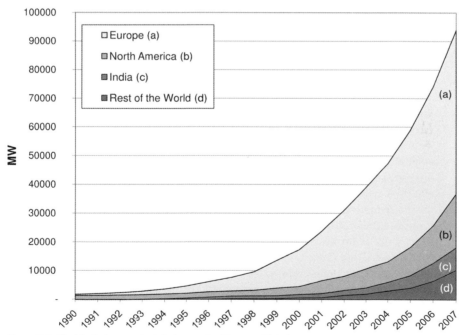

Figure 8.8 World wind power installed capacity. (Source: Global Wind Energy Council, European Wind Energy Association, IEA.)

Figure 8.9 Offshore wind farm, Denmark. (Courtesy: Elsam A/S, Denmark.)

speed can produce large changes in energy output. Therefore, the installation of new production capacity must be preceded by careful studies of wind flow conditions. Relatively constant winds, even with medium speeds, during extended periods of time are more suitable for reliable power generation than faster but more intermittent or unstable winds. Once the wind resource has been explored, wind turbines have to be constructed to transform the wind energy into useful electrical power. Modern wind turbines generally consist of a three-blade rotor connected to an electric generator, able to operate over a large range of wind speeds and fixed on the top of a hub. In most cases clusters containing up to several hundred wind turbines are concentrated in so-called wind farms (Figure 8.9). Over the past 20 years, the power of these turbines has risen dramatically as the rotor diameter increased from around 20 m, with generating capacities of 20–60 kW, to more than 100 m, with generating units over 2 MW in commercial applications, as well as prototypes of 126 m rotor diameter and 6 MW [110, 111]. Larger and taller turbines can also take better advantage of the fact that wind speed increases with increasing height above the ground level. These increases in size and technological know-how, coupled with the economy of scale from a fast-growing production, have greatly reduced the cost of wind power to the point that some of the most productive onshore wind farms are approaching price competitiveness with energy from fossil fuels. The exploitable onshore wind resource for the European Union is conservatively estimated at 600 TWh and the offshore wind resource at up to 3000 TWh [111]. Taken together, this may represent more than the present total electricity consumption of the European Union. Onshore wind potential is being developed first because of obvious lower installation costs compared to offshore facilities. To date, only Europe has installed wind capacity offshore, mainly in the North Sea, in relatively shallow waters and using large turbines to minimize the cost per unit of energy produced. However, because of the considerable potential of offshore facilities and often higher average wind speeds and lower turbulence at those sites, projects are under way all around the world.

One major problem with wind-generated electricity is that the resource is highly variable, changing on a daily, seasonal and annual basis. Therefore, power output prediction can be very challenging and additional generating capacity, as well as sufficient backup power, may be necessary in case of low wind speeds. Wind turbines being constructed in exposed, highly visible areas also raise public resistance because of the visual impact of wind farms as well as noise levels, which should also be taken into account by building them far from populated areas and sensitive landscapes. Although only limited land is necessary for the installation of each turbine, they must be placed at sufficient distances in order not to interfere with each other, and this can result in extensive land requirement for their establishment. However, as the footprint of the wind tower is small, much of the land comprised in wind farms can be used for other purposes such as agriculture or grazing livestock. On the other hand, wind farms also represent other environmental hazards by killing birds and thus interfering with ecosystems. Nonetheless, the environmental impacts can be greatly reduced or eliminated by the construction of offshore wind farms.

Based on the large and widespread resources, costs, maturity of the technology and relatively limited environmental impacts, wind power has a promising future in supplying an increasing share of electricity in countries willing to invest in it. The absence of greenhouse gas and pollutant emissions is also a strong incentive for electricity generation from wind. As with other alternative energy sources wind power-based electricity must, however, be fed directly into an electric grid as storage of electricity remains difficult. Alternatively, the electric power generated could be used to produce hydrogen and other products including methanol by recycling CO_2.

8.5
Solar Energy: Photovoltaic and Thermal

The sun is an enormous, effective and far-away (some 150 million km from earth) hydrogen-powered nuclear fusion reactor that can supply the earth with energy for some 4.5 billion years to come. Sunlight, or solar energy, emitted from the sun is the most abundant energy source. At any given time, sunshine delivers to earth as light and heat about 10 000 times more energy than what the entire world is consuming. Solar radiation, before entering the earth's atmosphere, has a power density of $1370 \, W \, m^{-2}$ [23]. Of all the sunlight that passes through the atmosphere annually, only about half reaches the earth's surface; the other half is scattered or reflected back to space by clouds and the atmosphere, or absorbed by atmospheric gases such as CO_2 and water vapor, the atmosphere and clouds. As 71% of our planet is covered with water, most of that energy that makes it to the surface is absorbed by the oceans. On a clear day, the radiation received on the earth's surface around noon is about $1000 \, W \, m^{-2}$ [100]. The amount of energy received on the surface (called insolation) is usually measured in $kWh \, m^{-2}$. Annual average insolation varies from maxima in hot desert areas to minima in polar regions. Seasonal variations are more important in regions far from the equator, as differences between winter and summer day lengths are more pronounced. In the extreme case, at the North Pole, insolation is near zero during the six months of polar winter when the sun never rises. In the United States, the highest insolations, with a daily average in excess of $6 \, kWh \, m^{-2}$, can be found in the dry and most of the time cloudless south-western states. These include states such as Arizona, New Mexico and Nevada, as well as Southern California and especially the Mojave Desert, where major solar energy technologies are tested. Using the current solar technology, an area of 160×160 km in this region could generate as much energy as the entire United States currently consumes [112]. The enormous flux of inexhaustible (at least on the human timescale) solar energy holds a tremendous potential for providing humanity with clean and sustainable energy. As with wind, real interest in utilizing the sun's energy arose in the aftermath of the oil crises of the 1970s. Today, the various ways in which solar energy can be used to generate electricity, provide hot water and heat or cool our buildings are at different stages of technological development and use.

8.5.1
Electricity from Photovoltaic Conversion

The conversion of daylight into electricity, called the photovoltaic effect, was first discovered by the French scientist Edmond Becquerel in 1839. The explanation for this effect was later provided by Albert Einstein, who received for this work (and not the theory of relativity) the 1921 Nobel Prize in Physics. However, the development of the first practical photovoltaic cell occurred only in 1954 at the Bell Telephone Laboratories, with the production of a silicon-based cell with 6% efficiency in converting light into electricity. The technology that was originally developed for space applications in the 1960s to power military and later commercial satellites has since been given much attention as a potential energy source for civilian uses. Photovoltaic (PV) systems that convert the energy of photons from sunlight directly into electricity using semiconductor devices are commonly known as solar cells. When photons enter the cell, electrons in the semiconducting material are freed, allowing them to flow and generate electricity. Solar cells are typically combined into modules in sizes from less than 1 W to 300 W to produce higher voltages and currents. If larger electricity production is required, these modules can be connected in series or parallel to form photovoltaic arrays. The modular nature of solar cells means that they can be used in applications ranging from a fraction of a watt, such as a solar wristwatch, to large multi-MW power plants containing millions of solar cells.

Solar cells are most commonly made from mono- or polycrystalline silicon. Their efficiency in converting sunlight into electricity is constantly improving, from around 15% in the mid-1970s to more than 35% today under laboratory conditions [113, 114]. Other materials such as cadmium telluride, gallium arsenide, copper-indium diselenide or amorphous silica, usually in the form of deposited thin films, have also been used to produce solar cells. In the past their large-scale utilization has met with limited success because of a shorter lifetime, low efficiency and/or toxicity issues in production and disposal. Recently, however, cadmium telluride-based thin film solar cells have attracted much attention. Companies such as First Solar developed a highly automated manufacturing process to produce these cells in high volumes and at a relatively low cost [113, 114, 116]. Large scale photovoltaic farms using this technology have been installed, notably in Germany and the US. On a former military air

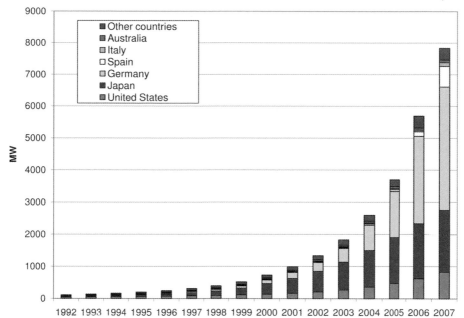

Figure 8.10 Cumulative installed photovoltaic (PV) power in reporting IEA countries. IEA-photovoltaic power systems programme.

field in Waldpolenz, Germany, for example, a 40 MW-capacity solar power plant has been constructed, covering $400\,000\,m^2$. However, tellurium is one of the rarest elements on earth, with only 147 tonnes produced worldwide in 2007 (about 15 times less than gold) [117]. This rarity could pose serious problems for the use of cadmium telluride-based photovoltaic technology on a large scale. Besides inorganic solar cells, photocells based on organic materials are now being designed and tested. Relatively inexpensive flexible plastic-embedded photovoltaic materials are currently being developed [118].

During the last decade, the cumulative PV capacity in the world has increased at a rate of more than 30% per year, reaching by the end of 2007 almost 8 GW, compared to less than 0.2 GW in 1992 (Figure 8.10) [114, 119]. This growth is impressive, but still equals only the production capacity of eight 1 GW nuclear reactors. Over 90% of the world capacity is installed in only four countries: Japan, Germany, the United States and Spain. Almost 50% of the installed capacity was located in Germany alone. Not surprisingly, Germany, Japan and Spain are major solar cell producers. Taiwan and China have also recently emerged as important solar cell manufacturers. From relative obscurity 5 years ago, China is expected to become the largest solar cell and module producer in the coming years.

PV technology has a wide range of applications. It can provide electricity to isolated systems far from power lines, such as telecommunication towers or water pumps in developing countries. Most of the future increase in solar power is, however, expected to come from the PV installed to generate all or part of the electricity needed in grid-connected buildings.

Solar cells have many advantages: they are silent, produce no emissions, have no moving parts, are easy to maintain, can be installed almost anywhere, are easily adapted to the customer's specifications and use no fuel.

The downside of solar cells is that energy generation is limited by available daylight. The sun does not shine during the night, and even during the day cloudy conditions also reduce the energy output to only 5–20% of that of full sunlight output. The intermittent nature of PV systems is an issue that must be addressed by the installation of supplemental conventional generating capacity or effective energy storage for cloudy or low-sunlight conditions. The main present challenge, however, is the cost of generating electricity from solar cells; this is in a range from \$0.2–\$0.8 per kWh, depending on the location [120, 121], and is an order of magnitude higher than electricity production from natural gas at some \$0.03 per kWh. As demand increases and larger quantities of PV are produced, the costs will decrease. However, further breakthroughs in solar-cell production are needed to decrease significantly their cost. Until then, the contribution of photovoltaic power to global electricity generation is likely to remain relatively limited and strongly dependent upon government incentives. Regardless, it is a promising and feasible alternate source of electricity generation.

8.5.2
Solar Thermal Power for Electricity Production

Solar thermal power systems use mirrors and optical devices to redirect, focus and concentrate the sun's rays onto a receiver, where heat is generated. The first application of this technique was recorded in 212 BC, when Archimedes is said to have used mirrors to burn Roman ships attacking Syracuse. Today, the thermal energy collected can be used to produce steam to drive generators for electricity production, or to promote chemical conversions such as methane reforming. Sunlight may be concentrated with different technologies, including parabolic troughs, solar power towers and parabolic dishes. These technologies differ in the way in which they collect solar radiation. Parabolic trough collector systems are commercially available and are the least expensive solar thermal technology. They use single-axis sun-tracking parabolic mirrors to focus sunlight onto a glass tube that runs in the trough and contains a heat-transfer fluid that reaches temperatures up to 400 °C and, via a heat exchanger, generates steam to drive an electric generator. Nine plants, ranging from 10 to 80 MW, have been installed in the Southern California Mojave Desert since 1991, providing a total capacity of 354 MW of peak electricity to the grid [122]. Nevada Solar One, a plant completed in 2008 close to Las Vegas, Nevada, has a total capacity of 64 MW, providing enough electricity for 14 000 households [123, 124]. Other projects are under way in several countries such as Spain, which has tested, at the Plataforma Solar de Almería, the possibility of using superheated steam directly as the heat-transfer fluid (Figure 8.11).

Figure 8.11 Two EuroDish-Dish/Stirling systems in operation at the Plataforma Solar de Almerìa, Spain. (Source: Schlaich Bergermann und Partner, World-Wide Information System for Renewable Energy (WIRE).)

One commercially less mature technology other than parabolic trough is that of solar towers, which use many large mirrors called heliostats that track the path of the sun throughout the day and focus the rays on the solar receiver used to heat a working fluid, typically molten salt, to generate steam and produce electricity. Part of the energy can be stored in the form of heat in the molten salt to produce electricity during the night. Because solar energy is concentrated on a point rather than a line in the case of parabolic trough, the temperature generated is higher; in the 500–1500 °C range. The Solar One and Solar Two projects, which were also operated in the Mojave Desert until 1999, demonstrated the technology at the 10-MW pilot scale.

Smaller solar towers have also been field tested in different countries, including Italy, France, Spain and Japan, not only for electricity generation but also other applications requiring high temperatures such as methane production and material testing [122]. Solar dishes, which resemble satellite dishes, use parabolic mirrors that track the sun in two axes, concentrate the sunlight at their focal point and generate temperatures similar to those of solar towers. In the most common approach, the concentrated solar energy is used to heat the hydrogen or helium working fluid utilized in Stirling engines [122]. The generated mechanical energy is then converted into electrical energy. On average, the dishes are between 8 and 10 m in diameter, though some can be much larger, like the "Big Dish" in Australia with a surface area of 400 m². Generally, units are small, producing between 10 and 25 kW. Like PV setups, they can be installed in large groups to serve utility needs or in smaller numbers for decentralized energy generation. Stirling Energy System recently announced its plan to construct the largest solar installation in the world, based on solar dishes, in the Mojave Desert in Southern California. When completed, the facility will consist of 20 000 25-kW solar dishes with a production capacity of 500 MW, comparable to a typical fossil fuel power plant [125, 126]. In addition, a 300-MW plant containing 12 000 solar dishes will also be constructed to provide electricity to the city of San Diego, California. Of all solar technologies, solar dish/engine systems have demonstrated the highest solar to electric conversion efficiency at 29%, compared to around 20% for other solar thermal technologies [122]. With receiver temperatures in excess of 1000 °C for solar tower or solar dish technology, processes to generate hydrogen thermochemically could also be exploited and are under investigation. The production cost of electricity with a parabolic trough, the most mature technology, is still high at $0.10–0.15 per kWh for the most efficient power plants [112]. To bring solar thermal power generation to the market, the key issue of cost reduction is currently being studied through research and technological development in numerous projects around the world. Solar thermal power plants seem currently the most promising to become the first large-scale solar electricity producers.

8.5.3
Electric Power from Saline Solar Ponds

A solar saline pond is a few meters in depth and artificially maintained so that the degree of its salinity, and consequently its density, is higher at the bottom than at the surface. The difference in salinity is created by dissolving large amounts of salt at the bottom of the pond and keeping the surface supplied with low-saline water to maintain the necessary salinity gradient. Because of the difference in salinity there is minimal mixing between the layers, and convection is prevented. Absorption of solar energy by the bottom of the pond heats the lower depths of water, which are prevented from rising by higher density relative to upper part of the pond. Under these conditions, water at the bottom of the pond can attain temperatures close to 90 °C. Research on saline solar ponds began in the 1950s in Israel, and resulted in the construction of a 25-hectare demonstration plant in Beit Ha'aravah near the Dead Sea in the early 1980s [127]. A 5-MW low-temperature turbine using a low-boiling working fluid was used to transform the temperature difference between the pond's water layers into electricity. The overall efficiency of such a relatively low-temperature system was only about 1%. On a continuous basis, the pond provided approximately 800 kW. The unique feature of the saline pond, however, is that it can store the energy of the sun in the form of heat and provide electricity even at night and during cloudy days. The Israeli solar saline pond power plant was operated until 1990. Since then, only one other plant using this technology has been built – in Texas to provide a food cannery with both electricity and heat. Owing to the relatively low efficiency of the system and the large area of land required, the potential of energy production from solar saline ponds is, however, limited.

8.5.4
Solar Thermal Energy for Heating

Solar thermal systems are widely used to provide heating and hot water for residential, commercial or industrial applications. Since their use took off in the 1970s as a result of high oil prices, millions of units for the production of hot water have been installed worldwide. In Tokyo alone there are 1.5 million buildings equipped with solar water heaters [100]. Some small countries such as Israel are also widely using solar thermal energy for homes and other installations [127]. They are the most widespread and easiest way to utilize solar power, as only moderate temperatures are needed and the technology is simple. Solar heating systems are composed of a collector in which a liquid, generally water, is heated and a pump to transfer the heat to living spaces or to a storage tank for later use. In the thermal collector, mounted on the roof facing the sun, a fluid running through a polymer or copper tubing is heated by sunlight. For some applications, solar water heating systems have become especially popular: in the western United States, heating of swimming pools accounted for more than 90% of the installations in the late 1990s [100]. Although government incentives and further cost reductions are still desirable, the technology is mature and already considered to be competitive in an increasing number of countries. In contrast, solar heating and cooling of buildings is not yet sufficiently competitive with conventional energy sources, and therefore is generally still rare.

8.5.5
Economic Limitations of Solar Energy

Today, the major problem facing solar energy, especially for the production of electricity, is its cost, which is still high when compared not only to fossil fuels and atomic energy but also to renewables such as wind or hydropower. Despite the free and inexhaustible solar power source, the up-front cost for the equipment to collect and store solar energy is high. Likewise, owing to the diffuse nature of sunshine, the collection surface area necessary to produce large amounts of solar energy is by necessity large. The absence of sunlight during the night and reduced insolation in cloudy conditions also necessitates back-up generators or expensive and limited-energy storage batteries [128]. Energy storage in the form of hot molten salt or pressurized air has also been proposed and tested. In the latter case, electricity produced from the sun can be used to compress air and pump it into underground caverns, aquifers, abandoned mines or depleted natural gas fields. When needed, the compressed air is released to move turbines, generating electricity. A 290-MW capacity compressed air energy storage (CAES) facility has been operated reliably in Huntorf, Germany for 30 years. In this plant off-peak electricity from the grid is stored and released back to the grid during peak periods. A similar 110 MW plant has been running since 1991 in McIntosh, Alabama. Compressed air storage facilities, like pumped hydro, are suitable for large-scale energy storage in the hundreds of MW for which batteries, flywheels, supercapacitors and other storage technologies are ill-suited and more expensive [129, 130]. The same compressed air storage technology could also be used for other intermittent and variable energy sources, especially wind energy [131–133]. Of course, energy storage will add to the cost of electricity produced from these renewable sources. With continued research and technological improvements, however, solar energy will, in the long run, certainly become an important part of our overall energy-mix. As mentioned, using solar energy to convert CO_2 with hydrogen generated by electrolysis of water is certainly a promising approach.

8.6
Bioenergy

Bioenergy or biomass energy refers to the use of a wide range of organic materials as fuels. These are produced by biological processes, and include forest products, agricultural residues, herbaceous and aquatic plants, and also municipal wastes. In principle, bioenergy is inexhaustible and renewable, provided that new plant life is grown to replace the ones harvested for energy. Biomass can either be burned to produce heat or electricity, or transformed into liquid fuels such as ethanol, methanol or biodiesel. Biomass has been our predominant source of fuel well into the nineteenth century. Whereas dried plants, plant oil, animal fat or dried dung were used for lighting and cooking needs, the most common biomass source was wood. Its dominance was progressively replaced by fossil fuels, first coal and then oil and gas during the nineteenth and twentieth centuries. Today, biomass supplies about 10% of the world's primary energy consumption. Indeed, in many poor countries it continues to be the most important source of energy for heating and cooking purposes, whereas in developed nations only a small fraction of the energy needs are covered by biomass.

8.6.1
Electricity from Biomass

The cheapest, most utilized and simplest way of using biomass to generate energy is to burn it. On a commercial scale, this process is carried out in a similar way to the one used for burning coal to produce electricity or heat. In these applications, wood, wood waste and municipal solid waste are the most utilized fuels. The average plant is generally small (around 20 MW) with an efficiency ranging from 15 to 30% for conversion into electricity [100]. With co-generation of electricity and heat, the total efficiency can reach 60%. Methane-rich biogas, if captured and collected from landfills, can also be used to generate sizeable amounts of energy.

New technologies that are commercially available for converting biomass into electricity include co-firing and gasification. Co-firing power plants use biomass as a supplementary energy source with a conventional fuel, typically coal. Gasification converts solid biomass, through partial oxidation at high temperature, into a combustible gas containing mainly carbon monoxide and hydrogen. The gas produced can then be burned in a gas turbine or internal combustion engine to generate electricity. Notably, during the Great Depression and World War II, small gasifiers were used for cars to convert wood and charcoal into gas that was then fed to the engine. These vehicles were not very efficient and needed extensive maintenance, but nevertheless functioned quite well.

The cost of biomass energy varies widely, depending on the fuel, its quality and the technology used. Electricity-generating costs are, however, generally higher than those for fossil-fueled plants because of lower efficiencies, higher capital and fuel costs. Most estimates for the fuel cost are in a range $150–250 per tonne, but this can be much lower in cases where the fuel is a by-product from some other process [100].

Table 8.1 Production of electricity from biomass and waste in 2006.

Country	Production (TWh)	Percentage of world electricity production from biomass	Percentage of the country's total electricty production
United States	58.7	29.3	1.5
Germany	19.7	9.9	3.4
Brazil	14.6	7.3	3.9
Finland	11.8	5.9	14.0
Japan	11.6	5.8	1.1
United Kingdom	9.3	4.6	2.5
Canada	9	4.5	1.6
Spain	8.2	4.1	3.1
Rest of the world	57.2	28.6	0.6
World	200.1	100	1.2

Data source: EDF and IEA key statistics.

Among the OECD countries (Organization for Economic Cooperation and Development, including most developed nations), electricity from bioenergy represented 1.5% of the total electricity generation in 2005 [121]. More than half was produced from solid products such as wood and agricultural residues, while waste also accounted for an important part of electricity produced by biomass. In 2006, in forest-rich Finland, electricity from bioenergy represented 14% of the total electricity production, but only 1.5% in the United States (which has the largest overall electricity-generation capacity from biomass, but also the highest electricity consumption) (Table 8.1).

8.6.2
Liquid Biofuels

Biofuels are liquid fuels produced from crops or biomass feedstock through different chemical or biological processes. Today, they are the only available renewable source for producing high-value liquid biofuels such as ethanol or biodiesel. These fuels can offer renewable alternatives to transportation fuels that presently are obtained almost exclusively from oil. Bio-ethanol, the most common biofuel, is produced by fermentation of annually grown crops (sugarcane, corn, grapes, etc.). In this process, starch or carbohydrates (sugars) are decomposed by microorganisms to produce ethanol. Ethanol can be produced from a wide variety of sugar or starch crops, including sugar beet and sugarcane and their by-products, potatoes and corn surplus. In Russia after the Bolshevik revolution, Lenin proposed the use of agricultural alcohol to produce industrial fuels and products. Because this process would have diverted the Russian people's beloved source of vodka to industrial use, the plan was soon abandoned. During World War II in Europe, blends of ethanol with gasoline were used, but only anhydrous ethanol is miscible with gasoline, phase separation otherwise causing

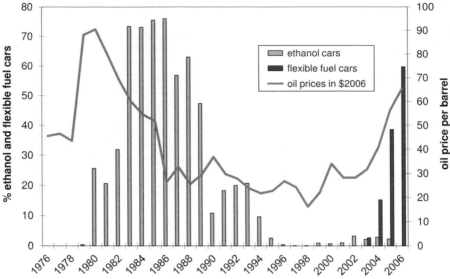

Figure 8.12 Percentage of ethanol-powered automobile produced
in Brazil compared to the country's total automobile production.
(Source: Associação Nacional dos Fabricantes de Veículos
Automotores ANFAVEA, *Brazilian Automotive Industry Yearbook
2007.*)

stalling of the engines. Ethanol has been promoted and used more recently and
extensively in Brazil and the United States as a response to the OPEC oil embargoes
and rising gasoline prices (but also to subsidize farmers). Initially, Brazil – one of the
largest sugarcane producers in the world – gave farmers financial incentives to switch
from sugar to ethanol production. This plan was implemented, and by the mid-1990s
about 4.5 million vehicles were running on pure ethanol, and the remainder on a
blend of gasoline containing up to 24% ethanol by volume. With falling oil prices and
the end of price subsidies for ethanol at the end of the 1990s, however, the sale of pure
ethanol cars fell to almost zero (Figure 8.12) [134, 135]. Increasing oil prices,
experienced more recently, have revived interest in ethanol. The production of ethanol
from sugarcane in Brazil reached some 22 million m³ per year in 2007 [136]. The
extraordinarily high productivity of sugarcane (up to 80 t cane per hectare, compared to
10–20 t for most plants cultivated under temperate climates), associated with low
wages, has contributed to a great extent to the competitiveness of ethanol in Brazil. So-
called flexible fuel vehicles (FFV), able to run on any mixture of gasoline and ethanol,
have also been introduced recently. In only a few years the market share for FFV has
increased to over 70% of the cars sold in Brazil. However, the amount of ethanol
produced annually in Brazil represents only the equivalent of some 10–11 million
tonnes of petroleum oil, less than the quantity consumed by the world in a single day.
Besides alcohol, sugarcane by-products – principally bagasse (sugarcane husk) – are
burned to supply Brazil's grid with about 600 MW of electricity [137]. This again, is less
than the output of a single large-scale fossil-fuel or atomic power plant.

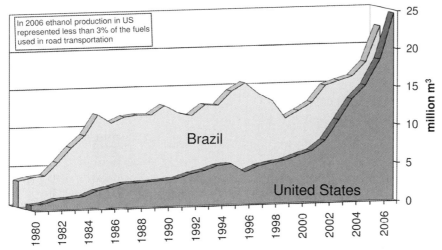

In 2006 ethanol production in US represented less than 3% of the fuels used in road transportation

Brazil

United States

Figure 8.13 Historic production of ethanol in the United States and Brazil. (Based on data from Renewable Fuel Association and Sao Paulo Agroindustry Union (UNICA).)

In the United States, ethanol produced from corn has been used in gasohol, a blend of 10% ethanol and 90% gasoline, as well as an oxygenated additive in gasoline since the early 1980s (Figure 8.13). E85, a blend containing 85% ethanol and 15% gasoline, is also available in a limited number of gas stations. However, ethanol has to be dry to be mixed with gasoline to avoid phase separation. Drying ethanol is an energy-intensive process, since ethanol readily forms azeotropic mixtures with water. Azeotropic mixures cannot be separated by simple distillation. Furthermore, ethanol is only economically competitive because of significant tax subsidies. Besides the current $0.51 per gallon federal subsidy for blending ethanol into gasoline there are also other direct and indirect federal and state subsidies directed to ethanol fuel. Total subsidies for ethanol fuel in the United States have been estimated at between $1.05 and $1.38 per gallon [138]. Growing corn for ethanol production is also very energy intensive because of the need and cost of fertilizing, harvesting and transporting the corn, as well as subsequent fermentation and distillation, which requires large amounts of energy that are generally provided from oil and natural gas. It should be emphasized that ethanol produced from corn produces at most only 25–35% more energy than was consumed in its production [139, 140]. Indeed, some claims state that the process is a net energy user [141–144]. In any case, corn used for ethanol production is far from an ideal feedstock, though dedicated energy crops and new genetically engineered crops could increase the energy efficiency. Utilization of the cellulose content of plants, which is resistant to fermentation, to produce ethanol has in the past only been possible by prior hydrolysis with sulfuric acid. Currently, the development of new strains of microorganisms capable of digesting cellulose directly is being explored, and this may allow for the use of other types of vegetation with lower production costs to be processed to ethanol, making the overall process cheaper and more efficient.

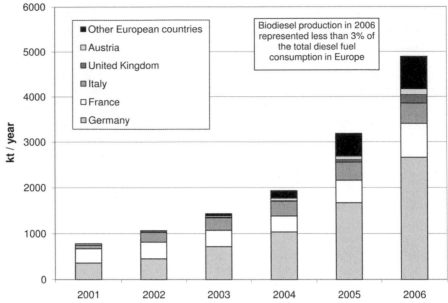

Figure 8.14 Production of biodiesel in Europe. (Source: European Biodiesel Board and Eurobserver.)

Biodiesel, processed from seed crops such as rape, sunflower and soy are currently produced mainly in Europe (Figure 8.14) and the United States on a limited scale. Market penetration is small and the production costs are relatively high, although interest is growing. In 2006, biodiesel represented only 1.8% of the 300 Mt fuel (gasoline and diesel fuel) consumed by road transport in Europe (EU 25) [145, 146]. The direct use of plant oils in diesel engines is not recommended as it will considerably reduce the engine's lifetime. This drawback has been mitigated by reacting these oils with methanol or ethanol in a so-called transesterification process to yield commercial biodiesel. Biodiesel can be blended without major problems with regular diesel oil in any proportion. The production of biodiesel is also less energy intensive than ethanol from corn because no fermentation and distillation is necessary. However, biodiesel from oil-seed crops requires up to five times more land per unit of energy produced than ethanol. A plant that is gaining wide interest as a biodiesel source is the Jatropha tree. The nuts of this tree contain a significant amount of oil but are not suitable for human consumption. As it can be planted even in degraded lands, the Jatropha tree does not compete with crops for arable land [147, 148]. India has identified 40 million hectares were Jatropha can be grown and hopes to use this resource to replace 20% of its diesel consumption in 2011. China, the Philippines, the US and many others are also pushing for the use of Jatropha.

In Brazil, biodiesel is also produced from sugarcane-based ethanol involving ethylene as an intermediate.

8.6.3
Biomethanol

Methanol was once produced from wood, and is therefore sometimes still referred to as wood alcohol. This process (called pyrolysis) was used well into the beginning of the twentieth century and involves the heating of wood in the absence of air to yield a mixture of solid, liquid and gaseous products from which methanol could be extracted in low yields. Today, methanol is predominantly produced by steam reforming of methane (natural gas) and subsequent reaction of the produced syn-gas to methanol. However, any source of carbonaceous material could be converted into syn-gas and thus methanol. These materials also include biomass, which could therefore become a source of methanol in the future.

8.6.4
Advantages and Limitation of Biofuels

Recently, considerable scrutiny and criticism has been directed to the production of ethanol and biodiesel from food crops. Their production has contributed to a significant extent to the sharp increases in food prices experienced in recent years [10]. Higher prices and subsidies have resulted in the conversion of forests, savannahs, grasslands and so on into cropland, especially in Southeast Asia and the Americas. The Brazilian Amazon is being converted into soybean cultivation and the Brazilian Cerrado Savannah into sugarcane and soybean production. In Malaysia, plantations of oil palms for biodiesel production have been held responsible for 87% of the deforestation. Growing demand for biofuels is only expected to increase the pace of destruction of rain forest in biologically rich ecosystems such as the islands of Sumatra and Borneo. Land clearing also releases large amounts of CO_2 as a result of fires set to clear the land and the decomposition of leaves and other plant material through microbiological processes. The vast amounts of CO_2 released from the land conversion is considered as a "carbon debt." Some studies suggest that the CO_2 emissions avoided by using current biofuels (ethanol from corn, palm oil and soybean biodiesel) compared to fossil fuel based gasoline would take between 35 and 450 years to repay this carbon debt [149, 150]. These results raise questions about the value of presently used biofuels based on agricultural crops to cut greenhouse gas emissions. In Europe these concerns have already prompted changes in future energy policies [151–154]. Besides green house gas emissions, biofuel production also has other environmental effects such as loss of biodiversity when large tracks of rainforest have to be cleared for sugarcane or palm mono-cultures, and utilization of vast amounts of water for irrigation [155–158].

Biomass as an energy source has, however, many advantages. It is renewable, provides a convenient way of storing energy (e.g., in the form of wood), which is not the case for wind or solar energy, and it can be found in different forms all over the world. It can reduce the energy dependence on foreign countries. Biomass is also versatile as it includes solid fuels such as wood or crop residues, liquid biofuels such as ethanol and biodiesel, and gaseous fuels in the form of biogas or syn-gas. Environmentally, biomass can help mitigate climate change, as the CO_2 released by use of bioenergy is captured from the atmosphere by the growing plants and should therefore result in no net CO_2 emissions (i.e., it is carbon neutral). However, the conversion of solar energy into biomass is only achieved at an efficiency of around 1% or less, which is very low even compared to the inefficient conversion of solar energy into electricity with solar cells (in excess of 10%). To generate bioenergy on a large scale, vast areas of land and water resources are necessary. Also, great care must be taken in choosing crops for energy production, as these should have a high photosynthetic efficiency, grow rapidly, use minimal amounts of fertilizers, herbicides and insecticides, and have limited water needs to minimize energy input for their cultivation. These "energy crops" should also preferably be grown on land not dedicated to food crops to avoid competition with food production. The use of land that has been degraded by human activities over historical times in particular should be promoted to stabilize erosion and related problems such as desertification and deterioration of watersheds. The vast expanses of the seas can also be used to grow algae, which can be utilized to produce bioenergy, and experimental facilities in the United States and Japan have explored this possibility. With the present technology, a large part of the world's agricultural land would have to be devoted to energy crops if they were to supply a substantial amount of our energy needs. Even if the use of bioenergy could be cost effective in certain cases for the production of heat and electricity, it has generally higher costs than conventional energy sources. An economical and sustainable large-scale use of this resource will therefore require technological advances or breakthroughs, especially in the bioengineering field, to design suitable high-yield energy crops. In the transportation sector, the production of cellulose-based ethanol and other liquid fuels, particularly methanol through biomass gasification, will allow a higher yield per unit of land [159]. Algae grown in ponds, tanks or the sea might also eventually extend the scope of bioenergy [160].

8.7
Ocean Energy: Tidal, Wave and Thermal Power

Energy contained in the oceans, which cover more than 70% of the earth's surface, exists in two forms: (i) mechanical energy in the waves, tides and marine currents; (ii) thermal energy absorbed from the sun's heat. Because ocean energy is abundant, all of these sources have been considered for energy production and are at various stages of development.

8.7.1
Tidal Energy

Tidal energy exploits the rise and fall of tides caused by the interaction of the gravitational fields of the moon and sun. Tidal movements are both periodic and predictable, and were already used before 1100 AD in tide mills in the United Kingdom and France to grind grain. Their modern version, the tidal power plant, which takes advantage of the difference in water levels between low and high tide, operates on the same principle as an ordinary hydroelectric plant. Behind a dam constructed across a bay or estuary, water flowing through sluices is stored during high tide and later released at low tide through turbines to generate electricity. To be exploitable for electricity generation, however, differences between high and low tides must be significant. Only a few locations worldwide, generally in river estuaries, are suitable. Today, the only large-scale tidal power station, with 240-MW generating capacity, is situated at the estuary of the Rance River, near Saint-Malo, France. It was built in the 1960s and has now completed more than 30 years of successful operation. Since then, the only other, much smaller, tidal power plants constructed have been an 18-MW project in Annapolis, Bay of Fundy, Canada, a 3.2-MW facility in China and a very small 0.4-MW unit near Murmansk, Russia [161]. In the European Union, the technically exploitable resource is estimated at about 100 TWh per year, with only around 50 TWh per year economically viable. Most possible sites (90%) are located in France and the United Kingdom. Beyond the EU, Canada, the states of the Former Soviet Union, Argentina, Western Australia and Korea have potential sites that have been investigated. Taking advantage of this resource, South Korea has recently announced its plans to construct the world's largest tidal plant (260 MW capacity). Tidal energy is a mature technology, but it requires high capital expenditure, long construction times and has low load factors, leading to long payback periods and high electricity generating costs. Governments, thus, are likely to remain the only entities to undertake such large-sized projects. One significant factor for the limited use of tidal power raised by ecologists has been the potential environmental impact of such projects. Studies conducted in the United Kingdom, however, concluded that this technology does not necessarily cause major environmental changes; rather, economic considerations are the real barrier to further development of tidal power. Nevertheless, even if all the potentially technically exploitable sites for tidal energy were to be developed, this alternative source of energy would still only represent a very small fraction of our energy needs.

Tides also drive marine currents. Although energy in marine currents is generally diffuse, it is concentrated in some locations near the coast where sea flows are channeled through constrained topographies such as straits and islands. Tidal turbines that resemble wind turbines, but are placed under water, can be used to generate electricity from these currents. Since the density of water is about 1000 times that of air, the power density of these currents is appreciably higher than that of wind, and therefore much smaller turbines are required. In contrast to wind, marine currents are also highly predictable, and they would also have considerably less environmental impact than the construction of a tidal dam. Tidal turbines, however, are for the most part still in the research and development phase with few prototypes currently in operation in the United Kingdom, Ireland, Norway and elsewhere [162]. Tidal turbines, with a total generating capacity of 200 kW have been installed in New York City [163, 164]. Recently, a commercial-scale tidal turbine with a 1.2-MW capacity was also installed close to Belfast in Northern Ireland [165].

8.7.2
Wave Power

Waves are generated by wind blowing across the sea surface. Wave energy is thus a concentrated form of solar energy. The amount of energy in waves depends on the wind speed and power. In particular, deep ocean waves with large amplitudes contain considerable amounts of kinetic energy. The strong winds blowing across the Atlantic Ocean, creating large waves, make the western coasts of Europe ideally suited for wave energy. Other wave-rich areas in the world include the coasts of Canada, northern United States, Southern Africa and Australia. Wave energy is a relatively new technology and is still in the research and development and demonstration stages. It was pursued vigorously in the 1970s and 1980s. A wide variety of designs have been proposed to harness wave energy with many different extraction methods. However, only a fraction of them have actually been realized and tested as demonstration units. Wave energy can be converted into electricity in both onshore and offshore systems. Several onshore systems have been built in Scotland, India, Norway, Japan and other countries using mainly three different designs: (i) systems that funnel waves into reservoirs; (ii) pendulor systems driving hydraulic pumps; and (iii) oscillating water columns that use waves to compress air within the systems [166]. The mechanical power created with these devices is then used directly or via a working fluid, usually water or air, to drive an electric turbine/generator unit. Shoreline systems have the advantage of being easier to maintain, but the energy potential is higher in offshore locations, where several systems have also been tested. Recently, construction of the first commercial wave farm was completed in Portugal [167], with electricity being generated by devices called Pelamis, named after a giant sea snake in Greek mythology. The name comes from their shape, and involves a series of connected cylindrical modules with a total length of 150 m and a diameter of only 3.5 m. In the initial phase of the project, three of these machines, each with a capacity of 750 kW, located 5 km off the coast will produce up to 2.25 MW. Eventually, an additional 25 Pelamis machines will be installed to generate up to 21 MW of power [168]. The global wave power resource is estimated to exceed 2 TW, with the potential to generate more than 2000 TWh annually (about 10% of the world's present electricity consumption). However, like tidal power, wave power – despite continuing cost reduction – is unlikely to be economically competitive with other power sources in the near future, except in special cases such as isolated coastal communities far from any electric grid.

8.7.3
Ocean Thermal Energy

Ocean thermal energy is also a potential source of energy. The oceans are the world's largest solar collectors, absorbing enormous amounts of solar energy in the form of thermal energy. A process called ocean thermal energy conversion (OTEC) uses the heat stored in the oceans to generate electricity. It exploits the difference in temperature between the sea's upper layer, which is warmed by the sun, and the colder deep water. The warmer water at the surface is used to vaporize a working fluid, or is converted into gas under vacuum, to run a turbine/generator system producing electricity. Cold water pumped typically from water depths of around 1000 m is then employed to recondense the vapor and close the cycle. To produce significant amounts of power, the temperature between the upper and lower water layer should differ by at least 20 °C. These conditions are encountered in tropical seas. Tapping the thermal energy of the oceans was initially proposed in 1881 by the French physicist Jacques Arsene d'Arsonval. It was, however, his student Georges Claude who built the first OTEC land-based system in Cuba in 1930, followed by a floating model off the coast of Brazil. Owing to poor location selection and technical difficulties these projects were abandoned before they actually produced more power than was necessary to run the systems. In the late 1970s, interest in OTEC rose again and a few experimental units were constructed in Hawaii, Japan and the Republic Island of Nauru in the Pacific Ocean. Besides electricity production, the spent cold deep seawater, rich in nutrients from an OTEC plant, could also be used for the culture of both marine and plant life near the shore or on land. In cases where water is used as the working fluid, fresh water obtained by vacuum distillation could be produced from seawater. High construction costs are, however, still an obstacle for this technology. The fact is that OTEC technology attempts to produce electricity from a small temperature difference that would be typically considered unusable for power generation. To compensate for this, massive amounts of water from the ocean's surface and their depths have to be pumped through the plant to generate reasonable quantities of energy. The challenge is to achieve this process economically. Thus, even if the potential resources are very large, OTEC is still in the research phase and is likely to remain there in the foreseeable future. If developed and eventually economically viable, the first market for this technology would most likely be in tropical island nations for the combined production of energy and desalinated water.

8.8
Nuclear Energy

In 2006, nuclear power, which is used almost exclusively to produce electricity in commercial applications, generated some 2800 TWh of electric power, representing about 15% of the world's electricity consumption (Figure 8.15). Its share of the total primary energy supply amounted to 6.2% [42]. Today, some 439 commercial nuclear reactors are operating in 30 countries with over 370 000 MW of total production capacity. Currently, the United States has 104 commercial nuclear power plants, accounting for 19% of its electricity generation. In Western Europe, nuclear energy generates around 35% of the electricity; more than from any other source. France and Belgium, for example, produce, respectively, 77% and 54% of their electricity through nuclear power. Other industrialized countries with limited or no fossil fuel resources, such as Japan and South Korea, also rely heavily on nuclear energy for their electricity supply.

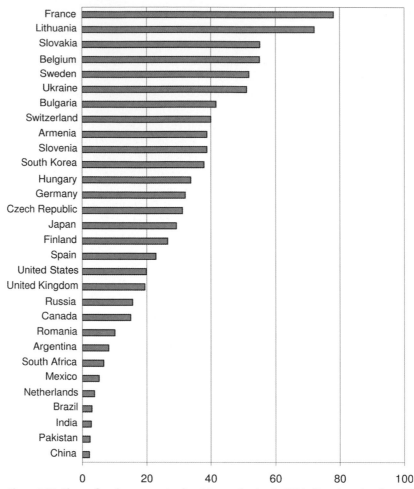

Figure 8.15 Share of nuclear energy in electricity production in 2004. (Based on data from IAEA.)

All commercial nuclear plants presently use uranium as fuel. Uranium is a slightly radioactive metal that occurs naturally throughout the earth's crust. It is about 500 times more abundant than gold, and as common as tin, tungsten and molybdenum. Uranium was originally formed in stars that, at the end of their life, exploded, with some of their shattered dust aggregating to form our planet. Uranium is present in most rocks in concentrations of 2–4 ppm. In phosphate rocks used as fertilizers, its concentration can be as high as 400 ppm, while some coal deposits have uranium concentrations in excess of 100 ppm. Uranium is also dissolved in sea water at a concentration of 3–4 ppm. There are, however, several areas where the concentration of uranium is much higher and economically exploitable. These deposits are particularly important in Australia and Canada, which are therefore currently the largest uranium producers in the world. Some Canadian deposits contain more than 100 kg of uranium per ton of raw ore. Like coal, uranium has to be mined in underground or surface mines depending on the depth at which the deposit is located. It is then sent to a mill where the ore is crushed to powder and leached with a strongly acidic or alkaline solution to extract the uranium from the rock. By precipitation from this solution, uranium oxide (U_2O_3) powder, referred to as "yellow cake" (because of its color), is obtained. Natural uranium consists of a mixture of two isotopes: uranium-235 (^{235}U) and uranium-238 (^{238}U). Only the isotope ^{235}U, which represents merely 0.7% of the natural uranium, can undergo fission, the process by which energy is generated in nuclear reactors. Even if some reactors are able to use natural uranium to produce energy, the vast majority of them require a higher concentration in ^{235}U. Thus, it is necessary to enrich uranium from its original ^{235}U concentration of 0.7% to typically 3–5%. For this enrichment process to occur, the uranium must be in a gaseous form, and this is carried out by conversion into uranium hexafluoride (UF_6), which is a gas at relatively modest temperature (solid UF_6 sublimes at 56 °C). UF_6 is the basis for the two enrichment processes used today on a large commercial scale: gaseous diffusion and gas centrifugation. Both of these take advantage of the difference in mass between ^{238}U and ^{235}U to separate the two isotopes. However, because the masses of the two isotopes are very close, their separation requires advanced technology. Once enriched to the desired level, UF_6 is converted into enriched uranium dioxide (UO_2, with a melting point of 2800 °C) powder that is pressed into small pellets inserted into long thin tubes, made from a zirconium alloy, to form fuel rods. The rods are then sealed and assembled into clusters to form fuel assemblies ready for use in nuclear fission reactors.

8.8.1
Energy from Nuclear Fission Reactions

Inside the nuclear reactor, the released energy comes from the fission of ^{235}U atoms (Figure 8.16). When a ^{235}U nucleus is struck by a slow neutron, it splits, releasing two smaller atoms and two or three neutrons in a process called "fission." The two new atoms may themselves undergo further radioactive decay, releasing beta or gamma radiation to achieve stability. The energy released by the fission results from the fact that the fission products and emitted neutrons together weigh less than the original ^{235}U atom. Following Einstein's famous equation, $E = mc^2$, the difference in weight is converted into energy. In fact, fission produces a tremendous amount of energy, as 1 kg of pure ^{235}U can generate over 2 million times more energy than 1 kg of coal (Table 8.2)! Because each fission event liberates two or three neutrons able to split other ^{235}U atoms, which themselves release two or three neutrons and so on, a rapidly multiplying sequence of fission events, known as a nuclear chain reaction, can occur and emit increasing quantities of energy. If controlled, this chain reaction can be used to produce a large and sustained amount of energy. This possibility was demonstrated in 1942, following Leo Szilard's original suggestion, by an outstanding team of scientists led by Enrico Fermi, who constructed the first nuclear reactor in a squash court under the stands of the University of Chicago's football field. At that time the project, funded by the US government, was directed to investigate the possibility of making an atomic bomb based on nuclear fission. This first nuclear reactor, called an

Nuclear fission chain reaction

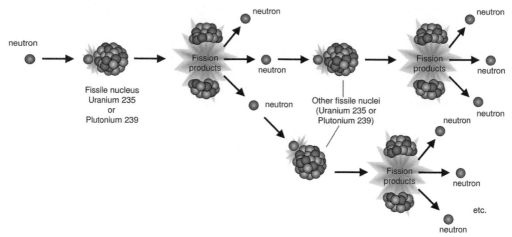

Figure 8.16 Nuclear fission chain reaction.

"atomic pile," consisted of highly purified graphite, uranium, uranium oxide and cadmium control rods, and was about the size of a two-car garage. To control the chain reaction, it was necessary to control the neutron flow. When a ^{235}U nucleus is broken by fission, the neutrons emitted have high kinetic energy, meaning that they have high speeds on the order of $20\,000\,\mathrm{km\,s^{-1}}$. These fast-moving neutrons will most probably be captured by ^{238}U atoms, which still constitute the major part of the uranium but do not undergo fission. Slower-moving neutrons have a much lower energy content. They are still able to split ^{235}U nuclei and sustain the chain reaction, but are less likely to undergo reactions with other kinds of atoms. To reduce the neutron's speed, so-called moderators are used. The best moderators are low-mass atoms such as deuterium, helium or carbon, which efficiently decrease, by successive collisions, the neutron's energy, keeping neutron losses at a minimum, and do not undergo fission themselves. Many early atomic reactors used graphite as a moderator because it is inexpensive and easy to handle. Water and heavy water (water in which the hydrogen atoms have been replaced by deuterium, D_2O), however, are more efficient moderators and are the most widely used for nuclear reactors today. To control the rate of the chain reaction, control rods, able to absorb neutrons

Table 8.2 Energy content of various fuels.

Fuel	Average energy content in 1 g (kcal)
Wood	3.5
Coal	7
Oil	10
LNG	11
Uranium (LWR, once through)	150 000

Source: Nuclear Energy Agency (Nuclear Energy Today, Nuclear Energy Agency, OECD publications, Paris 2003, Available from: http://www.nea.fr/html/pub/nuclearenergytoday/welcome.html).

without re-emitting them, are also necessary. These generally contain cadmium or boron, and can be moved in and out the reactor to regulate the flux of neutrons and thus the amount of energy produced. Fully inserted, the control rods will stop the chain reaction.

Using a combination of uranium, uranium oxide, graphite moderator and cadmium control rods, Fermi's first nuclear reactor, when successfully tested in December 1942, produced only a few watts of energy, but proved that the controlled use of atomic energy was possible.

Most importantly at the time, it also showed that the construction of an atomic bomb was achievable. It triggered the Manhattan project, which led to the explosion of the first atomic bomb in New Mexico in 1945 and put an end to World War II. In the early years, military applications dominated the use of nuclear energy. In 1954, for example, the first nuclear submarine, the USS *Nautilus*, was launched. It was able to stay underwater without refueling for long periods of time – an unimaginable feat before the advent of this new energy source.

Interestingly, Fermi's nuclear reactor was not really the first one on earth. As we now recognize, about 2 billion years ago, natural nuclear "reactors" existed in a rich deposit of uranium near what is now Oklo, Gabon. At the time, the concentration of ^{235}U in all natural uranium was above 3% instead of today's level of 0.7% (due to radioactive decay). Natural chain reactions started spontaneously in the presence of water, acting as moderator, and continued for about 2 million years before stopping. During that time, fission products as well as plutonium and other transuranic elements were naturally formed [169].

The world's first commercial-scale nuclear power plant opened in 1956 in the United Kingdom. It was equipped with a Magnox reactor using graphite as a moderator and CO_2 as coolant gas. It used, like in Fermi's reactor, non-enriched natural uranium containing only 0.7% ^{235}U. In the United States, the first commercial nuclear power plant began operation in 1957 in Shippingport, Pennsylvania. It was based on a so-called pressurized water reactor (PWR), which is still the technology used in 60% of the plants currently in operation worldwide. The heat generated by the fission reaction is used to heat water in which the fuel rods are immersed. Reaching temperatures of around 300 °C, the water, however, does not boil because it is kept under high pressure. The pressurized water serves both as a moderator and coolant. Via a heat exchanger it is used to boil water in a secondary loop, producing steam to propel a turbine that in turn spins a generator to produce electric power.

The second most common type of nuclear reactor is the so-called boiling water reactor (BWR), with more than 90 units operating worldwide. The design of the BWR has many similarities with the PWR, except that the water cooling the core is allowed to boil and the steam generated is used directly to drive turbines. After condensation, the water is returned to the reactor to close the cycle. This system has a simpler design than the PWR. As the water around the core is contaminated with traces of radioactive material, although generally of short half-life, the turbine must be shielded to avoid the escape of radiation. For safety reasons, PWRs are thus the preferred reactors in the Western World. In France, for example, all 59 nuclear reactors in operation are of the PWR type.

Both the PWR and BWR, representing together more than 80% of the commercial nuclear reactors worldwide, are light-water (H_2O) moderated and use uranium enriched at 3–5% in fissile ^{235}U isotope. Because light water not only slows neutrons but can also absorb them, it is not as selective a moderator as heavy water (D_2O) or graphite. Therefore, the CANDU (Canada deuterium uranium) pressurized water reactors, developed in Canada using natural uranium (0.7% ^{235}U), are moderated with heavy water (D_2O). The cost of uranium enrichment is avoided, but extensive amounts of expensive D_2O have to be employed. About 40 CANDU reactors are presently still operating in seven different countries, including Canada, India, South Korea and China. In the United Kingdom, advanced gas-cooled reactors (AGRs), derived from the earlier Magnox reactors using graphite as moderator, CO_2 as a coolant and uranium enriched to 2.5–3.5% in ^{235}U, are in operation.

The commercial reactors used today (PWR, BWR, CANDU, AGR, etc.), constructed between 1970s and 2000 (Figure 8.17), are considered second-generation reactors. Currently, the transition to a third generation of reactors is in progress. Two units have already been completed in Japan, and several others are under construction or planned in countries such as Taiwan, France, Finland and Korea. They are an evolution from the second-generation reactors, but feature enhanced safety systems and are less expensive to build, maintain and operate. At the same time, revolutionary designs known as generation IV systems, which have new and innovative reactors or fuel cycle systems, are well under development [170, 171]. Small scale (10 to 100 MW), sealed and inherently safe, modular nuclear reactors have also been proposed for distributed power and heat generation and are currently being developed by the Lawrence Livermore Laboratory (SSTAR: Small, Sealed, Transportable, Autonomous Reactor) Toshiba (4S: Super Safe Small & Simple Reactor), Hyperion Power Generation and others [172, 173].

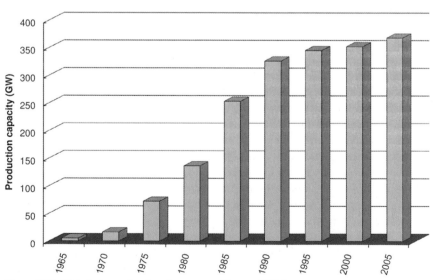

Figure 8.17 Historical growth of nuclear energy. (Source: IAEA.)

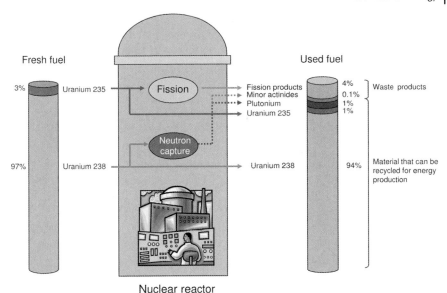

Figure 8.18 The nuclear fuel cycle in a once-through reactor.

Most commercial reactors currently in operation use enriched uranium in a once-through cycle (Figure 8.18); this means that the uranium is used only once and must then be disposed of. This cycle is the most uranium-reserve intensive, as only ^{235}U contributes to the production of energy. Uranium-238, which constitutes up to 97% of the fuel, and 99.7% of natural uranium, is left almost untouched. To solve this waste of resources a different fuel cycle has to be used. In the typical reactor, fast neutrons are slowed down by moderators to increase the probability of collision between these slow neutrons and the fissile ^{235}U nucleus, and thus increase the amount of energy generated by fission. Fast neutrons, however, can convert the ^{238}U isotope, which does not directly undergo fission, to plutonium 239 (^{239}Pu), a fissile material that can thus produce energy. Therefore, ^{238}U is referred to as a "fertile" isotope. Plutonium-239 was formed during the creation of the universe, but due to his its half-life of 24 110 years it disappeared a long time ago in Nature. In the existing fuel cycle, the enriched uranium, once used, still contains 1% ^{235}U but also 4% fission products, 0.1% of minor actinides and 1% ^{239}Pu. The spent fuel can be reprocessed and the useful ^{235}U sent back to the enrichment plant. Plutonium-239 can also be separated and mixed with uranium in the form of oxides to give a mixed-oxides fuel commonly known as MOx. MOx is currently in regular use to generate electricity in several countries, including Germany, Belgium and France. In the United States, the reprocessing of nuclear fuel was banned in 1977 by President Jimmy Carter. Despite a lifting of the ban in 1981, no reprocessing of spent nuclear fuel for commercial use has been carried out since. The use of ^{239}Pu in nuclear reactors to produce energy allows for the destruction of highly radioactive ^{239}Pu, which would otherwise have to be stockpiled, disposed of or used for the production of atomic weapons, and can also represent significant savings in ^{235}U. As plutonium is an essential material for the

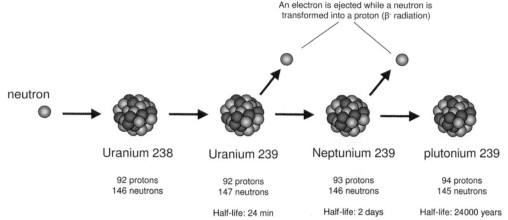

Figure 8.19 Uranium-238 fertilization by neutron capture.

production of some atomic bombs, its reprocessing for commercial use is, however, of great concern in regards to nuclear proliferation and possible acquisition by terrorist groups.

8.8.2
Breeder Reactors

Even more intriguing is the possibility of producing more fuel than is consumed, by using so-called "breeder reactors." In these reactors, based on fast neutrons, more fissile material is produced through the conversion of ^{238}U into ^{239}Pu than is consumed through fission (Figure 8.19). As fast neutrons are desired, no moderator is necessary in this type of reactor. In these reactors liquid sodium is typically used as the coolant, because it does not appreciably slow down neutrons and is an excellent heat-transfer material. Several fast breeder reactors have been constructed over the years in several countries. The best known are France's "Phénix" and "Superphénix" reactors. Phénix is an experimental reactor with a power capacity of 250 MW. Its successor, Superphénix, was a commercial scale reactor that was connected to the grid and had a capacity of 1300 MW, but it was closed down in 1997 after some technical problems, but mainly due to administrative and political concerns [174]. In Russia, Japan and India, breeder reactors are being developed with several units under construction. The United States, which pioneered this field with the construction of the first breeder reactor in 1951 at the Argonne National Laboratory [175], is now reconsidering this technology for future power plants. In fact, the United States initiated in 2000 the Generation IV International Forum (GIF) with nine other countries, aimed at determining and developing the most promising generation IV reactor designs that can provide for future worldwide electricity needs, and also produce hydrogen and derived products. Among the six selected systems, three are fast-neutron reactors which are all operated in a closed fuel cycle, meaning that the fuel is recycled [170].

About 4.6 million tonnes of proven uranium reserves are estimated to be exploitable at costs below $130 per kg. At a current worldwide demand of 60 000 t per year, this represents about 75 years of consumption. Based on geological surveys, potential resources exploitable at a cost of $130 per kg uranium are estimated at an equivalent of 280 years of consumption [176]. Despite an increase in uranium prices in recent years, the still-considerable reserves convey little incentive to open new mines or to start explorations in the quest for new uranium sources. However, if nuclear power were to supply a much larger part of our energy demand, in the long term the transition to breeder reactors is necessary. All the uranium could then be used as fuel, and not only the 0.7% of ^{235}U contained in natural uranium but also ^{238}U, multiplying the reserves by a factor of roughly 100 and fulfilling our energy needs for a least a thousand years. Eventually, the four billion tonnes of uranium contained in the oceans in the form of carbonates could also be exploited. Although very dilute (3–4 mg m^{-3}) and with an estimated extraction price of up to $1000 per kg, it could become a viable source because the cost of uranium is only a minor component of the price of electricity generated by breeder reactors. This would leave us with an almost unlimited source of energy [177].

Besides ^{238}U, another "fertile" isotope that is susceptible to be transformed into fissionable atoms exists in nature, namely thorium-232 (^{232}Th). Almost 30 isotopes of thorium are known, but only ^{232}Th (with a half-life of 14 billion years) is present naturally. When bombarded with neutrons, ^{232}Th becomes ^{233}Th, which eventually decays to ^{233}U. Uranium-233 is a fissionable material with similar properties to ^{235}U, and can be used as nuclear fuel. Thorium is considered to be about three times more abundant than uranium, and so the potential energy available from its exploitation is therefore tremendous. The use of thorium for electricity generation has been studied and has been demonstrated successfully in several reactor prototypes, including the high-temperature gas-cooled reactor (HTGR) and the molten salt reactor (MSR). The MSR was first considered by the United States in the early 1950s for aircraft propulsion. Using a ^{232}Th–^{233}U fuel cycle, MSR is now one of the six systems selected by the Generation IV International Forum on future nuclear reactors. Countries such as India, with limited deposits of uranium but large thorium resources, are particularly interested in such a technology. The use of thorium fuels, compared to uranium fuels, also produces much less plutonium and other actinides, so that induced radiotoxicity is considerably reduced. This reduced radiotoxicity is due to the fact that ^{232}Th has two protons and four neutrons less than ^{238}U, which makes it less probable for ^{232}Th, through a succession of neutron captures, to be transformed into long-lived toxic transuranic actinides such as neptunium, americium, curium or plutonium. From the viewpoint of decreased nuclear waste production, the thorium-based fuel cycle is thus more desirable.

8.8.3
The Need for Nuclear Power

Since the construction of the first nuclear reactor (pile) by Fermi and his team more than 65 years ago, the history of nuclear energy has been a varied one. In the early years, after World War II, the concept of nuclear power holding the promise of cheap and abundant energy was overwhelmingly supported by public opinion. On the other hand, since nuclear energy was first used for the development of the atomic bomb, many still associate atomic energy with destruction and killing. Over the years, following the Three Mile Island incident in 1979 and, more importantly, after the Chernobyl accident in 1986, serious public concerns emerged and public opinion increasingly turned against nuclear power, forcing many governments to drastically reconsider their policies for electricity generation from atomic energy. In the United States, after the Three Mile Island incident (which, incidentally, did not lead to any casualties), anti-nuclear activism successfully slowed down and eventually stopped the construction of any new nuclear power plant ordered after 1973. By filing innumerable lawsuits against utility companies, construction was halted at the cost of millions, if not billions, of dollars to power companies. It effectively scared any corporation that planned to build a new nuclear power plant. The actual cost of the Shoreham nuclear power plant in New York, for example, was first estimated at $240 million but finally rose to $4 billion at its completion. Seabrook, in New Hampshire, which was estimated to cost around $1 billion, was abandoned before completion after years of delays in construction and with $6 billion already spent. Similarly, in Europe, countries such as Sweden, Spain and Germany, fearing possible accidents involving nuclear reactors, have put a hold on their construction. Other nations with no or limited fossil-fuel resources such as France, Japan and South Korea, however, are actively developing and expanding their nuclear power sector. France's decision to launch a large nuclear program dates back to the 1970s under the presidency of Georges Pompidou, and was aimed at minimizing the country's dependence on imported oil. Following the popular saying: "no oil, no gas, no coal, no choice" the French authorities of the time quickly came to the conclusion that nuclear power was not a choice but a necessity if their country was to keep its energy independence. Both benefits as well as risks were efficiently explained to the public, which understood that life would be very difficult without nuclear energy. The fact that Japan, the only country ever to experience a nuclear bomb attack, also strongly favors the use of atomic power should tell us much about its safe and peaceful use. In recent years, a less passionate, more rational, facts-based view of nuclear energy, coupled with increased safety controls and enhanced need for energy and reduced carbon dioxide emissions, has begun to emerge, especially in Asia where numerous nuclear power-plants are under construction (Table 8.3). The United States is currently also starting to slowly reconsider the construction of new nuclear power plants.

8.8.4
Economics

Surprisingly, in some ways anti-nuclear activism had a positive long-range influence on nuclear energy. It forced energy companies and governments under much higher public scrutiny to make nuclear power plants even safer and more productive and reliable than almost any other power industry. At the same time, in the United States, the average power-load capacity factor of nuclear power plants increased from 58% in 1980 to 70% in 1990, and to around 90% since 2001. The increased capacity factor resulted in lower generation costs and an increased electricity production using nuclear power. The increase in electricity production during 1990–2003 was the

Table 8.3 Nuclear power reactors under construction.

Country	Units	Total capacity (MW)
China	9	8220
Russian Federation	8	5809
India	6	2910
Korea	5	5180
Japan	2	2191
Ukraine	2	1900
Bulgaria	2	1906
Taiwan	2	2600
Argentina	1	692
Finland	1	1600
France	1	1600
Iran	1	915
Pakistan	1	300
United States	1	1165
Total	42	36 988

Source: IAEA (as of December 2008).

equivalent of adding more than 20 new nuclear reactors to the US capacity. The cost of producing electricity at US nuclear power plants – including fuel, operation and maintenance – has been declining over the past decade, from more than $0.03 per kWh in 1990 to less than $0.02 per kWh since 2004 (Figure 8.20). This price is

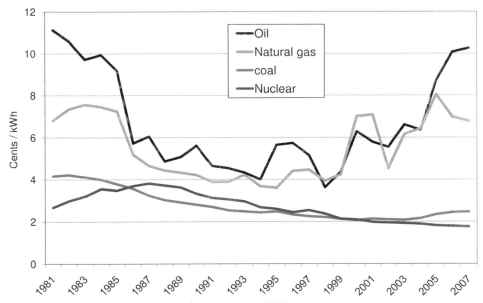

Figure 8.20 United States electricity production costs (in 2007 cents per kWh). Costs include fuel, operation and maintenance. (Based on data from NEI.)

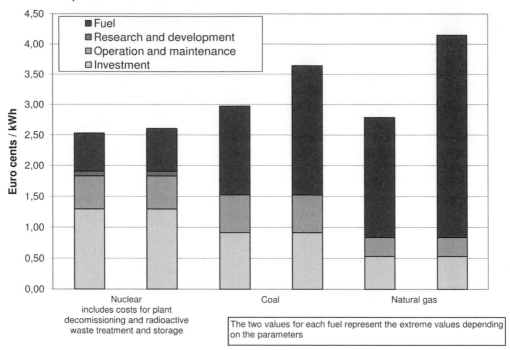

Figure 8.21 Electricity production costs in France (€ per kWh). (Based on data from CEA.)

comparable to the cost of electricity production from coal, but much lower than that from oil and gas. Even including construction capital, as well as the decommission cost of old power plants and the treatment and storage of nuclear waste, the cost of electricity from nuclear origin is estimated in France to be less than €0.03 per kWh (Figure 8.21). This low-generation cost, due in large part to the homogeneous and standardized nature of the French reactors, with all 59 reactors based on a PWR design, allows the country to be the largest electricity exporter in the world [42]. Often-cited arguments against nuclear energy based on high costs have thus become baseless.

8.8.5
Safety

Nuclear power plants are closely regulated by national and international agencies, which provide rigorous oversight of the operation and maintenance of these plants. The safety record of nuclear power plants has been exemplary over the years and was achieved through improved plant designs, high-quality construction, regular staff training, safe operation and careful emergency planning. Diverse and redundant systems prevent accidents from occurring, and multiple safety barriers are in place to mitigate the effect of accidents in the highly unlikely event they occur. The safety of the nuclear industry has been significantly improved since the Three Mile Island incident in 1979, which, although it was a serious nuclear accident, did not lead to any casualties.

The Chernobyl disaster was on a completely different scale. It was a consequence of human error, lack of safety measures and poor construction and design. The power plant had no reactor-containment building to prevent radioactivity from escaping to the atmosphere, and would never have been licensed and allowed to operate in any Western country. The explosion that occurred in Chernobyl, however, was not nuclear, but chemical. When the reactor went out of control, huge amounts of heat generated by nuclear fission instantaneously evaporated the water used under normal conditions to cool the fuel rods. The steam produced reacted with extremely hot zirconium to produce hydrogen, as well as with the graphite, used as the moderator, to form carbon monoxide and hydrogen. The pressure increased dramatically and blew off the 2200-ton concrete lid of the reactor pressure vessel. The remaining graphite then ignited, with flames carrying, in the absence of any containment, highly radioactive material into the atmosphere. Thirty-five people who tried to extinguish the fire died shortly after the accident from the direct effects of radiation. According to the latest and most comprehensive study from the United Nations, 20 years after the accident, fewer than 50 deaths have been directly attributed to radiation from the disaster [178]. Over time, however, a total of up to 4000 people could eventually die from radiation exposure from the Chernobyl accident. Poverty and "lifestyle" diseases, rampant in the former Soviet Union, pose far greater threat to local communities than the exposure to radiation resulting from the accident. Although Chernobyl was the most serious nuclear accident, it should also be compared with other energy-related losses of life such as coal mining accidents, which draw much less public attention but nevertheless cause thousands, if not tens of thousands, of deaths every year. The Chernobyl accident is considered as the archetype of the worst conceivable civilian nuclear disaster. The probability for a disaster of this magnitude to occur today in a Western World-built reactor has been estimated by nuclear safety experts to be on the order of one-millionth per reactor per year of operation in currently used reactors, and even less in the next generation of reactors with advanced safety features. Other potential accidents, such as a dam rupture or the explosion of a LNG tanker, have a much larger probability of causing significant casualties. It is also important to point out that, contrary to widespread belief, it is impossible for a civilian nuclear reactor to undergo a nuclear explosion of the kind generated by nuclear bombs. Nuclear fuel used in commercial units contains at most 5% ^{235}U, whereas for nuclear bombs the uranium used must contain at least 90% ^{235}U, and be contained in specifically designed and shaped devices before a nuclear explosion can occur.

8.8.6
Radiation Hazards

Unstable atomic nuclei can split to form other particles, simultaneously ejecting different types of radiation (alpha, beta, gamma and X-ray) in a process called radioactive decay, discovered by Henri Becquerel in 1896. These radiations are able to penetrate matter and can disrupt biological systems and thus essential processes in human body cells. The degree of penetration will depend on the energy of the

Table 8.4 Radiation exposure from different activities.

Activity	Exposure (mSv year^{-1})
Natural background radiation	2.4
Working at a nuclear power plant	1.15
One diagnostic X-ray	0.2
Living in a stone, brick or concrete building	0.07
One round-trip flight Paris–New York	0.05
Living at the gate of a nuclear power plant	0.03
Watching television	0.015
Luminous wrist watch	0.0006
Coal fired power plant, average within 80 km	0.0003
Average radiation from nuclear power production	0.0002
Smoke detector	0.00 008

Source: UNSCEAR, NEI and R. Morris, *The Environmental Case for Nuclear Power. Economical, Medical and Political Considerations*, 2000 [184].

radiation, with gamma and X-rays being the most energetic. Radioactivity is a part of our daily life, and is present everywhere from various natural sources: cosmic rays, uranium and thorium contained in the earth's crust, granite used as a construction material, radon gas produced by the natural decay of uranium, potassium in fertilizers and food, and so on. The average natural irradiation to which a human is exposed in a year is around 2.4 millisieverts (mSv, the unit that quantifies the biological effect of radiation in our body).

Besides natural radioactivity, populations in developed countries are also exposed to artificial sources of radiation amounting to less than 1 mSv, essentially for medical purposes (X-rays) but also daily activities such as watching television (0.015 mSv) or taking an airplane trip (0.05 mSv for a Paris to New York round trip) (Table 8.4; Figure 8.22).

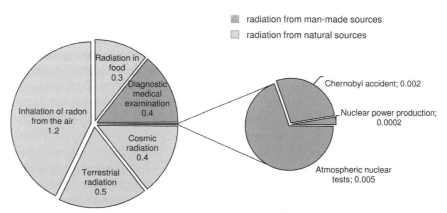

Figure 8.22 Average radiation dose to the public (in mSv year^{-1}). (Based on data from UNSCEAR, Sources and effects of ionizing radiation (New York, 2000).)

Nuclear power plants are responsible for emissions of around 0.0002 mSv per year, similar to the commonly used smoke detector, which contains americium, but 10 000 times less than naturally occurring radiation. Thus, they present no risk in terms of radiation in normal operation. Nuclear power plants de facto are even emitting less radiation than coal-burning power plants! A part of the radioactive contaminants in coal, including uranium and thorium, pass up the chimney stack and are released to the atmosphere, while another part stays in the ashes [179, 180]. It has been calculated that about 13 000 t thorium and 5000 t uranium are released yearly into the environment as a result of coal combustion [181].

8.8.7
Nuclear By-Products, Waste and Their Management

Nuclear wastes are classified according to their activity and lifetime. Some radioactive elements have a half-life of only a few seconds, while others have half-lives of millions or even billions of years; the danger of nuclear waste is inversely proportional to its lifetime (Table 8.5). The longer the half-life of a nucleus the lower the number of disintegration events per unit of time.

With regard to nuclear waste management, most of the public attention has been focused on high-activity spent nuclear fuels. Nuclear fuel rods are generally present in a commercial reactor for three to four years, during which time they are progressively used up. Subsequently, they have to be replaced by new rods to keep the electricity output constant. Used nuclear fuel is not a waste, however, as it still contains very large amounts of compounds with high energy value: 95% uranium (of which about 1% is ^{235}U) and 1% plutonium. It also contains about 0.1% actinides (neptunium, americium, curium, etc.) that could eventually be used to produce energy, and 4% of fission products that have no energy potential and are thus considered as waste that must be disposed. Treatment of the used fuel in special facilities in Europe and Japan enables the recycling of unused ^{235}U and plutonium. It also reduces the amount of highly radioactive waste produced and allows a better use of natural resources. For political reasons, based on fears of nuclear proliferation, the

Table 8.5 Half-life (years) of some radioactive elements.

Uranium-238	4 470 000 000
Uranium-235	704 000 000
Neptunium-237	210 000
Plutonium-239	24 000
Americium-243	7400
Carbon-14	5730
Radium-226	1600
Caesium-137	30
Strontium-90	29
Cobalt-60	5.3
Phosphorus-30	2.55 min

Source: Nuclear Energy Agency (Nuclear Energy Today, Nuclear Energy Agency, OECD publications, Paris 2003, Available from: http://www.nea.fr/html/pub/nuclearenergytoday/welcome.html).

reprocessing of nuclear fuel was banned in the United States in 1977. Despite a lifting of the ban some years later, the reprocessing of nuclear fuel for commercial uses is just recently being reconsidered. Ironically, reprocessing is now the preferred option to destroy stocks of military-grade plutonium by transforming them into usable MOx fuel for nuclear reactors. This reprocessing enables the generation of energy while simultaneously reducing the amount and toxicity of spent fuels. It was also found to be cheaper than considering plutonium as waste and trying to dispose of it in depositories, such as under the Yucca Mountain in Utah [182]. This facility was designed and built to store nuclear waste safely for at least an assured 10 000 years. Following pressure from groups opposing the storage of nuclear waste, standards were subsequently increased to 100 000 years and more recently to 1 000 000 years, guaranteed to protect the next 25 000 generations of residents living near the site [183]. This time frame is hardly a reasonable requirement. The underground storage of used nuclear fuel without reprocessing is clearly also a waste of energy potential. It is thus not the best approach for nuclear waste management. Regardless, the storage of nuclear materials in specifically chosen underground facilities is feasible and safe for at least many thousands of years, and does not involve unsolvable technical difficulties. Furthermore, the quantities of used nuclear fuel generated each year from a commercial power plant are on the order of only a few tonnes and thus should be easily manageable. Technical solutions also exist for the treatment of all other nuclear wastes. The disposal of nuclear waste has never been an "unsolvable problem," as some would like us to believe. Rather than technical, the radioactive by-products and waste problem is more often than not ethical and political and should ultimately be solvable.

To reduce the radiotoxicity of used nuclear fuels and make much better use of current uranium resources, reprocessing and the construction of breeder reactors to convert not only ^{235}U but also most of ^{238}U into energy is a preferable solution. The fast neutrons in breeder reactors can also break down the highly active minor actinides, such as neptunium, americium and curium, produced during fission reactions, which contribute (besides plutonium) to most of the radioactivity of the spent nuclear fuel.

8.8.8
Emissions

Of all the large-scale energy sources used today, nuclear energy has probably the lowest impact on the environment. In contrast to fossil fuel-powered plants, nuclear power plants do not emit any greenhouse gases or air pollutants such as SO_2, NO_x and particulates. Thus, the use of nuclear energy helps to keep the air clean, to mitigate earth's climate changes, avoid ground-level ozone formation and prevent acid rains. Although not emphasized, in 2007 the US nuclear power plants saved us from the emission of 3.1 million tonnes of SO_2, 1 million tonnes of NO_x and 690 million tonnes of CO_2 entering the earth's atmosphere [184]. Globally, nuclear energy avoids the emission of more than 2 billion tonnes of CO_2 per year. Air pollution, especially from SO_2 emissions and sulfate aerosols produced in large quantities in coal-burning power plants, is estimated to be the cause of approximately 40 000 deaths per year in the US alone [185] – a number that should be compared with no casualties due to the operation of commercial nuclear reactors in the US over a period of almost 50 years. Coal also contains mercury, cadmium, arsenic and selenium and, as mentioned above, even radioactive elements such as uranium and thorium, all of which are released into the atmosphere or are present in coal ashes. Because there is virtually no regulation regarding the millions of tonnes of coal ash produced every year, these can be dumped almost anywhere without much concern for public health [186]. To generate electricity, coal is clearly much more harmful to use than nuclear power. Even natural gas – the cleanest fossil fuel – still generates large quantities of CO_2 and NO_x. In the context of increased concern about the greenhouse effect and global climate changes due in large part to CO_2 emissions, "clean" nuclear energy therefore has an increasing role to play.

8.8.9
Nuclear Fusion

The sun, like innumerable other stars, is a giant thermonuclear reactor that obtains its energy from nuclear fusion of small nuclei, primarily hydrogen, to produce larger ones, mostly the transformation of hydrogen into helium, which subsequently leads to the formation of heavier elements. The mass of the fused reaction products being less than that of the initial particles, the difference is converted into energy in accord with Einstein's equation, $E = mc^2$. This mass difference represents a tremendous amount of energy – about ten times greater than a typical fission reaction for the same mass of nuclear fuel. Since the positive electric charges of the nuclei provide strong repulsive forces, the energy of the particles, and consequently the temperature, must be very high for the fusion reaction to occur at a sufficient and sustained rate. In the center of stars this is achieved by intense gravitational forces in a high-temperature plasma (about 15 million °C). Plasma is (beside solid, liquid and gas) the fourth state of matter, where the atoms' nuclei and electrons are separated and move independently. On earth, instead of natural gravitational confinement, different technologies must be used to contain very energetic particles and prevent them from escaping before reacting. The two techniques presently considered to achieve this goal are: (i) magnetic confinement, in which a very high-temperature plasma is contained by a strong magnetic field for suitable extended periods of time, and (ii) inertial confinement, where fusion is realized in a small concentrated volume of plasma heated and compressed extremely rapidly with high energy lasers. For electricity production, magnetic confinement is presently the most advanced and favored option. Experimental studies have been conducted in several countries since the 1950s, with a breakthrough in 1968 when Russian scientists obtained, for the first time, a 10-million-°C plasma in a so-called Tokamak. Since then, Tokamaks have become the standard, leading to the construction of large-scale fusion experiments in the 1980s: the Japanese JT-60, American TFTR and European JET. A Tokamak is basically a donut-shaped cylinder in which a strong magnetic field is created. The plasma, being a combination of independent, electrically charged electrons and ions, is trapped in this magnetic field and circulates in a circle in the Tokamak. The temperature necessary to maintain a fusion reaction in such systems is on the order of 100×10^6 °C. To achieve such staggering temperatures, the plasma is heated by the injection of highly energetic neutral particles, which also serve as fuel, and with electromagnetic waves at specific frequencies, which transfer their energy to the plasma through antennas placed in the confinement chamber. The magnetic field contains the hot plasma far enough from the inner Tokamak's wall to avoid cooling and allow fusion to occur. The heat generated by the reaction is removed through heat exchangers placed in the reactor's wall. This heat is then used to produce steam and finally electricity via a turbine/generator system, much like in today's fossil-fuel and nuclear power plants.

To be viable from an energetic standpoint, and to reach the so-called "break-even" point, the energy produced by the fusion reaction must be at least equal to the amount of energy furnished to heat the system. Beyond break-even, the important task will be to create much more energy than is injected into the system and to establish the feasibility of controlled nuclear fusion as an economically viable source of energy. The hydrogen bomb is a fusion devise initiated by the explosion of a fission bomb, but not in a controlled fashion.

Currently, the best ratio of energy produced over energy injected into the system, named gain and represented by the symbol Q, has been obtained at the Joint European Torus (JET) with a $Q = 0.65$, close to break-even. The ITER (International Thermonuclear Experimental Reactor), an international collaboration between Europe, Japan, Russia, China, India, the United States and South Korea on fusion reaction, is expected to yield a gain between 5 and 10 (Figure 8.23), with a fusion power of 500 MW. With fusion reactions being more efficient the larger the installation, this next generation of Tokamak, which will be built in Cadarache, France, will be much bigger than its predecessor and constitutes the preliminary step to the construction of commercial reactors of even larger size with Q values >30.

On earth, the most accessible and practical nuclear fusion reaction, on which most efforts have been concentrated, is the one between deuterium (D) and

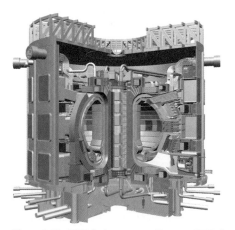

Figure 8.23 ITER fusion reactor (Source: ITER, http://www.iter.org.)

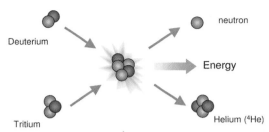

Figure 8.24 Typical fusion reaction.

tritium (T), the two heavier isotopes of hydrogen, which offers by far the highest gain (Figure 8.24). Deuterium is a stable element and abundant in water (33 g m^{-3}), from which it can be extracted. Existing deuterium resources represent more than 10 billion years of annual world energy consumption! Tritium, however, is a radioactive element that does not exist in nature and has to be prepared from lithium by neutron bombardment. Lithium resources are estimated at 2000 years, but can be extended to several million years if it is extracted from sea water. In existing experimental reactors, tritium is generated outside and subsequently injected into the plasma. In future systems, however, tritium will be produced directly inside the thermonuclear reactor. The inner walls will be made of lithium-containing materials which, under bombardment of neutrons from the fusion reaction, will be transformed to tritium.

Fusion reactors are intrinsically safe because at any time only small amounts of fuel are present in the plasma chamber, and any uncontrolled perturbation will immediately induce a temperature drop, leading to a cessation of fusion reactions. The risk for potential runaway reactions is therefore eliminated. Like fission, fusion does not produce any air pollutant contributing to acid rains or greenhouse gases responsible for global climate change, and could thus mitigate the environmental risks associated with fossil-fuel burning. None of the basic fuels for fusion, deuterium and lithium, or the reaction product helium are radioactive or toxic. Tritium, which will be entirely produced on site, decomposes to helium by emission of a low-energy beta ray and has a relatively short half-life (12.3 years). Its radioactivity is low, but reactors will have to take into account the permeation of this gas through matter. As in every system under intense high-energy particle flow, the materials constituting the reactor will be activated. However, a judicious choice of materials with rapidly decreasing activity will allow a minimization of the amounts of radioactive waste to short-lived, low-activity materials. The deuterium–deuterium fusion reaction involves lower radioactivity but is less practical as it requires about an order of magnitude higher plasma pressure to produce the same power as the deuterium–tritium reaction. The reaction of ^{3}He with deuterium generates even less radioactivity. Helium-3, however, is not available on earth and must be prepared. Interestingly, some have proposed to mine ^{3}He from the moon's surface, where it is present in minute quantities [187]. There, ^{3}He was deposited over millions of years by solar winds, a stream of particles emitted by the sun, striking the powdery lunar soil. Unlike earth, the moon has no magnetic field able to deflect the wind-charged particles. Eventually, with increased knowledge and advanced technology, both the deuterium–deuterium and ^{3}He–deuterium fusion reactions could be used in fusion reactors.

Nuclear fusion is still an energy source of the future. Much progress has already been accomplished, but extensive research and development are still needed. With mega-projects such as the ITER and its followers, however, it is very likely that this technology will be made practical during the twenty-first century. Fusion, when compared to present sources of energy, offers numerous advantages: the fuels, deuterium and lithium, are widespread and virtually inexhaustible, obliterating energy security problems and resource-based conflicts. Fusion also has large-scale power-generating capacity with minimal environmental impacts, and is inherently safe. Water used as working fluid in heat exchangers to cool the reactor and produce electricity could also be replaced by liquid metals or helium to achieve higher operating temperatures (1000 °C) and allow, besides electric power generation, the production of hydrogen by thermochemical splitting of water, as has already been proposed with nuclear fission (Chapter 9). Because of the high cost and complex technology of eventual fusion energy plants, there will most likely initially be only few such large installations serving major power needs, with smaller and more dispersed atomic fission plants providing more decentralized electricity production. Advances in superconductive power transmission lines could, however, substantially help and alleviate present electric transmission limitations.

8.8.10
Nuclear Power: An Energy Source for the Future

Nuclear power already generates a significant part of the world's electricity and has much to offer for the future, not only for electric power generation but also for other applications such as hydrogen production (Chapter 9) and water desalination. Over the years, the safety record in Western-built reactors has been remarkable, and advanced reactor designs promise to be even safer. With near-zero emissions, nuclear power has a clear advantage over fossil fuels with regard to air pollution and growing concerns about global climate change. Moreover, with most resources of uranium and thorium being located in politically and economically stable areas of the world, stability in prices and production would also be greatly improved. The progressive introduction of breeder reactors should allow a better usage of the these resources and secure the world's energy supply far into the future, until better energy-generating technologies – eventually even controlled nuclear fusion – emerge.

8.9
Future Outlook

Alternative energy sources to fossil fuels are numerous, but have their drawbacks and limitations. Hydropower has been used on a large scale for over a hundred years for electricity generation, but the installation of new capacity is becoming more limited because the best sites are already developed and environmental and socio-economic considerations must increasingly be taken into account before flooding large areas for reservoirs. Energy from geothermal wells can play an important role on a local scale for some countries, but resources are limited on the global scale. Solar energy, photovoltaic and thermal, is still too expensive and intermittent. Suitable energy storage or other supplementary energy sources are needed, as there is only limited solar energy production under cloudy conditions, and none during the night. Regardless, its use is increasing in suitable areas, particularly deserts. Wind power is also reliant on the intermittent power and speed of winds, but is presently much cheaper than solar energy and has therefore presently a greater potential. Its increasing use in Europe, the US and other countries is a significant aspect of the growing use of renewable energies. Biomass can also provide a significant but nevertheless limited amount of energy that on its own is inadequate to sustain our modern society's needs. Ocean power in the form of tides, waves and thermal energy is unlikely to represent a significant share of the global energy production in the foreseeable future. Taken together, these renewable energy sources must – and will – certainly play an increasing role in our future global energy mix. However, they will be unable to replace by themselves the bulk of the energy obtained from fossil fuels for many decades to come. With decreasing petroleum and gas reserves, we could rely for some time more heavily on the larger resources of coal. This shift to coal, however, is only a temporary solution until coal production eventually starts to decline. With present-day technologies, it would also imply much larger air pollutants and greenhouse-gas emissions, with severe health consequences and contributions to global climate change. The chemical recycling of carbon dioxide emissions from fossil fuels, as proposed in the Methanol Economy is therefore a significant technological advance to mitigate the global warming problem (see following chapters).

By considering environmental, energy security and long-term stability viewpoints, it is clear that nuclear power – albeit made even safer and with problems of radioactive by-product reprocessing and storage solved – is a major energy source of choice based on current knowledge. Indeed, for the foreseeable future it may produce a large part of the vast and increasing amounts of energy needed by humanity. Advanced breeder reactors and new fusion technologies could provide for our energy needs for centuries or millennia to come.

Eventually, it may be possible to find more efficient, as-yet unknown, ways of using the sun's energy, our major and inexhaustible energy source, but this problem is for future generations. Even nuclear energy should not be viewed as an essentially human invention. Rather, it is a naturally occurring event throughout the universe, and humankind has only relatively recently succeeded in controlling and harnessing the energy of the atom. Without thermonuclear reactions in the sun and nuclear decay inside our own earth, there would be no renewable energies such as solar, wind, biomass, geothermal or hydro energy, or even fossil fuels on our planet. Regardless of how we produce energy, it still must be stored, transported and provided in a suitable form for subsequent use. This remains a major challenge. In addition, we must still find new solutions to provide convenient hydrocarbon-based fuels for transportation and household needs, as well as the various products and materials from renewable and sustainable sources. These processes all require substantial energy, which can be provided by any of the discussed sources. The "Methanol Economy," to be discussed in the following chapters, offers a feasible new approach to achieve these goals.

9

The Hydrogen Economy and its Limitations

Inexhaustible and non-polluting, hydrogen is considered by many as the fuel of choice for our future energy needs. Being involved in some way in the so-called "Hydrogen Economy" seems these days to be quite obligatory for governments and any large energy-related company, automobile manufacturers and other industries. The idea sounds rather simple: take hydrogen, one of the most plentiful elements on earth and in the cosmos, and use it as a clean-burning fuel or in fuel cells to power cars, heat houses and offices, generate electricity and so on. It produces only water as a by-product and none of the CO_2 and other pollutants formed by burning fossil fuels as in the current carbon (fossil fuel)-based economy. As some 60% of our oil consumption is used in transportation, numerous programs have been launched around the globe for the development of hydrogen-based fuel cell-powered cars. In 2003, the US announced a five-year, $1.2 billion budget for the Hydrogen Fuel Initiative aimed at funding hydrogen research and commercializing hydrogen-powered cars by 2020 [188, 189]. The European Union launched a €2.8 billion public–private partnership over a 10-year period to 2011 to develop hydrogen fuel cells. In Japan, the New Hydrogen Project, which began in 2003, focuses on the commercialization of hydrogen fuel cell technologies. The governmental budget for this project has been growing constantly to reach some ¥34 billion in 2006 [189]. Other nations such as China and Canada have also increased their efforts in this field. Most automobile manufacturers have already invested large amounts of money in the development of hydrogen fuel cell-powered cars, and major energy and oil companies are testing ways to provide and refuel these new vehicles with hydrogen. Despite all of these efforts, the challenges that lie in the way towards to the Hydrogen Economy seem prohibitive. Fundamental problems will have to be solved if hydrogen gas is ever to become a practical, everyday fuel that can be filled into the tanks of our motor cars or delivered to our homes as easily and safely as gasoline or natural gas is today.

9.1
Hydrogen and its Properties

Hydrogen is the lightest element of the periodic table. Hydrogen was first formed as the universe began to cool down after the Big Bang, and represents 90% of the atoms

Beyond Oil and Gas: The Methanol Economy, Second updated and enlarged edition
George A. Olah, Alain Goeppert, and G. K. Surya Prakash
Copyright © 2009 WILEY-VCH Verlag GmbH & Co. KGaA, Weinheim
ISBN: 978-3-527-32422-4

Figure 9.1 The sun is converting 600 million tonnes of hydrogen into helium every second (Photo source: NOAA.)

present in the cosmos, the rest of it being mostly helium. By fusion reactions in the stars, hydrogen subsequently formed the heavier elements, and can thus be considered as their common precursor. Hydrogen is the fuel of the stars. Every second, 600 million tonnes of hydrogen are converted into helium in our sun alone by nuclear fusion, releasing enormous amounts of energy as well as providing light and heat, which makes life on earth possible (Figure 9.1).

Hydrogen is also one of the most widespread and plentiful elements on earth. Owing to its high reactivity, however, hydrogen combines with other elements. As our atmosphere contains 20% oxygen, molecular hydrogen (H_2) is not present, except in small amounts in the upper atmosphere. In nature, hydrogen is nearly always found combined with other elements. In every water molecule (H_2O), covering 70% of the earth's surface, two hydrogen atoms are attached to an oxygen atom. Hydrogen is also found in hydrocarbons and their derivatives as well as in every living organism, plants and vegetation.

Hydrogen as a distinct element was first identified, and some of its properties described, in 1766 by the English scientist Henry Cavendish, who called it "inflammable air" (Table 9.1). By applying a spark to hydrogen, water was produced. This later led the French chemist Antoine Lavoisier to name the gas hydrogen from the Greek *hydro* and *genes* meaning "water" and "born of." Shortly after the French revolution (during which Lavoisier was beheaded as a tax collector on the guillotine, despite his fellow scientists pleading for his life), the first practical use for hydrogen was found in the military for reconnaissance balloons filled with hydrogen gas able to fly high above enemy lines. The large quantities of hydrogen gas needed were initially produced by passing steam at high temperature over iron filings. The possibility to generate hydrogen and oxygen gases by water electrolysis was discovered in the early 1800s by Englishmen William Nicholson and Anthony Carlisle. William Grove, in 1839, found a way to reverse the electrolysis process and generate electricity by combining hydrogen and oxygen to form water in what would be later called a fuel cell.

Table 9.1 Properties of hydrogen.

Chemical formula	H_2
Molecular weight	2.0159
Appearance	Colorless and odorless gas
Melting point	$-259.1\,°C$
Boiling point	$-252.9\,°C$
Density at $0\,°C$	$0.09\ kg\ m^{-3}$
Density as a liquid at $-253\,°C$	$70.8\ kg\ m^{-3}$
Energy content	$28\ 670\ kcal\ kg^{-1}$
	$57.7\ kcal\ mol^{-1}$
Octane number	$130+$
Autoignition temperature	$520\,°C$
Flammability limits in air	4–74%
Explosive limits in air	15–59%
Ignition energy	0.005 mcal (milli-calorie)

As mentioned, unlike fuels such as wood, coal, oil or natural gas, hydrogen is not found in its free form on earth and thus cannot be collected as such for combustive energy production. A significant amount of energy must first be expended to produce hydrogen, which is bound to other elements such as in water or hydrocarbons, to be able to use it as a fuel. Hydrogen is thus not a primary energy source but only an energy carrier. Some of its physical and chemical characteristics, however, are not well-suited for this purpose, especially as a transportation fuel. Despite its physical properties there is presently great interest for such applications. The lightness of hydrogen (indeed, it is the lightest of all elements) represents a handicap for its storage, transmission and use in its gaseous form. The small H_2 molecules also diffuse through most materials and makes steel brittle, especially at high pressure or/ and temperature. Being a volatile gas, it can only be condensed to a liquid at a very low temperature of $-253\,°C$, only $20\,°C$ above absolute zero. According to Avogadro's law a mole (2 g) of hydrogen occupies 22.4 liter at atmospheric pressure and $0\,°C$. Thus its storage and handling generally involves the use of low temperature and/or high pressure. Hydrogen can also ignite or explode in contact with air and should thus be handled with substantial care.

9.2
Development of Hydrogen Energy

From the early nineteenth century onwards, hydrogen obtained from the incomplete combustion of coal, and together with carbon monoxide in a mixture called "town gas," was widely used to heat and light homes, apartments, businesses and to provide street lighting in increasingly industrialized cities. However, with the advent of electricity, and the development of naturally occurring oil and natural gas that could be used directly without prior processing, the importance of hydrogen as a fuel rapidly declined. Today, the use of hydrogen as a fuel is limited to niche markets, principally as a rocket propellant and to specific experimental development as a transportation fuel. Nevertheless, since the nineteenth century, the unique properties of hydrogen has fascinated generations of scientists, futurists and even science fiction writers. As early as 1874, Jules Verne in one of his visionary books, *The Mysterious Island*, described in a discussion between his characters what would happen to America's commerce and industry when the world runs out of coal several centuries later. Cyrus Harding, the engineer of the group, explains that one will then turn to another fuel, proposing to the astonishment of his companions that water would be the fuel of the future. Or more precisely, ". . . water decomposed into its primitive elements and decomposed doubtless, by electricity which will then have become a powerful and manageable force".

> Yes my friends, I believe that water will one day be employed as fuel, that
> hydrogen and oxygen which constitute it, used singly or together,
> will furnish an inexhaustible source of heat and light, of an intensity of
> which coal is not capable. Some day the coalrooms of steamers and the
> tenders of locomotives will, instead of coal, be stored with these two
> condensed gases, which will burn in the furnaces with enormous
> calorific power. . . . I believe, then, that when the deposits of coal are
> exhausted we shall heat and warm ourselves with water. Water will be the
> coal of the future.

To our knowledge this is probably the earliest reference to a "Hydrogen Economy." Verne, however, never mentioned where the primary energy necessary to produce the needed hydrogen from water electrolysis would come from.

In the 1920s, Canada's Electrolyser Corporation Ltd. opened the way to commercial-scale hydrogen production through water electrolysis. This technology allowed hydroelectric power plants to utilize their excess capacity to produce hydrogen and oxygen. The generated gases were used mainly for non-fuel-related applications such as steel cutting and synthesis of nitrogen fertilizers. At about the same time, German engineers, especially Rudolf Erren, experimented with hydrogen as a fuel for trucks, automobiles, trains, buses and other internal combustion driven devices [190]. In aviation, hydrogen was first exploited in the German Zeppelin airships (Germany having no helium sources), which offered regular transatlantic flights years before airplanes did (Figure 9.2). During these trips, liquid fuels carried aboard were consumed for propulsion, gradually reducing the weight of the airship. To maintain proper buoyancy, part of the hydrogen that kept the vessel afloat in the air was used as extra fuel instead of being simply blown off. The catastrophic fire of the airship *The Hindenburg* in 1937, however, ended the era of the hydrogen-filled Zeppelins. During World War II, hydrogen fuel attracted some interest for submarines and trackless torpedoes. After the war, however, and during the era of cheap oil and gas the potential use of hydrogen as a fuel (except for space and military applications; Figure 9.3) was widely ignored. It only resurfaced with the oil crises of the 1970s and the necessity of finding alternatives to petroleum oil. Also aided by a growing public awareness of pollution problems, suggestions of using clean hydrogen fuel flourished, considering, however, only its combustion and not necessarily its production. This was also the time when the term "Hydrogen Economy" was introduced and the

Figure 9.2 Zeppelin LZ-129 "Hindenburg" flying over New York.

International Association for Hydrogen Energy was created. Interest by governments and private companies, however, lasted only as long as the cost of oil remained high. With sharply declining oil prices in the 1980s, funding for the development of hydrogen energy and alternative energy sources was significantly reduced. For example, the budget for renewable energy in the United States was cut by almost 80% in the early 1980s. Regardless, in the former Soviet Union, a Tupolev 155 experimental airplane tested the use of liquid hydrogen and natural gas as alternatives to jet fuel in 1988 [191]. The use of cryo-fuels, however, was found to be impractical and technically too challenging for regular operation. The large insulated spherical tanks needed to keep the gases liquid were too voluminous and had only enough capacity for relatively short flights. Liquid hydrogen was also deemed too expensive when compared with kerosene fuel. Interest in hydrogen fuel began to rise again in the 1990s, based on concerns about decreasing petroleum and gas reserves, and reports on increasing CO_2 emissions that were considered to be a major cause of global climate change. At the same time, considerable advances in the development

Figure 9.3 Space shuttle launch at Cape Canaveral, Florida.

of fuel cells, and especially proton exchange membrane (PEM) fuel cells, have made a commercial hydrogen fuel cell-powered motor car potentially feasible. This technological advance resulted in the transportation sector making by far the most extensive hydrogen-related development investments. Many major carmakers, including Daimler-Chrysler, Honda, Toyota, General Motors and Ford, have built prototype fuel-cell cars, buses or trucks, and consequently the term "Hydrogen Economy" became commonplace and attracted much public attention. Today, hydrogen-powered vehicles are still attracting interest and wide media coverage. Many organizations, industries, media, public figures and others promote hydrogen fuel via publications, television, meetings and exhibitions. As indicated previously, some governments in the industrialized world themselves have also pledged and provided significant funding for the development of the Hydrogen Economy.

Hydrogen as a fuel undoubtedly has many advantages. Its oxidative conversion to produce electricity or heat is clean, producing only water and generating no pollutants. In special areas, such as the space industry, it plays a significant, well-established role. However, the crucial question remains as how to generate and handle, economically and safely, the large quantities of hydrogen needed. If a significant part of this hydrogen is to be produced by reforming of fossil fuels, as is the case today, it would only displace and possibly even increase – but not solve – the problem of pollution and greenhouse gas emissions. However, if hydrogen is to be generated by the electrolysis of inexhaustible water sources using nuclear and renewable energy, it could become a low emission or emission-free fuel and energy storage medium. However, because of its unfavorable physical properties and high reactivity, hydrogen storage, transportation and use pose major challenges.

9.3
Production and Uses of Hydrogen

Today, hydrogen is used on a large scale as a feedstock in the chemical and petrochemical industry to produce principally ammonia, refined petroleum products and a wide variety of chemicals. It is also used in metallurgic, electronic and pharmaceutical industries (Figure 9.4). Except as a propellant for rockets and space shuttles, hydrogen is still rarely used today as a fuel. Industrial facilities often build their own hydrogen production unit to ensure secure supply and safety, and also avoid transportation difficulties. Consequently, the hydrogen market is presently mainly a "captive" market. To cover the present needs, about 50 million tonnes of hydrogen are produced yearly worldwide, representing some 140 million tonnes oil equivalent (toe), or less than 2% of the world's primary energy demand. Using hydrogen as the main energy source would thus imply enormous investments to increase the production capacity and to establish the needed infrastructure for storage and distribution.

As mentioned earlier, hydrogen is not a primary energy source, but rather an energy carrier, that is, it must first be manufactured before it can be used as a fuel. Currently, almost 96% of the world's hydrogen needs are produced from fossil fuels [192], with almost half being generated by the steam reforming of methane

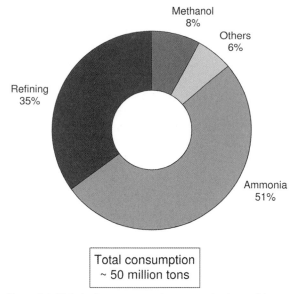

Figure 9.4 Main hydrogen-consuming sectors in the world.

(Figure 9.5). Although relatively inexpensive, this process relies on diminishing natural gas (or oil) resources and emits large amounts of CO_2. At present, water electrolysis is much costlier, and represents only 4% of the production. It is preferentially used when high-purity hydrogen is needed. At the same time, as mentioned, it uses inexhaustible water resources and the energy required can come from any source, including atomic and non-fossil fuel-based alternative energy sources.

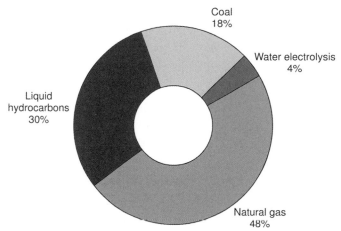

Figure 9.5 Sources for current worldwide hydrogen production.

9.3.1
Hydrogen from Fossil Fuels

Compared with other fossil fuels, natural gas is the most suitable feedstock for hydrogen production because of its wide availability and ease of handling. It also has the highest hydrogen-to-carbon ratio (4 to 1), which minimizes the amount of CO_2 produced as a by-product. Methane can be converted into hydrogen by steam reforming or partial oxidation with oxygen, or by combining both (autothermal reforming). Steam reforming is presently the preferred method, accounting for 50% of the hydrogen produced worldwide, and for more than 90% in the United States. In this process, natural gas reacts with steam over a metal catalyst in a reactor at high temperatures and pressures to form a mixture of carbon monoxide (CO) and hydrogen. In a second step, the reaction of CO with steam (water-gas shift reaction) produces additional hydrogen and CO_2. After purification, hydrogen is recovered, while the CO_2 by-product is generally vented into the atmosphere. In the future, however, it will need to be captured and sequestered if stricter measures to mitigate climate change are imposed. Methane steam reforming can also be performed on a small scale using varied converters. Hydrogen could thus be directly produced locally, including at filling stations. These decentralized units would have higher hydrogen production costs and lower efficiency than larger industrial ones, but could avoid the cumbersome (and potentially dangerous) transportation of hydrogen from distant centralized production centers. In this case, however, the cost of CO_2 capture and disposal would be high. Purification from even small amounts of carbon monoxide would be essential, particularly when used in fuel cells where it poisons the catalysts. Partial oxidation and autothermal reforming of methane are more efficient than simple steam reforming, but require oxygen, whose separation from air at low cost and its safe use in methane oxidation is still technically difficult. Producing hydrogen from petroleum oil and natural gas, albeit established, is not attractive for the long run because it would not solve our energy dependence on diminishing oil and gas reserves and related environmental problems.

With the largest reserves of all fossil fuels, coal could supply significant amounts of hydrogen well into the next century. The current technology to achieve this goal is the so-called integrated gasification combined cycle (IGCC). This "clean coal" technology would allow the co-generation of hydrogen and electricity and therefore significantly improve the overall energy efficiency compared to current commercial plants. In this process, much as in methane reforming, coal is gasified by partial oxidation with oxygen and steam at high temperature and pressure. The created synthesis gas, a mixture containing mostly CO and H_2 (but also CO_2), can be further treated with steam to use CO to increase the hydrogen yield by the water-gas shift reaction. The gas can then be cleaned to recover hydrogen. However, because coal has a low hydrogen/carbon ratio, it also releases much more CO_2 per unit of hydrogen or electricity produced than natural gas or even petroleum oil. In current projects, this issue is considered to be addressed by capturing and sequestering the emitted CO_2 into suitable geological formations or depleted oil and gas reservoirs. Based on this technology, Vision 21 and other programs have been funded by the United States

Government through the Department of Energy (DOE) [193, 194] to promote the use of coal and other domestic resources for the production of hydrogen. The billion-dollar FutureGen program, started in 2003, originally had the goal to co-produce hydrogen and electricity in a 275-MW prototype zero-emission coal-fueled facility [195, 196]. Since then, this program has been restructured. Although it still includes the CO_2 capture and sequestration part, it no longer considers concomitant hydrogen production. Nevertheless, considering the large amounts of coal reserves still available in the US, the DOEs approach is sensible in its goal to diminish reliance on foreign energy sources. Coal gasification is, however, a less mature technology than other hydrogen-generation processes, although the cost of hydrogen production by this method is among the lowest available. In large centralized plants, the current cost of producing hydrogen is estimated to be just above $1 per kg [192], with substantial potential for improvement and further price reduction. However, the planned sequestration of CO_2 produced in large quantities from coal and other fossil fuel burning power plants will be technologically and economically very challenging. None of the existing CO_2 separation and capture technologies has yet been adapted for a large-scale power plant, and costs are uncertain. Carbon dioxide sequestration has so far only been tested on a relatively small scale. The environmental impacts of large-scale sequestration must be carefully assessed before we start pumping billions of tons of CO_2 into subterranean cavities and under the seas. Sequestering is in many ways only a temporary solution to CO_2 emissions. The energy needed for the capture and sequestration processes will also reduce the overall efficiency of the power plants by as much as 14% [197]. Although coal, as long as it is readily available, is a viable alternative to generate hydrogen in large centralized plants, it is not suited for decentralized hydrogen production.

The use of fossil fuels to produce hydrogen necessitates that the CO_2 emitted as a by-product is, according to present plans, captured and sequestered to reduce greenhouse gas emissions and to mitigate global climate change. Such storage, although possible, is not without dangers, as the earth's movements, such as earthquakes or volcanic eruptions, could lead to the catastrophic release of large amounts of CO_2. The proposed chemical recycling of CO_2 to methanol would avoid this problem (*vide infra*). Regardless, the use of fossil fuels would only be a temporary solution as oil, natural gas and eventually coal will all be depleted. Thus, more sustainable methods are needed, such as generating hydrogen from biomass or from water by electrolysis using renewable as well as atomic energy sources.

9.3.2
Hydrogen from Biomass

Biomass could potentially become an important source of hydrogen [198]. Biomass includes a large variety of materials such as agricultural residues from farming and wood processing, dedicated bioenergy crops such as switchgrass, and even algae in the sea. As mentioned in Chapter 8, biomass can be used for the production of such liquid fuels as ethanol, biodiesel and even methanol. Like fossil hydrocarbons, biomass can also be converted into hydrogen by gasification or pyrolysis coupled with steam reforming. This approach can greatly benefit from the extensive knowledge accumulated over the years in the field of fossil-fuel transformation and refining. Gasification plants designed for biomass are generally limited to small or midsize-scale operations due to the high cost of gathering and transporting the usually dispersed and relatively limited amounts of available biomass. Currently, such plants operate only at some 26% efficiency with estimated hydrogen production costs over $7 per kg H_2 [192]. Although with technology improvements and increased efficiencies lower hydrogen prices can be expected, the cost is expected to remain at or above $3 per kg. An interesting alternative, which has been demonstrated commercially [13], is to gasify biomass together with coal, with mixtures containing up to 25% biomass. The construction of biomass-specific gasification units would be unnecessary, and in case the biomass feedstock is seasonal, the plant could also operate on coal alone.

With regard to greenhouse gas emissions, biomass combustion releases CO_2 that was itself captured from the atmosphere by Nature's photosynthetic cycle, so that the net CO_2 emission is near zero. However, the cultivation of the needed crops requires much land and water as well as fertilizers (that use hydrogen in the form of ammonia), and significant amounts of energy for their cultivation, harvesting and transportation. All of these factors, together with their environmental impacts on soil, water supply and biodiversity, must be taken into account and the possible consequences of a large-scale intensive energy-related crop farming carefully assessed. Dedicated high-yield energy crops such as switchgrass, which can be grown with minimal energy input, would be preferable, although biomass for energy would still have to compete for huge areas of land with other agricultural products. Algae grown in the vast expanses of the sea could, however, remedy this picture in the future. In any case, biomass could only be expected to supply a part of the large quantities of hydrogen required.

9.3.3
Photobiological Water Cleavage

Besides biomass transformation processes, a technology aimed at producing hydrogen by the direct cleavage of water with microorganisms without first producing biomass is emerging. Photobiological processes using the sun's energy and suitable microorganisms such as green algae and cyanobacteria could potentially be several-fold more efficient than current biomass gasification. The microbes used at present, however, split water at a rate much to slow to be used for efficient hydrogen production. Photobiological water splitting is still in the research stage and significant efforts and breakthroughs will be needed if it is to become a potential, practical hydrogen source [199, 200].

9.3.4
Water Electrolysis

Electrolysis, the process of cleaving water into hydrogen and oxygen using electricity, is an energy-intensive but well-proven method of producing hydrogen:

$$H_2O \xrightarrow{e^-} H_2 + \tfrac{1}{2}O_2 \quad \Delta H_{298K} = 68.3 \, \text{kcal mol}^{-1}$$

In most locations it is currently about three to four times more expensive than the production of hydrogen from natural gas reforming, which explains its present small share in global hydrogen production. However, it is potentially the cleanest method of producing hydrogen with respect to greenhouse gas emissions, as long as the electricity needed comes from renewable or nuclear energy sources and not from fossil fuels. In this regard, one should always bear in mind that hydrogen energy is only as clean and environmentally friendly as the process used to produce it. Commercial electrolysis is a mature technology that has been around for over a century to produce high-purity hydrogen. Oxygen produced concomitantly is a valuable by-product that can be used in industrial applications such as the smelting of iron ore to steel and metal cutting and welding, and in modern fossil fuel-burning power plants and syn-gas production.

Electrolysis of water is presently used to a significant extent only in locations where cheap electricity sources exist, such as hydropower in Canada and Norway. The power consumption at 100% theoretical efficiency is 39.4 kWh per kg of hydrogen; however, in practice it is closer to 50–65 kWh kg^{-1} [201]. A typical commercial electrolyzer system thus has an efficiency of about 70–80% [202, 203], but a higher efficiency can be obtained with more elevated temperature water or steam electrolysis. Since the efficiency of the electrolysis reaction is independent of the size of the cell or cell stack, electrolyzers allow both centralized and also decentralized hydrogen production, such as in local service stations. The absence of moving parts requires low maintenance, and electrolyzers are well suited for use with intermittent and variable power sources, such as wind or solar. Furthermore, any excess electricity generated during off-peak periods could be stored in the form of hydrogen, which then could be used to produce additional power during peak demand.

Basically, any energy source that produces electricity can be used to produce hydrogen by water electrolysis. Today, more than 60% of the electricity in the world is still produced by fossil fuel-burning power plants. It would, however, be unreasonable to use fossil fuels to generate electricity and then use the electricity to generate hydrogen. As each transformation involves energy loss, the overall efficiency would be lowered and much more CO_2 would be emitted than if fossil fuels had been used directly or transformed into hydrogen by reforming. To be sustainable and environmentally friendly in the long term, electricity for water electrolysis should be preferably derived from renewable or nuclear energy sources that do not emit CO_2 and air pollutants such as SO_2 and NO_x.

Hydropower, which is by far the largest renewable electricity source today, is clearly well suited to produce hydrogen, although its availability, as discussed in Chapter 8, is limited.

Electricity from geothermal energy is a feasible approach in some geothermally active areas, such as Iceland, the Philippines and Italy, to produce hydrogen by electrolysis. However, even if economically viable in specific areas, it would only be of relatively minor importance on the global scale, as the number of geothermal sources with the quality of thermal energy required for electricity generation is limited.

Resource limitations, a lack of mature technologies and difficulties of exploitation or environmental concerns also explain the limited development of energy extraction from oceans under various forms: tidal power, wave power or thermal energy, which are thus not expected to play any significant role in hydrogen production.

Considering its enormous potential, *wind power*, compared to all the other renewable energy sources, has probably the greatest possibilities for the production of pollution-free hydrogen at a reasonable cost in the foreseeable future. Electricity from wind is already competitive with power from fossil fuels in some areas, and constant developments and improvements in turbine technologies are expected to further reduce significantly its cost. However, the intermittent nature of wind energy, with capacity factors of only about 30%, is a serious drawback, resulting in coupled electrolyzers for hydrogen production operating at full capacity only for limited periods of time, and requiring considerable hydrogen storage capacity to offset the lack of production when the wind ceases to blow. With significant optimization of wind-coupled electrolysis and hydrogen storage systems, however, costs for hydrogen generation could fall from current estimations of $6–7 per kg to less than $3 per kg [192].

Another source of energy that could potentially meet all our energy needs in the future and be used to generate hydrogen is solar energy. Like wind, solar energy is a non-polluting and plentiful source of energy but, being also an intermittent source, it suffers from the same drawbacks as wind energy. Unlike wind energy, however, it is still a very expensive way to generate electricity. With current technology, the production cost of hydrogen generated by photovoltaic systems is estimated to be as high as $28 per kg [192] – an order of magnitude higher than that based on fossil fuels, and also considerably more expensive than that based on other renewable energy sources. Even with further development, including improved efficiency and the use of thin-film technology instead of crystalline silicon solar cells, the cost is estimated to remain above $5–6 per kg of hydrogen [192]. Important technological breakthroughs will be needed to make any significant reduction in the costs of solar electricity and thus hydrogen produced by photovoltaic cells. Currently, there are promising new concepts at the research stage based on conductive organic polymers or nanostructured films, which could possibly be mass produced at lower costs than silicon-based photovoltaic cells. By avoiding the need to couple a photovoltaic device with an electrolyzer, the possibility of producing hydrogen directly from sunlight and water in a so-called photoelectrolysis (PE) device is also under development [204]. Besides photovoltaics, electricity from thermal solar power plants could be used to produce hydrogen, but for the foreseeable future this also is too expensive. Experiments have also been conducted in France, Canada, Israel and other countries to thermally split water into oxygen and hydrogen at high temperature (2000–2500 °C) using solar furnaces. Although there has been limited success, no practical applications are in sight. Thermochemical water splitting using solar energy is another possibility that must be further explored [205]. As solar energy is clearly a potentially low-cost option for the sustainable production of hydrogen in the future, further research is warranted.

9.3.5
Hydrogen Production Using Nuclear Energy

As with renewable energy sources, nuclear power reactors for electricity generation do not emit any CO_2 or other pollutant gases into the atmosphere. Using off-peak periods, hydrogen could be produced by nuclear power through electrolysis, enabling an extended and more efficient utilization of these plants. As mentioned earlier, higher temperatures improve both the thermodynamics and kinetics of the process: hydrogen can be generated more efficiently in less time. Most of the new generation IV reactors are planned to operate at more elevated temperatures (700–1000 °C) than existing reactors, which operate between 300 and 400 °C [170]. These new reactors are thus well suited for high-temperature electrolysis of steam for hydrogen production. The direct thermal decomposition of water is impractical as it requires temperatures in excess of 2000 °C, but thermochemical water splitting into hydrogen and oxygen can be achieved efficiently at of 800–1000 °C by using chemical cycles [206]. Among various processes, the currently most studied is the so-called iodine–sulfur cycle, in which SO_2 and iodine are added to water in an exothermic reaction to form sulfuric acid and hydrogen iodide (Figure 9.6) [192]. Above 350 °C, HI decomposes to hydrogen and iodine, the latter being recycled. Sulfuric acid decomposes at temperatures in excess of 850 °C into SO_2 (which is also recycled), water and oxygen. With SO_2 and iodine being continuously recycled, the only feeds to be used up in the process are water and high-temperature heat, giving the products hydrogen, oxygen and low-grade heat. Considering its near-zero emission characteristics, nuclear power is particularly suited for the generation of hydrogen, and the development of such processes is being conducted at several locations, including the Japan Atomic Energy Research Institute (JAERI), the Oak Ridge National Laboratory and the Commisariat à l'Energie Atomique (CEA). High temperatures generated by nuclear plants could also be used in other energy-intensive industrial applications and for processes related to methanol

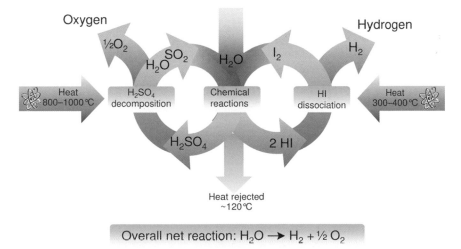

Figure 9.6 Sulfur–iodine thermochemical cycle for the production of hydrogen.

production. If used for natural gas (or methane) steam reforming or bi-reforming, the heat provided by a high-temperature nuclear reactor could significantly increase the efficiency in synthesis gas production and reduce CO_2 emissions.

Eventually, if and when fusion reactors become available the heat generated could also be used in hydrogen production and other high-temperature applications.

9.4
The Challenge of Hydrogen Storage

Producing hydrogen is only the first step in the envisioned Hydrogen Economy (Figure 9.7). The next challenge is the storage of the generated hydrogen in a form that should be economical, practical, safe and user friendly. Because hydrogen is a very light gas, it contains much less energy per unit volume than conventional liquid fuels under the same pressure. Under normal conditions, hydrogen requires about 3000 times more space than gasoline for an equivalent amount of energy. Thus, hydrogen must be compressed, liquefied or absorbed on a solid material to be of any practical use for energy storage.

Depending on the use for stationary or mobile applications, hydrogen will have very different storage characteristics. In stationary applications, including heating and air-conditioning of homes and buildings, electricity generation and varied industrial uses, hydrogen storage systems can occupy a relatively large space and their weight is not a major factor. In contrast, hydrogen storage in transportation such as in cars is limited by volume and weight which, to provide a reasonable driving range of some 500 km, must remain minimal. Hydrogen storage is therefore a key factor for the successful introduction of hydrogen as a transportation fuel, the presently considered major application for the Hydrogen Economy. The challenges faced by hydrogen storage in transportation are great and far from resolved. With a driving range requirement of 500 km, a storage capacity of 5–10 kg of hydrogen will be needed even for a fuel cell-propelled vehicle, which also requires high purity hydrogen free of carbon monoxide. At the same time, refueling should take less than 5 minutes and should be as easy and safe as with hydrocarbon fuels today. Finally, the system should also be economically affordable.

Current hydrogen storage technologies involve either physical or chemical storage. Physical storage is effected in insulated or high-pressure containers in which hydrogen is stored as a liquefied or compressed gas. Chemical storage involves metals and other materials that absorb or contain hydrogen and can subsequently readily release it for use. Each of these methods has its advantages, but at the same time has serious drawbacks.

9.4.1
Liquid Hydrogen

On a weight basis, hydrogen has the highest energy content of any known fuel (almost three times that of gasoline). However, being the lightest gas, the density of

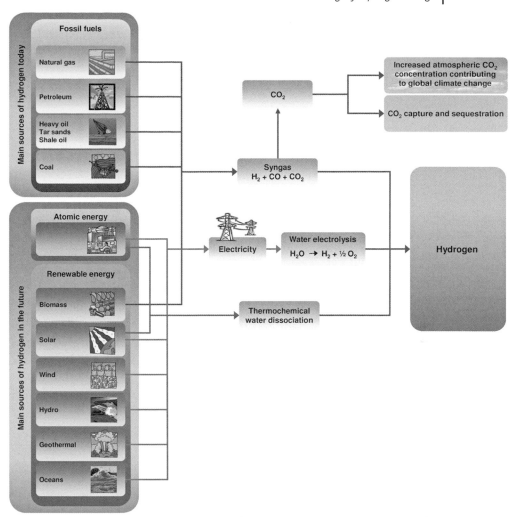

Figure 9.7 Different routes for the production of hydrogen.

liquid hydrogen is only $70.8 \, \text{kg m}^{-3}$, corresponding on a volume basis to an energy content about a factor of three less than gasoline. Nevertheless, liquid hydrogen is a compact form of hydrogen, making it in principle an attractive candidate for hydrogen storage, especially in transportation. In fact, it is in this form that hydrogen is used as a propellant for space vehicles. However, hydrogen, which has a boiling point of $-253\,^\circ\text{C}$, is, after helium, the most difficult gas to liquefy. Complex and expensive multi-stage cooling systems are necessary to obtain liquid hydrogen. Typically, in the first step, hydrogen is pre-cooled with liquid ammonia to $-40\,^\circ\text{C}$ and then to $-196\,^\circ\text{C}$ using liquid nitrogen. In the following step, helium is used in a multi-stage compression-expansion system to obtain liquid hydrogen at $-253\,^\circ\text{C}$. The efficiency of this complex process increases with the plant's size and is thus more adapted for centralized production. The process is not only complex and expensive but also very energy intensive: about 30–40% of the energy content of the hydrogen is required for its liquefaction [207]. Moreover, liquid hydrogen storage systems inevitably lose hydrogen gas over time by evaporation or "boil off." The rate of loss is dependent on the amount stored and the tank's insulation, and is generally lower for larger quantities of liquid hydrogen. In the case of automobile tanks with small capacities, the result is that 1–5% of the hydrogen content would be released to the atmosphere each day to avoid pressure build-up and possible explosion [13]. Given the cost and energy invested into producing liquid hydrogen, this is unacceptable both from an economic and also from an environmental viewpoint. Although suitable indicators could be installed, guessing how much fuel is still in the storage tank after some days is certainly not something that most people would like to worry about! In addition to the expensive cost of cryogenic storage, liquid hydrogen, because of its extremely low temperature, must be handled with great care. Hydrogen leaks also represent major safety hazards.

9.4.2
Compressed Hydrogen

To store sufficient amounts of energy in a given space, hydrogen compression is currently the preferred solution used in most hydrogen fuel cell-powered prototype cars (*vide infra*). Because the same quantity of hydrogen can be stored in smaller tanks with increased pressure, containers (tanks) were developed over the years that are able to withstand increasingly high pressures. Hydrogen can now be held under 350 or even 700 atm in tanks made from new lightweight materials, such as carbon-fiber-reinforced composites. However, even under these conditions, hydrogen still has a much lower energy content per volume than gasoline (4.6 times less than gasoline at 700 bar of H_2; Figure 9.8) and thus requires several-fold more voluminous tanks. In contrast to liquid fuel tanks, which can adopt varied shapes and easily be adapted to any vehicle, compressed hydrogen tanks have a fixed cylindrical shape necessary to ensure their integrity under high pressure. Designers and engineers will need to pay great attention to how and where to integrate the pressure tanks in their vehicles. Although hydrogen compression is less energy intensive than liquefaction, depending on the pressure, it still uses the equivalent of 10–15% of the energy contained in the hydrogen fuel [207]. Because of its small size, as mentioned, hydrogen can diffuse through many materials, including metals. During prolonged exposure to hydrogen, some metals can also become brittle. Consequently, because many parts of the fuel system in contact with hydrogen will be metallic, it is necessary to prevent material failure that could have grave consequences, especially under high pressures. The risk of leaks is a major safety hazard as hydrogen is a highly flammable and explosive gas when exposed to air. This is of course an even greater concern in accidents involving collision. Although the high-pressure hydrogen storage systems made from high-tech materials are also complex and presently very expensive, technological improvements and larger-scale production will certainly reduce the costs.

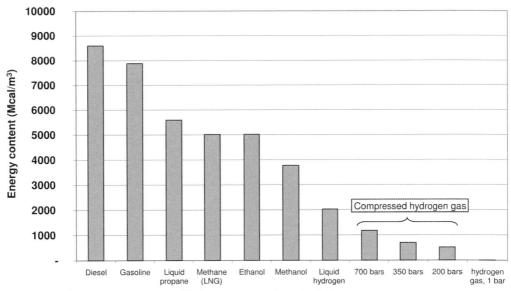

Figure 9.8 Volumetric energy content of hydrogen compared to other fuels.

9.4.3
Metal Hydrides and Solid Absorbents

An alternative to liquefaction and compression is to store hydrogen in solids, either physically absorbed or chemically bound. Most research in this field has concentrated on metals and metallic alloys that have the ability to absorb hydrogen, like a sponge, to form hydrides. In these materials, the metal matrix is expanded and filled with hydrogen. The process, depending on the nature of bonding, can be either reversible or irreversible. Hydrogen stored irreversibly in some materials (including chemical hydrides) can only be released by the chemical reaction of these compounds with another substance, such as water, producing by-products that must be collected and reprocessed before they can be used again to store hydrogen. Thus, for practical purposes, using irreversibly formed hydrides is not an attractive means of hydrogen storage.

Reversible hydrides are generally solids that can release the hydrogen contained in them in the form of molecular hydrogen, H_2. The uptake and release of hydrogen is typically controlled by temperature and pressure, and is different for different hydrides. Some metals absorb hydrogen rapidly but release it slowly, while others require higher temperatures before release is possible. The release of the absorbed hydrogen may also be only partial. The storage and release of hydrogen in any case should take place at temperatures and on a time scale needed for practical applications. Volume and temperature increases due to hydride formation should also to be taken into account for storage vessel design. Conventional metal hydrides, as well as their hydrogen-storage capabilities, have been well characterized, with most containing relatively heavy elements: $TiFe$, $ZrMn_2$, $LaNi_5$ and so on. As only a few hydrogen atoms can be bound by each metal atom, this explains why typically only 1–3% by weight of these metal hydrides is actually usable hydrogen. In other words, to store 5 kg of hydrogen a tank will need to weigh 200 kg or more. So, unless compensated for by the use of lightweight materials in other parts of the vehicle, the additional weight of the storage tank would reduce fuel efficiency, which is one of the main goals of developing hydrogen-powered vehicles. Metal hydrides, however, do have the advantage of being compact, requiring less space than compressed hydrogen for an equal amount of stored energy. Being only under moderate pressure, hydride tanks can also be shaped more freely, and this facilitates their integration into the vehicle body. Today, research into metal hydrides is focused mainly on lighter compounds such as $NaAlH_4$, Na_3AlH_6, $LiAlH_4$, $NaBH_4$, $LiBH_4$ and MgH_2, which offer higher hydrogen contents per unit of mass.

Besides metal hydrides, other solid-absorbing materials for potential hydrogen storage are under investigation. Fullerenes and carbon nanotubes have attracted much attention, but they are still a long way from finding their way into fuel tanks, partially because of their still exorbitant price and unproven potential. Metal-organic frameworks (MOFs), a new class of highly porous materials with high surface areas, have also been recently discovered and show promise as high-capacity hydrogen-storage media [208, 209].

9.4.4
Other Means of Hydrogen Storage

In seeking to overcome the problems associated with hydrogen storage and distribution, many approaches are being pursued to use liquids that are rich in hydrogen, such as gasoline or methanol, as a hydrogen source. In contrast to pure hydrogen, these are compact (they contain, on a volume basis, more hydrogen than even liquid hydrogen), and are easy to store and handle. The possibility of generating hydrogen with more than 80% efficiency by reforming gasoline on-board of vehicles has been demonstrated. As the reforming produces a mixture of H_2 and CO, the later must be separated and disposed of as CO_2, adding to atmospheric levels of greenhouse gases. Also, as CO will poison the fuel-cell catalysts, every trace must be removed. The reforming process is also expensive and challenging, because it involves high temperatures and needs considerable time to reach steady-state operational conditions. The advantage is that the distribution network for hydrocarbon fuels already exists, though this would not solve the problems of diminishing oil and gas resources.

Using methanol reformers operating at much lower temperatures (250–350 °C) [13], though still expensive, is more adaptable for vehicle on-board applications. The absence of any C—C bonds in methanol also greatly facilitates its reforming to hydrogen, and was the fuel used to generate hydrogen in Daimler's Necar 5 fuel cell demonstration vehicle. As a liquid, methanol – much like gasoline – can be distributed through the existing infrastructure, including filling stations, with only minor modifications. Furthermore, methanol can be made from different sources, including the recycling of CO_2 (Chapter 12). The use of direct methanol fuel cells (DMFCs) would, however, eliminate the need to first reform methanol to hydrogen (*vide infra*).

9.5
Centralized or Decentralized Distribution of Hydrogen?

If hydrogen is produced from gasoline or diesel fuel on-board vehicles via reformers, then the existing distribution network could be used with essentially no modifications. However, even if the reformer technology can be made economically viable, it would not solve our dependence on decreasing oil resources and would still produce considerably increased amounts of CO_2.

For the hydrogen gas to become the energy carrier of the future, it would need to be easily available anywhere at an affordable price, and its distribution should be safe and user friendly – similar to today's hydrocarbon-based fuels. These criteria imply the creation of an entirely new infrastructure specifically designed for the transportation, storage and distribution of hydrogen. Hydrogen could, however, also be produced directly in local fueling stations by natural gas reforming or via water electrolysis using electricity. Decentralized production would not require a nationwide delivery system of hydrogen involving trucks or pipelines, but it would be very expensive and energy consuming. Only some 100 hydrogen-producing and refueling stations are currently operating worldwide, namely, in Germany, the United States, Japan and other countries, providing hydrogen for a small number of automobiles and buses [210]. California, in its Hydrogen Highway program, was planning to construct numerous such fuel stations within a decade. The original plan presented in 2004 called for hundreds of hydrogen stations to be built all across the California highway system by 2010 to power thousands of cars, trucks and buses [211]. As of 2009, however, only some 20 hydrogen stations have been built with only a handful accessible by the public. Locally distributed hydrogen generation may be the preferred option as long as the number of hydrogen-powered vehicles and the demand for hydrogen remain low. However, there are drawbacks: hydrogen being generated in limited amounts has a high cost, and on-site production by reforming of natural gas emits CO_2 which, considering the scattered locations and small scale, could be expensive to capture and recycle. Electrolysis is only emission-free if the electricity used comes from renewable or atomic energy sources. In most countries, if electricity is taken from the existing grid, a significant part of it will have been generated by fossil fuel-burning power plants, thereby eliminating any possible benefits expected from the use of hydrogen. The local manufacture of leak-prone and explosive hydrogen in populated areas would also raise serious safety issues.

The alternate solution would be to produce hydrogen in large quantities in centralized plants and then to transport it to the local stations by road, rail or pipelines. In such plants, the production costs would be lower, the efficiency higher and CO_2 capture and sequestration (or recycling) easier to implement. Furthermore, hydrogen generation is not limited only to methane (natural gas) reforming or water electrolysis, but can also involve other technologies such as coal and biomass gasification or thermal splitting of water in high-temperature nuclear reactors. However, such means of hydrogen production would require the establishment of an extensive storage and delivery system to service customers. Currently, the feasible options taken in consideration for delivering hydrogen are by road transport using trailers containing hydrogen in its cryogenic liquid or high-pressure form, as well as hydrogen gas pipelines.

The transportation by truck of hydrogen as a cryogenic liquid is today commonly used when delivering hydrogen to industries with limited needs, and where on-site generation would be uneconomical. It is by this means, for example, that the hydrogen required as propellant for Space Shuttle launches is transported from Louisiana to the NASA launch pads in Florida. Although liquid hydrogen has a density about ten-fold lower than that of gasoline or diesel oil, it still has the advantage of being relatively compact compared to compressed hydrogen. Commercially available trailer-trucks can transport some 3500 kg hydrogen in liquid form, and this is energetically equivalent to about 13 000 L of gasoline [212]. Because the cryogenic trailers used must be double hulled and vacuum insulated to avoid excessive blow-off they are expensive. It has been suggested that transportation of hydrogen in its liquid form from centralized production centers to local distribution stations could play an important role in the initial transition phase to a hydrogen infrastructure. On the large scale necessary, however, this solution is not suitable. As mentioned earlier, up to 40% of the energy content of the shipped hydrogen is consumed by its liquefaction, making the process far too expensive from both energetic and economic viewpoints. Hydrogen compression requires less energy than liquefaction. The steel cylinders and other on-board equipment needed for the safe handling of highly pressurized hydrogen is both heavy and expensive. Moreover, with road transport being limited by weight, and considering the extremely low density of hydrogen, this means that the current tube trailers used to transport hydrogen under high pressure (200 atm) can each deliver only about 300 kg of hydrogen. Even taking into account any expected future technical advances, a 40 000-kg truck would enable the delivery of only 400 kg of hydrogen, or about 1% of its dead weight [207]. In comparison, a similar truck could deliver some 26 tonnes of gasoline, containing more than twenty times more energy than the compressed hydrogen truck. So, instead of one driver and truck, more than twenty would be needed to deliver the equivalent amount of energy as gasoline; this in turn would generate higher costs and increase traffic congestion. The introduction of lightweight materials for high-pressure hydrogen storage, such as those currently under development for use in motor cars, could potentially be utilized, but the transportation capacity is expected to remain modest. Besides economic and energetic considerations, a substantial increase in the transportation of highly flammable and explosive hydrogen, both in liquid and compressed form, by road would imply considerable safety issues and risks.

The most commonly used system for the transmission of hydrogen in large quantities for the chemical and petrochemical industries is by pipelines. To date, worldwide, these have a combined length of only about 2500 km, of which 1500 km are located in Europe and 900 km in the United States [213]. Transport by pipeline allows a direct connection to be made between main hydrogen producers and users. Hydrogen pipelines, however, due to the ability of hydrogen to diffuse through or react with common materials, require the use of special steels or metals, seals and pumps, and they are also expensive to build, maintain and operate. Although there are many hundreds of thousands of kilometers of existing pipelines for natural gas, oil and other hydrocarbon products around the world, these are not well suited for hydrogen transport. Most of the metal pipelines, seals, pumps and other equipment, when exposed to hydrogen, would allow hydrogen to diffuse through and would become brittle over time. Owing to the small size of hydrogen, leakage – especially during transportation over long distances – is likely to occur and should therefore be carefully controlled to minimize significant losses and explosion hazards. The cost of hydrogen transport is also at least about 50% higher than that for natural gas for the same volume [214], while hydrogen contains three times less energy than natural gas. Pipeline shipment and dispensing of hydrogen is estimated to cost some $1 per kg with current technology (about $0.7 per kg expected by future improvements). It is thus much more expensive than the $0.19 per gallon currently paid to ship and distribute gasoline [192]. Nonetheless, the transport of hydrogen by pipelines may be the best solution to date, though the installation of a large hydrogen pipeline infrastructure would be highly capital intensive. Indeed, it would only be an option for the long term, when the numbers of cars and other fuel cell-driven devices running on hydrogen would be sufficient to support the vast investments needed.

9.6
Hydrogen Safety

As mentioned earlier, the chemical and petrochemical industries have been using hydrogen in their operations for many years, and the space industry has used it for several decades. Within industrial settings, the production, storage and transportation infrastructures have been developed for the safe use of hydrogen. However, it must again be understood that hydrogen is a volatile, dangerous and explosive gas. Perhaps the most vivid image of this fact is the aforementioned fire that destroyed *The Hindenburg* Zeppelin in 1937 while landing in Lakehurst, New Jersey. Initially, hydrogen used to keep the airship aloft was blamed for the disaster, but later investigations showed the real cause of the accident was the extremely flammable lacquer (it had similar properties to rocket fuel) that was painted onto the outer hull of the airship, which ignited due to electrostatic discharges. The ensuing fire, helped by hydrogen, burned the entire airship in only about 30 seconds.

In the past, hydrogen has also been widely used in homes for cooking and heating (the owners being most often unaware of it) in the form of town gas, which contained up to 60% hydrogen in a mixture with carbon monoxide. It was due in part to the high toxicity of CO, and not only because of the properties of hydrogen, that town gas has been replaced by natural gas. Owing to its unique physical properties compared to liquid and gaseous hydrocarbon fuels, the safety issues associated with the use of hydrogen are, however, also significant and quite specific. Being small and light, hydrogen is a most leak-prone gas. Hydrogen itself is non-toxic, but it is explosive and flammable. Moreover, being colorless, odorless and tasteless, it is difficult to detect leaks. In the case of natural gas, which is also odorless, colorless and tasteless, volatile sulfur compounds are added to make leaks readily detectable, but the addition of such odorants is impractical in the case of hydrogen. In addition, the odorants would leak at different rates compared to the extremely small hydrogen molecules. Consequently, it is necessary to use sensors for hydrogen detection, though even these have been found to be relatively ineffective. Additives could also contaminate and poison the hydrogen fuel cells. Hydrogen is flammable over a wide range of concentrations in air (4–75%), and the minimum energy necessary for its ignition (0.005 mcal) is about 20-fold lower than that for natural gas and gasoline. Common electronic devices such as cell phones or even the friction of sliding over a motor car seat could cause ignition if the correct concentration of hydrogen in air is present [190, 13]. Hydrogen burns with a scarcely, almost invisible, slightly bluish flame, which means that a person could actually step unknowingly into hydrogen flames. Hydrogen, as mentioned earlier, can also cause many metals (including steel) to become brittle over time, raising the risk of cracks and fractures that would result in failures with possible catastrophic consequences, especially in high-pressure systems. Hence, specialized materials or/and liners would be necessary for hydrogen storage.

Until now, the good safety record of hydrogen use in industry has been largely due to the numerous precautions, codes and standards required for hydrogen handling by trained professionals. It is also related to the fact that most hydrogen is produced on-site and so is not transported over long distances in large quantities. However, if hydrogen were to be handled by the wider public as a transportation fuel by people with no formal training or awareness of its potential danger, then it would be vital that strict new safeguards be introduced. Such safety measures would most likely be very costly and public compliance difficult to ensure.

9.7
Hydrogen as a Transportation Fuel

The development of a hydrogen infrastructure, besides the difficulties discussed earlier, is facing the "chicken and egg dilemma." As long as there is no adequate hydrogen distribution infrastructure, there will only be a limited demand for hydrogen-powered cars and other applications, despite its obvious attractiveness as an energy storage material. On the other hand, there is no real incentive for investing hundreds of billions of dollars in a hydrogen infrastructure unless there is a solid and sustained demand for it. The questions to be answered are: is hydrogen a feasible, economic and safe fuel for the future and if so how will the needed production, transportation and distribution be stimulated and advanced? Today, the fate of the widespread use of hydrogen seems to be closely connected with the development of hydrogen fuel cells, which offer the promise of very efficient and zero-emission vehicles. However, questions about generation, handling and distribution of hydrogen are often neglected. Owing to significant technical and economic challenges, the road to commercialization still has a long way to go. Currently, most prototype hydrogen vehicles are very expensive, and even using optimistic assumptions the US Department of Energy has estimated that future hydrogen fuel cell vehicles (FCV) would likely be 40–60% more expensive than conventional ones [13]. Thus, it may take decades before hydrogen FCVs begin to seriously replace internal combustion engine (ICE)-powered cars and trucks. In aiming to accelerate the transition to a Hydrogen Economy, the use of hydrogen as a fuel in conventional ICE vehicles has also been suggested. Except for replacing the fuel tank with a hydrogen tank, only minor and relatively inexpensive changes would be necessary to run an ICE vehicle on hydrogen. Although safety precautions must be extensive, several automobile makers are considering this pathway. BMW, in particular, has been conducting research on hydrogen-powered engines since 1978, testing the first prototype motor car a year later. The company's sixth-generation hydrogen-powered car, the BMW Hydrogen 7, has been produced in limited numbers. Beginning in 2006 a fleet of 100 of these cars has been leased to influential people and celebrities around the world. The Hydrogen 7 has a hybrid propulsion system that can run either on hydrogen or gasoline. This allows the car to use today's road network, where hydrogen filling stations are still few and far between. The liquid hydrogen tank, with a capacity of 170 L, can hold 8 kg of hydrogen, allowing the car to cover a distance of about 200 km. A secondary gasoline tank with a 60 liter capacity extends this range to 680 km (Figure 9.9). In addition to BMW, Ford (hydrogen-powered model U concept car) and Mazda are also working on hydrogen ICE vehicles. Thirty hydrogen powered ICE shuttle buses built by Ford have been used since 2007 in the United States and Canada, including at some major airports, in the city of Las Vegas and SeaWorld in Orlando. Mazda is road testing a hydrogen-gasoline version of its RX-8 model, equipped with a rotary engine and storing hydrogen on-board under high pressure. Thirty of these RX-8 Hydrogen RE vehicles were given to the Hydrogen Road of Norway (HyNor), to be initially used mainly on a stretch of highway between Oslo and Stavanger (580 km). In 2007 HyNor had already purchased 15 Toyota Prius hybrid vehicles modified by Quantum to run on hydrogen. All these vehicles have the advantage of producing almost no pollutants (except for small amounts of NO_x) and could be introduced relatively fast in the commercial market compared to FCVs. To

Figure 9.9 Hydrogen ICE car from BMW. (Courtesy of ©BMW AG.)

significantly increase efficiency, the hydrogen ICE could also be coupled with an electric hybrid system, as found in Toyota's Prius (an ICE engine running on gasoline also charges batteries, which take over to provide an electric drive in slow city traffic). Regretfully at the time of writing Toyota indefinitely delayed plans for the production of its Prius cars in the US due to strongly decreasing demand as a consequence of economic slowdown and lower gasoline prices [215]. The overall efficiency, however, is expected to remain lower than for fuel cell vehicles. Despite some advantages of hydrogen ICEs, the problem of on-board hydrogen storage, which presently limits the driving range, also remains. Besides fueling cars with hydrogen produced at central locations and in delocalized small units using electrolysis of water at filling stations, it should also be possible to fuel these cars with hydrogen produced by on-board reforming of various hydrocarbon fuels. This would, however, not result in any advantage over conventional hydrocarbon-burning ICEs. Carbon monoxide must be carefully separated from the generated syn-gas, so as not to poison the fuel cells. If the CO would then be oxidized the produced CO_2, emission to the atmosphere would only be relocated to the hydrogen-generating facilities or devices, but not eliminated.

9.8
Fuel Cells

9.8.1
History

Fuel cells are devices that convert the chemical energy of a fuel directly into electrical energy by electrochemical reactions. Fuel cells are considered to be one of the main solutions for the efficient utilization of fossil fuel-derived fuels. The concept of fuel cells was discovered by William R. Grove, a Welshman, during the late 1830s. Grove discovered that by arranging two platinum electrodes with one end of each immersed in a container of sulfuric acid and the other ends separately sealed in containers of hydrogen and oxygen, a constant current would flow between the electrodes. The sealed containers held water as well as the gases, and Grove noted that the water level rose in both tubes as the current flowed. In 1800, William Nicholson and Anthony Carlisle in England had described the process of using electricity to decompose water into hydrogen and oxygen (the electrolysis of water). But combining hydrogen and oxygen to produce electricity and water was, according to Grove, "... a step further that any hitherto recorded." Grove realized that by combining several sets of these electrodes in a series circuit he might "... effect the decomposition of water by means of its composition." His device, which he named a "gas battery," was the first ever fuel cell.

The device remained a curiosity, however, with no practical application in sight until, more than a century later, in 1953, Sir Francis T. Bacon constructed the first fuel-cell prototype with a power output in the kW range. Bacon began experimenting with alkali electrolytes in the late 1930s, settling on potassium hydroxide (KOH) instead of using the acid electrolytes known since Grove's early discoveries. Potassium hydroxide performed as well as acid electrolytes and was not as corrosive to the electrodes. Bacon's cell used porous "gas-diffusion nickel electrodes" rather than solid electrodes as Grove had used. Gas-diffusion electrodes increased the surface area in which the reaction between the electrode, the electrolyte and the fuel occurred. Bacon also used pressurized gases to keep the electrolyte from "flooding" the tiny pores in the electrodes. Over the course of the following 20 years, Bacon made enough progress with the alkali cell to present large-scale fuel cell demonstration units. The US space agency (NASA) selected alkali fuel cells for the Space Shuttle fleet, as well as for the Apollo program, mainly because of power-generating efficiencies that approach 70%. Importantly, alkali cells also provided clean drinking water for the astronauts. The cells use platinum catalysts that are expensive (probably too expensive for large scale commercial applications), but several companies are examining ways to reduce costs and improve the cells' versatility by using less-expensive cobalt catalysts. Most of these alkali fuel cells are currently being designed for transport applications.

9.8.2
Fuel Cell Efficiency

In contrast to heat engines (gasoline and diesel engines), the fuel cell does not involve conversion of heat into mechanical energy and the overall thermodynamic efficiencies can be very high (Figure 9.10).

The thermodynamic derivation of the Carnot cycle of a heat engine states that all of the heat supplied to it cannot be converted into mechanical energy, and thus some of the heat is rejected. The heat is accepted from a source at higher temperature (T_H in Kelvin), where part of it is converted into mechanical energy, and the remainder is rejected into a heat sink at lower temperature (T_S in Kelvin). The greater the temperature difference between the source and the sink, the greater the efficiency. The Carnot efficiency of a heat engine is given by Equation 9.1. In contrast, the fuel cell

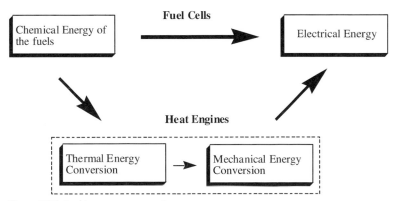

Figure 9.10 Fuel to energy conversions.

efficiency is related to the ratio of two thermodynamic properties, Gibbs free energy (ΔG°) and the total heat energy or enthalpy (ΔH°) (Equation 9.2):

$$\text{Maximum efficiency (Carnot),} \quad \eta_{\text{Carnot}} = (T_H - T_S)/T_H \tag{9.1}$$

$$\text{Fuel cell efficiency,} \quad \eta_{\text{Fuel cell}} = \Delta G^\circ/\Delta H^\circ \tag{9.2}$$

The theoretical thermodynamic efficiency of a hydrogen-oxygen fuel cell is $\sim$83% at ambient temperature. To achieve acceptable efficiencies, an internal combustion engine under ideal conditions must operate at a very high temperature. Figure 9.11 shows the variation of a hydrogen fuel cell's theoretical efficiency versus the corresponding Carnot efficiency of a heat engine.

The ambient temperature maximum thermodynamic intrinsic fuel cell efficiencies of different fuels can be very high. Table 9.2 lists these data, along with reversible cell potentials of selected fuels.

Fuel cells, therefore, are considered as very efficient electrical energy-producing devices with high power densities at relatively low temperatures. The possible applications of fuel cells are numerous, from micro fuel cells, producing only a few watts needed in cell phones, to on-board fuel cells for the automobile sector, and large units able to produce several MW to provide buildings with electricity. Major drawbacks to the widespread commercialization of fuel cells are mainly technological (reliability issues, material durability, catalyst utilization, mass transport, etc.) and cost related. Different types of fuel-cell designs exist, with some being more suited to certain applications than others. However, they all function on the same electrochemical principle. Fuel cells, in principle, can be built based on any exothermic chemical reaction.

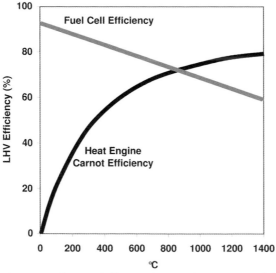

Figure 9.11 Theoretical efficiency change with temperature of a hydrogen fuel cell and a heat engine.

Table 9.2 Theoretical reversible cell potentials (E_{rev}^{o}) and maximum intrinsic efficiencies for fuel cell reactions under standard at 25 °C.

Fuel	Reaction	n	$-\Delta H^{\circ}$ (kJ mol^{-1})	$-\Delta G^{\circ}$ (kJ mol^{-1})	E_{rev}^{0} (V)	E (%)
Hydrogen	$H_2 + 0.5O_2 \rightarrow H_2O$ (l)	2	286.0	237.3	1.229	82.97
Methane	$CH_4 + 2O_2 \rightarrow CO_2 + 2H_2O$ (l)	8	890.8	818.4	1.060	91.87
Methanol	$CH_3OH + 1.5O_2 \rightarrow CO_2 + 2H_2O$ (l)	6	726.6	702.5	1.214	96.68
Formic acid	$HCOOH + 0.5O_2 \rightarrow CO_2 + H_2O$ (l)	2	270.3	285.5	1.480	105.62
Ammonia	$NH_3 + 0.75O_2 \rightarrow 0.5N_2 + 1.5H_2O$ (l)	3	382.8	338.2	1.170	88.36

9.8.3
Hydrogen-Based Fuel Cells

Hydrogen-based fuel cells produce electricity, heat and water by catalytically combining hydrogen with oxygen. They are composed of two electrodes, an anode (negatively charged) and a cathode (positively charged), separated by an electrolyte. This electrolyte can be made of various materials, from polymers to ceramics, which are in general ion (H^+, OH^-, CO_3^{2-}, O^{2-}, etc.) conductors. The nature of the electrolyte determines many of the fuel cell's properties, including the temperature of operation, and this is therefore used to categorize the different fuel cell types.

In a proton-exchange membrane (PEM) fuel cell (*vide infra*), hydrogen entering the fuel cell is split with the help of a catalyst (generally platinum) on the anode side into electrons and protons (H^+). The electrons move along an external circuit to power an electric device, while the protons migrate through the electrolyte. At the cathode, by action of a catalyst, protons and electrons are recombined with oxygen of the air to produce water. The device is shown in Figure 9.12. Since every cell produces less than 1 V, many cells must be stacked together to produce higher voltages.

In an alkaline-based fuel cell, instead of protons, hydroxide ions (OH^-) move from cathode to anode (Figure 9.13). Although, the alkaline fuel cells are utilized in space applications, their commercial use is hampered by their sensitivity to CO_2, which reacts with the alkali.

The most studied fuel-cells designs include phosphoric acid fuel cells (PAFCs), molten carbonate fuel cells (MCFCs), solid oxide fuel cells (SOFCs), proton exchange membrane (PEM) fuel cells and direct methanol fuel cells (DMFCs). The latter is discussed in more detail in Chapter 11.

PAFC, MCFC and SOFC are generally designed to be used in stationary applications because they are heavy, bulky and require operating temperatures from around 200 °C for PAFC to about 650–1000 °C for MCFC and SOFC. PAFC, the most mature fuel cell technology, is commercially available from United Technologies Corp. Close to 300 units have been installed worldwide. As the name indicates, these fuel cells use liquid polyphosphoric acid as the electrolyte, the electrodes being made of carbon coated with finely dispersed platinum catalyst. The hydrogen required is obtained by

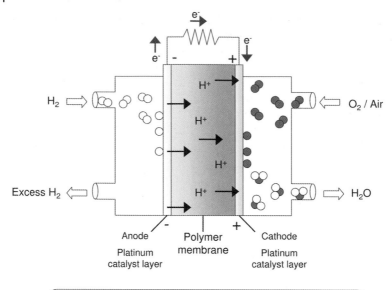

Anode	$H_2 \longrightarrow 2H^+ + 2e^-$	$E^0 = 0.00$ V
Cathode	$\frac{1}{2}O_2 + 2H^+ + 2e^- \longrightarrow H_2O$	$E^0 = 1.23$ V
Overall	$H_2 + \frac{1}{2}O_2 \longrightarrow H_2O$	$E_{cell} = 1.23$ V

Figure 9.12 Proton exchange membrane (PEM) hydrogen fuel cell.

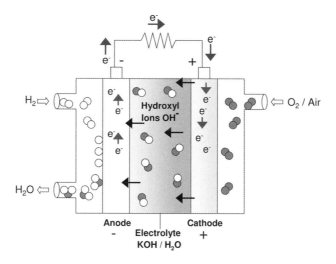

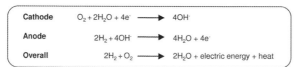

Cathode	$O_2 + 2H_2O + 4e^- \longrightarrow 4OH^-$
Anode	$2H_2 + 4OH^- \longrightarrow 4H_2O + 4e^-$
Overall	$2H_2 + O_2 \longrightarrow 2H_2O + \text{electric energy} + \text{heat}$

Figure 9.13 Alkaline fuel cell.

methane (natural gas) reforming, and the overall efficiency from methane to electricity is 37–42%. With co-generation, a heat efficiency approaching 80% can be achieved, which is comparable to conventional systems burning natural gas. At around $4500 per kW capacity [216], PAFCs also remain expensive compared to conventional fossil fuel-based technologies, with costs of less than $1000 per kW capacity. However, because fuel cells have no moving parts they are generally very reliable and require low operation and maintenance costs. This explains why PAFCs have only found their way into "niche markets," mainly for consumers in need of a very stable, reliable and clean on-site electricity source such as banks, airports, hospitals or military bases.

In MCFCs, the electrolyte is made of lithium and potassium carbonate salts heated to about 650 °C (Figure 9.14) [216]. At this high temperature the molten carbonate salts act as electrolytes and CO_2 formed by the reaction of carbonate with hydrogen is transported between the anode and cathode through carbonate ion. Because they operate at high temperature, natural gas (or other fuels, including methanol and ethanol) can be converted into hydrogen-rich gases directly inside the fuel cell in a process called internal reforming. Without the need for an external reformer, MCFCs can reach fuel to electricity efficiencies above 50% – much higher than the 37–42% obtained with PAFCs. The higher temperature also allows nickel to be used as a catalyst instead of expensive platinum, which is much more active than nickel at lower temperature. The first commercial unit was delivered by FuelCell Energy, Inc.

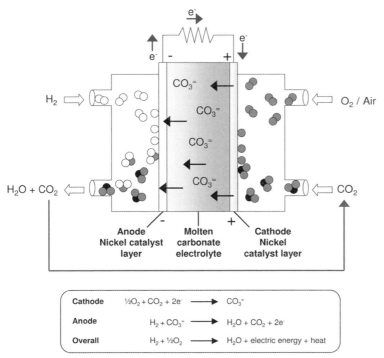

Figure 9.14 Molten carbonate fuel cell (MCFC).

to a brewery in Japan in 2003, and today more than 50 units from that company and others are operating worldwide. Their price is in the same range as PAFCs. Molten carbonates are, however, highly corrosive, and this raises some concerns about the fuel cell's lifetime. FuelCell Energy, Inc., in collaboration with the US Department of Energy, is also developing a hybrid system combining a MCFC with a gas turbine that could eventually lead to power plants with fuel to electricity efficiencies approaching 75%.

SOFC is the technology that currently attracts the most attention for stationary applications. These cells operate at high temperature (800–1000 °C) and thus, like MCFCs, do not require a reformer. However, in contrast to PAFCs and MCFCs, the SOFC uses a solid ceramic (usually Y_2O_3-stabilized ZrO_2) instead of a liquid as an electrolyte. O_2^- ions are transported from cathode to anode, the latter being made of Co-ZrO_2 or Ni-ZrO_2 (Figure 9.15). This feature allows the electrolyte to adopt different shapes, such as tubes or flat plates, giving a greater freedom in fuel cell design and also avoiding problems associated with the use of corrosive liquids. The efficiency of SOFCs is in the 50–60% range. The high temperature of the exhaust gases produced is ideal for co-generation and combined cycle electric power plants. In combination with gas turbines, efficiencies of 70% or more could thus be achieved. In co-generation units, the use of waste heat could bring overall fuel efficiencies to 80–85%. In 2000, the US Department of Energy formed the Solid State Energy Conversion Alliance (SECA), involving companies, universities and national laboratories, with the goal of producing a highly efficient SOFC system that would cost

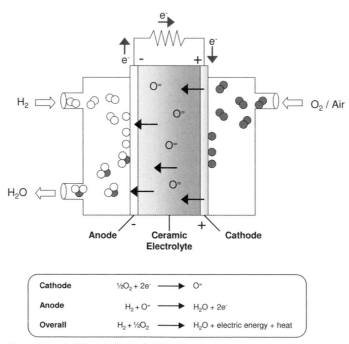

Figure 9.15 Solid oxide fuel cell (SOFC).

about \$400 per kW by 2010, allowing this technology to compete with diesel generators and natural gas turbines and to rapidly gain widespread market acceptance. The mass production of standardized basic ceramic modules, using manufacturing technologies similar to those developed for the production of electronic components, is believed to be the key to this ambitious cost reduction [217].

9.8.4
PEM Fuel Cells for Transportation

For transportation, PEM fuel cells are at present the most promising option to replace the current internal combustion engine. They operate like a refuelable battery, generating electricity for as long as they are supplied with hydrogen fuel and oxygen (from air). At the core of these fuel cells, a thin polymer film constitutes the electrolyte. Until now, this PEM has been based predominantly on a fluorocarbon polymer produced by DuPont and known under the commercial name of Nafion; this polymer is permeable to protons when saturated with water, but does not conduct electrons. It is sandwiched between two platinum-impregnated porous carbon (graphite) electrodes. PEM fuel cells can, in theory, convert more than 50% of the fuel into usable energy. They have a high power density, which translates into low weight and volume requirements. In addition, they operate at low temperature (generally 80 °C), which allows for fast start-ups, and they can very rapidly change their power output as a function of demand. Furthermore, they are safe, quiet and easy to operate and maintain. For these reasons, PEM fuel cells are seen as the most suitable candidate for automotive applications; they are also considered and being developed for small power applications. However, before millions of PEM fuel cell-powered cars leave the assembly lines, numerous problems will have to be solved. The price of a prototype PEM fuel cell is today on the order of hundreds if not thousands of dollars per kW. A fuel cell with about 65–70 kW output (equivalent to about 90 horse power) is needed to power a small car. Thus, dramatic cost reductions are necessary to reach a price of less than \$50 per kW, which would begin to make the PEM fuel cell competitive with ICEs.

The presently used perfluorinated membranes are far too expensive, representing about one-third of the cost of a fuel cell stack. New and inexpensive, efficient materials, such as hydrocarbon membranes, will have to be developed to replace them. At the same time, they should be chemically and mechanically stable, have a great durability and a high tolerance to fuel impurities or reaction by-products, such as carbon monoxide (CO).

Another key element for proper PEM fuel cell operation is the thin layer of platinum catalyst coated with electrically conductive graphite on both sides of the membrane. The use of platinum is necessary because of the low temperature of operation. Platinum, however, is an expensive precious metal, and so extensive research is being carried out to reduce its content. These efforts include ways of increasing the catalyst's activity (e.g., by using nano-sized dispersed metals) so that less can be used without sacrificing power output, and seeking alternative and cheaper catalysts. At a low operating temperature, platinum also has the disadvantage of binding strongly with CO, a common impurity in hydrogen obtained by the reforming of fossil fuels. This binding reduces the platinum's availability for hydrogen chemisorption and electro-oxidation. At 80 °C, the platinum catalyst can tolerate only a few ppm of CO in hydrogen before its activity begins to subside. Thus, the hydrogen fuel used must be of very high purity, requiring additional purification steps such as oxidation of CO over gold or other catalysts – all of which will add to the cost of hydrogen obtained from hydrocarbons. To increase the catalyst's tolerance to CO, new binary catalysts, such as those based on platinum/ruthenium, are currently being developed for use in fuel cells. PEM fuel cells that can potentially operate at temperatures of 120 °C or higher will also alleviate CO poisoning. However, high-temperature membranes with adequate water content and proton conductivity need to be developed to achieve this lofty goal.

Today, numerous PEM fuel cell-powered prototype vehicles are being tested worldwide by the major automobile companies, including Ford, General Motors, Honda, Toyota, Renault and Volkswagen. Daimler, for example, has been working intensively on fuel cell technology since the early 1990s, and is now leasing a group of 60 of its latest hydrogen fuel cell car, the F-Cell, to be tested for their performance under everyday driving conditions by customers in the US, Europe and Asia. In addition, the company has also developed both a light-duty vehicle and a bus that run on hydrogen. Small fleets of these buses have been successfully driven on a daily basis for public transportation in Amsterdam, Luxembourg, Reykjavik and other cities. Fleets of transit buses are ideal for the introduction of alternative fuels because they travel only short distances and generally refuel at a central depot. Furthermore, the installation of voluminous fuel tanks, which can be easily accommodated in these larger vehicles, is less of a problem than in a passenger car. In fact, transit buses are one area where alternative fuels have had the most success. Diesel buses in many cities have been progressively replaced by much less-polluting compressed natural gas (CNG) buses. To make fuel cell-powered vehicles affordable, however, major technological breakthroughs are clearly needed to significantly lower production costs and mitigate safety issues. If hydrogen fuel has to be used to fuel cars, the on-board hydrogen storage capacity must also be substantially improved, and a massive and very expensive infrastructure for delivering needed hydrogen to the users must be built from scratch. Therefore, we need to look more carefully at other options. ICEs, for instance, have been continuously improved for more than a century, and are becoming increasingly efficient and ever less pollutant-emitting. Their combination with batteries and electric motors in hybrid vehicles increases their efficiencies even further. To compare various vehicles with different fuels, engines, electric motors and drive trains, a well-to-wheel (WTW) analysis, representing the overall efficiency of an energy source from the time it is extracted from earth or any other resource to when it actually turns the wheels of the vehicle, is generally conducted. In the case of a gasoline vehicle operating with a regular ICE, the WTW gives an overall efficiency of only 14%. A gasoline ICE hybrid vehicle, however, can achieve a WTW efficiency of 28%, which is comparable to the present day 29% obtained with a fuel cell vehicle using compressed hydrogen generated from natural gas. The efficiency of a diesel hybrid ICE vehicle was found to be even higher than that for a fuel cell motor car [198]. Fuel cells may be more efficient in converting hydrogen into electricity than through mechanical conversion, but the energy required to produce, handle and store hydrogen lowers the overall efficiency considerably. Considering greenhouse gas emissions, a WTW analysis shows that hydrogen fuel cell vehicles offer only a slight advantage over hydrocarbon-based hybrid vehicles if the hydrogen used in the fuel cell is derived from fossil sources.

The ICE is a proven and reliable technology, and its combination with a battery/electric motor has already been applied today in hybrid vehicles (Toyota Prius, Honda Insight, etc.). These are slightly more expensive than regular cars, but they allow a considerable reduction in fuel consumption and lower emissions. Fuel cell vehicles, on the other hand, are still in the developmental phase, with developers striving to bring down manufacturing costs and improve the lifetime and reliability of the fuel cell stacks. To reduce the consumption of petroleum and reduce CO_2 emissions, deploying hybrid cars on a large scale seems to be a better and more reasonable solution in the short term than to rely on the still uncertain future advances in fuel cell technology, which, however, might become economically viable for the transportation sector in the long term.

9.8.5
Regenerative Fuel Cells

If a hydrogen-oxygen fuel cell is designed to operate in a reverse fashion as an electrolyzer, then electricity can be used to convert water back into hydrogen and oxygen. Such a dual-function system, known as a regenerative fuel cell (also called unitized regenerative fuel cell, URFC), is lighter than a separate electrolyzer and generator and is an excellent energy source in situations where weight is a major concern [218, 219].

Scientists at AeroVironment of Monrovia, California and NASA have developed a propeller-driven aircraft called Helios, to be used for high-altitude surveillance, communications and atmospheric testing. Helios was a $15 million dollar, solar-electric project. The unmanned aircraft has a wingspan of 75 m and has been described by some as more like a flying wing than a conventional plane. In 2001, in a test flight, Helios reached an altitude of almost 29.4 km, an altitude considered by NASA to be a record for a propeller-powered, winged aircraft. Helios was designed for atmospheric science and imaging missions, as well as relaying telecommunications up to 30 km. The 5-kW prototype was powered by solar cells during the day and by fuel cells at night (a URFC device). Regretfully, in another test flight in 2003, Helios crashed.

For automotive applications, the Livermore National Laboratory and the Hamilton Standard Division of United Technologies have studied URFCs in great detail and found that, compared with battery-powered systems, the URFC is lighter and provides a driving range comparable to that of gasoline-powered vehicles [220]. Over the life of a vehicle, the URFC was found to be more cost effective as it does not require replacement. In the electrolysis (charging) mode, electrical power from a residential or commercial charging source supplies energy to produce hydrogen by electrolyzing water. The URFC-powered motor car can also recover hydrogen and oxygen when the driver brakes or descends a hill. Such regenerative braking feature increases the vehicle's range by about 10%, and could replenish a low-pressure (about 14 atm) oxygen tank, the size of a regular football.

In the fuel-cell (discharge) mode, stored hydrogen is combined with air to generate electrical power. The URFC can also be supercharged by operating from an oxygen tank instead of atmospheric oxygen to allow peak power demands such as entering a freeway. Supercharging permits the driver to accelerate the vehicle at a rate comparable to that of a vehicle powered by a regular ICE.

The URFC, in a motor car, must produce at least ten times the power of the Helios aircraft prototype, or about 50 kW. A car idling requires just a few kilowatts, highway cruising about 10 kW, and climbing a hill about 40 kW, but acceleration onto a highway or passing another vehicle demands short bursts of 60 to 100 kW. For this burst of energy, the URFCs supercharging feature can supply the additional power. A URFC-powered vehicle must be able to store hydrogen fuel on board, but existing tank systems are relatively heavy, reducing the car's efficiency and range. Under the Partnership for a New Generation of Vehicles – a government–industry consortium dedicated to developing vehicles with very low fuel consumption – the Ford Corporation provided funding to Lawrence Livermore National Laboratory, EDO Corporation and Aero Tec Laboratories to develop a light-weight hydrogen storage tank (a pressure vessel). The team combined a carbon-fiber tank with a laminated, metalized, polymeric bladder to produce a hydrogen pressure vessel that was lighter and less expensive than conventional hydrogen tanks. Moreover, its performance factor – a function of burst pressure, internal volume and tank weight – was about 30% higher than that of comparable carbon-fiber hydrogen storage tanks. In tests where cars with pressurized carbon-fiber storage tanks were dropped from heights or crashed at high speeds, the cars generally were demolished while the tanks still held all of their pressure – an effective indicator of tank safety. Apart from other hydrogen-fueled vehicles in which refueling needs depend entirely on commercial suppliers, the URFC-powered vehicle carries most of its hydrogen infrastructure on board. Unfortunately, even a highly efficient URFC-powered vehicle needs periodic refueling, and until a network of commercial hydrogen suppliers is developed, an overnight recharge of a small motor car at home would generate enough energy for a driving range of about 240 km (150 miles), exceeding the range of present-day electric vehicles. With the infrastructure in place, a 5-min fill up of a 350 atm (5000 psi) hydrogen tank would give a range of 580 km (360 miles). The commercial development of the URFC for use in automobiles is, however, at least five to ten years away.

Utilities are also looking at large-scale energy storage systems employing regenerative fuel cells. The proposed systems store or release electrical power through an electrochemical reaction between two liquid electrolytes such as sodium bromide and sodium polysulfide stored in separate tanks (Figure 9.16) [130]. Inside the cell, the two electrolytes are separated by a thin, ion-selective membrane. Inside this big rechargeable battery-like device, when subjected to current during charging, bromine is produced at the anode. The bromide ions in the electrolyte combine with bromine to give perbromide ions. In the discharge cycle, perbromide is converted into bromide ions, producing at the same time electric energy [221].

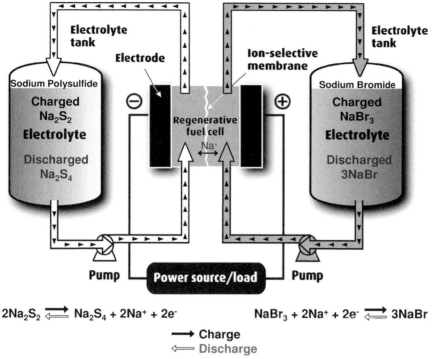

Figure 9.16 Schematic of a polysulfide bromide battery. (Source: EPRI [130] and Electricity Storage Association.)

Systems based on vanadium salts, zinc/bromine or sodium/sulfur are also being developed and are commercially available [130, 222, 223]. Such regenerative fuel cell storage systems are expected to store up to a whopping 500 MW of energy for up to 12 h [224].

9.9
Outlook

Although hydrogen is widespread and abundant on earth, its extraction from water or hydrocarbons requires much energy. Today, most hydrogen production is based on natural gas and coal, because they are still the most inexpensive sources for hydrogen. One of the goals of the envisioned Hydrogen Economy is the mitigation of greenhouse gas emissions, but this mitigation would imply capture and sequestration of CO_2 on a very large scale. Even if this might become technologically and economically possible, the consequences of storing huge amounts of CO_2 underground or at the bottom of the seas are, at best, uncertain. The chemical recycling of carbon dioxide to methanol would mitigate this problem and also transform hydrogen into a suitable liquid product (methanol) for storage, transportation and use. In the long term, however, considering the finite amounts of fossil fuels, hydrogen will have to be produced increasingly from water, using not only renewable energy sources such as wind and solar, but also nuclear energy. Once generated, the physical and chemical properties of hydrogen make its storage, transportation and safe handling difficult and potentially hazardous. Whether the production is centralized or decentralized, due to its unique properties, a totally new and expensive infrastructure would have to be built to supply consumers with hydrogen. For vehicles, on-board hydrogen storage is likely to remain voluminous and costly, while the use of hydrogen as an automotive fuel from an energy efficiency and emissions viewpoint should be preferably used only in combination with fuel cells. Efficient, reliable and affordable hydrogen fuel cells could become a reality, but this will take time. Unless an efficient and safe metal hydride or other storage system is developed it is also questionable to what degree people would feel safe, knowing that they are driving cars with a high-pressure tank filled with an explosive and highly flammable gas under their seats. Realizing the numerous technical difficulties associated with the use of hydrogen as a fuel, the US Energy Secretary Steven Chu announced in May 2009 that the DOE was putting the brakes on research into hydrogen fuel cell for automotive applications by cutting its funding to the development of the technology [225].

Other static applications of hydrogen fuel are feasible in suitable cases, and will be developed in time. We have expressed serious but realistic questions about the storage and transportation of energy in the form of hydrogen, which, as discussed here, has serious drawbacks and problems. Hydrogen will indeed be an essential part of the future for providing fuels to the transportation, industrial and residential sectors. However, instead of using hydrogen in its highly volatile and inconvenient pure form, a feasible alternative for energy storage involves its conversion into a convenient and safe liquid, namely, methanol. The proposed "Methanol Economy" (Chapters 10 and 14) will achieve the chemical recycling of CO_2 with hydrogen to methanol rather than its sequestration. This will provide an inexhaustible fuel source as well as a carbon source for synthetic hydrocarbons and their products while mitigating global warming caused by the generation of excess CO_2 from burning fossil fuels. To achieve its goals, the Methanol Economy will also require the production of hydrogen on a massive scale through water electrolysis and other water-splitting methods, using electricity generated from any non-fossil sources (renewable energy, and also atomic energy). In this way, energy will be stored, not as volatile hydrogen gas, but by its conversion with CO_2 into convenient and easy-to-handle liquid methanol and used as such subsequently.

10
The "Methanol Economy": General Aspects

Oil and natural gas, the main fossil fuels besides coal, are not only still our major energy sources and fuels but also the raw materials for a great variety of derived hydrocarbon materials and products. These range from gasoline and diesel oil to varied petrochemical and chemical products, including synthetic materials, plastics and pharmaceuticals. However, what Nature provided as a gift, formed over the eons, is being used up rather rapidly. The expanding world population (now approaching 7 billion and projected to reach 8–10 billion during the twenty-first century), and the increasing standard of living and demands for energy in rapidly developing countries such as China and India, is putting growing pressure on our diminishing fossil fuel resources and making them increasingly costly. Whereas coal reserves may last for another two or three centuries, readily accessible oil and gas reserves – even considering new discoveries, improved technologies, savings and unconventional resources (such as heavy oil deposits, oil shale, tar sands, methane hydrates, coalbed methane, etc.) – may not last much beyond the twenty-first century. Once fossil fuels are combusted or used they form CO_2 and H_2O and therefore are not naturally renewable on the human timescale.

To satisfy humankind's ever-increasing energy needs, all feasible alternative and renewable energy sources must be considered and used. These include biomass, hydro and geothermal energy as well as the energy of the sun, wind, waves, tides of the seas and others. Nuclear energy will also have to be further developed and utilized. Our discussion here, however, is not dealing with the question of energy generation, which we believe humankind must and will solve (as in the final analysis most of our energies are derived from the sun, an enormous and lasting energy source), but rather with the challenges of how to store, transport and best use energy in an environmentally friendly way.

Most of our energy sources are used to provide primarily heat and electricity. Electricity is generally a good way to transport energy over relatively short distances where a suitable grid exists. It is, however, very difficult to store electricity on a large scale; batteries, for example, are still inefficient, expensive and bulky. Its transfer over long distance is still difficult and wasteful and the construction of transmission lines across oceans is not possible with present technologies. Besides finding new energy

Beyond Oil and Gas: The Methanol Economy, Second updated and enlarged edition
George A. Olah, Alain Goeppert, and G. K. Surya Prakash
Copyright © 2009 WILEY-VCH Verlag GmbH & Co. KGaA, Weinheim
ISBN: 978-3-527-32422-4

sources, it is therefore necessary to identify and develop new and efficient ways to store and distribute energy from whichever source it is derived from.

One approach that has been proposed and widely discussed recently is the use of hydrogen. This would be generated in the long term by water electrolysis using any available energy source and subsequently used as a clean fuel in the frame-work of the so-called "Hydrogen Economy." Hydrogen is clean in its combustion, producing only water, although its generation is less clean if the requisite energy is derived from fossil fuels with their attendant polluting effects. As we have seen in Chapter 9, hydrogen has some desirable attributes for energy storage and as a fuel but, due to its extreme volatility and explosive nature, many difficult issues will need to be resolved for it to be used on a practical scale as an everyday energy source and fuel. Hydrogen, the lightest element, has serious limitations in terms of its storage, transportation and delivery of energy. The handling of volatile and potentially explosive hydrogen gas necessitates special conditions, including cryogenic and/or high pressure technologies, special materials to minimize diffusion and leakage, as well as strict adherence to extensive safety precautions, all of which makes its use difficult and very costly. In addition to these difficulties a currently non-existent infrastructure must be developed for the "Hydrogen Economy," which seems, at least presently, economically prohibitive. In addition, hydrogen itself cannot solve our continuing need for hydrocarbons and their products. For this problem, new practical synthetic methods must be developed that use renewable or regenerative resources and new technologies such as the hydrogenative chemical recycling of carbon dioxide (*vide infra*).

Today, the transportation field is one of the major consumers of oil. Oil's main products such as gasoline, diesel and jet fuel and synthetic hydrocarbons are produced on a large scale. They are easy and relatively safe to handle, to transport and to distribute, mainly because a vast dedicated infrastructure already exists. Consequently, the preferred alternative fuels for transportation should be compatible with existing infrastructure and uses.

Some years ago, one of us (G. A. O.) suggested a new, viable alternative approach of how to use more efficiently our still available oil and natural gas resources and, eventually, to free humankind from its dependence on fossil fuels. This approach is based on methanol and derived dimethyl ether (DME), as well as their varied hydrocarbon products, which form the basis of the "Methanol Economy" [226–229]. Methanol (CH_3OH) is the simplest, safest and easiest to store and transport liquid oxygenated hydrocarbon. At present, it is prepared almost exclusively from synthesis gas (syn-gas, a mixture of CO and H_2) obtained from the incomplete combustion and reforming of fossil fuels, mainly natural gas or coal, with steam and in some cases carbon dioxide. Methanol can also be prepared from biomass, including wood, agricultural by-products and municipal waste, but these play only a relatively minor role. As discussed in Chapter 12, the production of methanol and dimethyl ether is also possible by the direct oxidative conversion of methane, from natural gas or any other source, avoiding the initial preparation of syn-gas. Most importantly, they can also be produced by reductive hydrogenative recycling of CO_2 from natural or industrial sources, including fossil fuel-burning power plants, cement plants, and so on, and eventually from the atmosphere itself. The hydrogen required for this process can be generated from water using any energy source, and is thus stored in the form of a safe and easily transportable liquid. The chemical recycling of CO_2 would also help to mitigate or eliminate the major human activity cause of climate change due, in a significant part, to excessive burning of fossil fuels.

Methanol is an excellent fuel in its own right, with an octane number of 100, and it can be blended with gasoline as an oxygenated additive. Alternatively, even neat methanol can be used in today's ICEs with only minor modifications. Methanol can also be efficiently used to generate electricity in fuel cells. This was initially achieved by first catalytically reforming methanol into H_2 and CO. The H_2, after separation from CO, is then fed into a hydrogen fuel cell. However, methanol can also react directly with air in the direct methanol fuel cell (DMFC), without the need for reforming. The DMFC greatly simplifies fuel cell technology, making it readily available to a wide range of applications such as portable electronic devices (e.g., cell phones, laptops), as well as for motor scooters and, soon, for cars. It can also be used for electricity generators and emergency back-up systems and in areas of the world where electricity is still not available from a grid. DME, readily produced from methanol, is itself an excellent substitute for diesel fuel as well as household gas for cooking and heating.

Another potentially significant application is the conversion of natural gas (methane) into methanol for its ready and safe transportation when pipelines are not feasible or available. Today, LNG is transported across oceans under cryogenic conditions by using supertankers (>200 000 tonnes). The LNG is unloaded at terminals and fed into pipelines to satisfy increasing needs, or to serve as a substitute for diminishing local natural gas sources. LNG is potentially hazardous, however, due to possible accidents or terrorist acts. A single supertanker exploding close to densely populated area might have the devastating effect of a small atomic bomb. Whilst hoping that such a situation will never occur, realistically we must be prepared to find alternative, safer methods for transporting natural gas. In this respect, its conversion into methanol close to the source is a feasible alternative. Natural gas can also be converted into liquid methanol using existing technologies as well as new and improved ones, such as direct methane oxidation without prior production of syn-gas (Chapter 12). Methanol produced close to the natural gas sources can then be easily and safely transported.

Besides its use for energy storage and fuel, methanol and DME serve as convenient starting materials for basic chemicals such as formaldehyde and acetic acid and a wide variety of other products, including polymers, paints, adhesives, construction materials, synthetic chemicals, pharmaceuticals and even single-cell proteins. Methanol and DME can also be conveniently converted via a simple catalytic step into ethylene and/or propylene (the methanol-to-olefin, MTO, process), which serve as building blocks in the production of synthetic hydrocarbons and their compounds (Chapter 13). Practically all hydrocarbon fuels and products currently obtained from fossil fuels can be efficiently obtained from methanol, which in turn, as emphasized, can be produced by the chemical recycling of natural or industrial CO_2 sources (Chapter 12). Methanol, therefore, has the ability to liberate humankind from its dependence on fossil fuels for transportation and hydrocarbon products by allowing these to be produced by a technical hydrogenative recycling of CO_2, supplementing and in many aspects eliminating limitations of nature's photosynthetic cycle.

The concept of the "Methanol Economy" has broad advantages and possibilities. Methanol, together with derived dimethyl ether, can be used as: (i) a convenient energy storage medium; (ii) a readily transportable and dispensable fuel for ICE, compression ignition engines and fuel cells; and (iii) as a feedstock for synthetic hydrocarbons and their products, including varied fuels, polymers and even single-cell proteins (for animal feed and/or human consumption). The regenerative carbon source for methanol will eventually be the CO_2 contained in the air, which is available to all on earth and practically inexhaustible. The required energy for the regeneration of CO_2 into methanol can be obtained from any available energy source.

It should be emphasized that there is no preference for any particular energy source needed for the production of methanol. All energy sources, including renewable sources or atomic energy, can be used in the most economical, safe and environmentally acceptable ways. The ready conversion of methanol into synthetic hydrocarbons and their products will ensure that future generations will have unlimited and inexhaustible access to the essential synthetic products and materials that today form an integral part of our life. At the same time, the "Methanol Economy," by recycling excess atmospheric CO_2, will mitigate one of the major adverse effects on the earth's climate caused by human activity, namely global warming.

The concept of the "Methanol Economy" has been developed over the past decade, with the use of methanol as a fuel and gasoline additive attracting increasing interest, particularly during times of shortages and higher fossil fuel prices. When discussing "alcohol fuels," the public generally means ethanol and its availability from agricultural sources, including fermenting corn (in the United States), sugarcane (in Brazil), or by bio-converting various other natural materials. The production and use of bio-ethanol as a transportation fuel was discussed in Chapter 8 together with its limitations. Although the use of ethanol is feasible in some countries under certain conditions (e.g., Brazil, the United States), it will at best satisfy only a relatively small part of our overall fuel requirements and have a significant effect on food availability and prices as long it is produced from corn, wheat, sugarcane and other food crops.

Although methanol and ethanol are chemically closely related (ethanol is the next higher homolog of methanol, CH_3CH_2OH vs. CH_3OH), when the public consider "alcohols" as fuels they usually fail to realize the significant differences between the two. The fermentation of agricultural or natural products can be used to produce both alcohols. Alcohols can also be obtained from cellulosic sources, primarily wood, although methanol is presently nearly exclusively produced by synthetic processes. Industrial synthetic ethanol, which can be readily produced by hydration of ethylene, is rarely mentioned. In contrast, bio-ethanol prepared by fermentation of natural sources is a renewable, non-fossil fuel-based fuel, but vast areas of land are required to grow the sugarcane, corn, wheat or other crops from which it is produced. The agricultural production of ethanol is also highly energy demanding, and currently most of this energy comes from fossil fuels. The fundamental difference between the production of bio-ethanol and methanol is that the latter does not only rely on agricultural resources but can be readily produced from varied feedstocks, including fossil fuels and most significantly by chemical recycling of carbon dioxide from natural or industrial sources.

Eventually, methanol will be made by chemical conversion of atmospheric CO_2, with hydrogen generated by the electrolysis of water, using any form of energy, including renewable non-fossil fuel based energies and atomic energy. In this way, humankind will produce methanol by chemically recycling CO_2 from the air, which is accessible to all, together with water – both inexhaustible resources on earth. In the interim, still-available fossil fuels can be more efficiently converted into methanol.

The Olah group has long been involved in the study of various new aspects of methanol chemistry. *Inter alia*, beginning in the 1970s, the selective oxidation of methane to methanol in superacidic media, preventing further oxidation, was studied. During the 1980s, it was followed by the discovery of the bifunctional acid–base-catalyzed conversion of methanol or dimethyl ether into ethylene and/or propylene, and through them to both gasoline-range aliphatic as well as aromatic hydrocarbons. These studies were conducted independently of the zeolite-catalyzed chemistry developed by Mobil (now ExxonMobil), UOP and others for the conversion of syn-gas-based methanol into hydrocarbons (Chapter 13).

Based on the general realization in recent times that our non-renewable fossil fuel resources are diminishing, increasing efforts have been directed to finding solutions to counteract their depletion. Their more efficient and economic use has been explored together with increasing use of non-fossil fuel alternative energy sources and safer atomic energy. It should be emphasized, however, that in addition to finding new solutions to satisfy our overall energy needs there will still be in the post fossil-fuel era continued need for convenient and safe energy storage, transportation fuels and for a multitude of synthetic hydrocarbon products that will necessitate the creation of vast quantities of synthetic hydrocarbons. Whilst the "Hydrogen Economy," as discussed, cannot itself fulfill these needs it appears that hydrogen produced and combined with the chemical recycling of carbon dioxide in the proposed "Methanol Economy" represents a viable alternative.

In summary, the goals of the "Methanol Economy" encompass:

- New and more efficient ways of producing methanol and/or derived dimethyl ether from fossil fuel sources, primarily natural gas, by their oxidative conversion, without prior production of syn-gas.
- Production of methanol by hydrogenative recycling of CO_2 from natural and industrial sources, and eventually from the air itself as an inexhaustible carbon source.
- The use of methanol and derived dimethyl ether as convenient energy storage media for transportation fuels for both combustion engines as well as new generations of fuel cells, including DMFC.
- The use of methanol as the raw material for producing ethylene and/or propylene as the basis for synthetic hydrocarbons and their products.

The "Methanol Economy" offers a feasible way to liberate humankind from its dependence on diminishing oil and gas resources, while simultaneously utilizing and storing all sources of energies (renewable and atomic). At the same time, by chemically recycling CO_2, one of the major man-made causes of climate change – global warming – will be mitigated while rendering carbon-containing fuels and materials regenerative. These points are discussed in more detail in Chapters 11–14.

11

Methanol and Dimethyl Ether as Fuels and Energy Carriers

11.1
Background and Properties

Methanol, also called methyl alcohol or wood alcohol, is a colorless, water-soluble liquid with a mild alcoholic odor. It freezes at $-97.6\,°C$, boils at $64.6\,°C$ and has a density of 791 kg m^{-3} at 20 °C (Table 11.1). Methanol in its (relatively) pure form was first isolated in 1661 [230] by Robert Boyle, who called it "spirit of the box" because he produced it through the distillation of boxwood. Its chemical identity or elemental composition, CH_3OH, was described in 1834 by Jean-Baptiste Dumas and Eugene Peligot. They also introduced the word methylene to organic chemistry, from the Greek words *methu* and *hyle*, meaning respectively wine and wood. The term methyl, derived from this word, was then applied to describe methyl alcohol, which was later given the systematic name methanol. Containing only one carbon atom, methanol is the simplest of all alcohols. Methanol is commonly referred to as wood alcohol because it was first produced as a minor by-product of charcoal manufacturing, by destructive distillation of wood. In this process, one ton of wood generated, along with other products, only about 10–20 L of methanol. Beginning in the 1830s, methanol produced in this way was used for lighting, cooking and heating purposes, but was later replaced in these applications by cheaper fuels, especially kerosene. Up until the 1920s wood was the only source for methanol, which was needed in increasing quantities in the chemical industry. As hard as it may be to believe today, all the methanol required during World War I, along with acetone and other essential chemicals, was derived from charcoal furnaces [231]. With the industrial revolution, wood was largely replaced by coal in many applications. Coal and coke gasification processes through the action of steam and heat were developed, by which gases containing carbon monoxide and hydrogen could be obtained to supply cities with "town gas." Using hydrogen produced with this technology, Fritz Haber and Carl Bosch developed the technical hydrogenation of molecular nitrogen, N_2, to ammonia at high temperature and pressure. This breakthrough resulted in the development of several other chemical processes, necessitating similar severe conditions and feedstock, including methanol synthesis. In fact, from the earliest days, the synthesis of methanol and ammonia were so interrelated that they are often produced in the same

Beyond Oil and Gas: The Methanol Economy, Second updated and enlarged edition
George A. Olah, Alain Goeppert, and G. K. Surya Prakash
Copyright © 2009 WILEY-VCH Verlag GmbH & Co. KGaA, Weinheim
ISBN: 978-3-527-32422-4

Table 11.1 Properties of methanol.

Synonyms	Methyl alcohol, wood alcohol
Chemical formula	CH_3OH
Molecular weight	32.04
Chemical composition	
Carbon	37.5%
Hydrogen	12.5%
Oxygen	50%
Melting point	$-97.6\,^{\circ}C$
Boiling point	$64.6\,^{\circ}C$
Density at $20\,^{\circ}C$	$791\,kg\,m^{-3}$
Energy content	$5420\,kcal\,kg^{-1}$
	$173.6\,kcal\,mol^{-1}$
Energy of vaporization	$9.2\,kcal\,mol^{-1}$
Flash point	$11\,^{\circ}C$
Autoignition temperature	$455\,^{\circ}C$
Explosive limits in air	7–36%

plant. The synthetic route to methanol production, by reacting carbon monoxide with hydrogen, was first suggested in 1905 by the French chemist Paul Sabatier [232]. In 1913, the Badische Anilin und Soda Fabrik (BASF), based on the investigations of A. Mittasch and C. Schneider, patented a process to synthesize methanol from syngas, produced from coal, over a zinc/chromium oxide catalyst at 300–400 °C and 250–350 atm [232, 233]. After World War I, BASF resumed its research into synthetic methanol and in 1923 built the first commercial high-pressure synthetic methanol plant in Leuna, Germany. Between 1923 and 1926, F. Fischer and H. Tropsch reported from the Mühlheim Coal Research Laboratory their extensive studies of the production of hydrocarbons, including that of methanol, from syn-gas, a mixture of carbon monoxide and hydrogen, which was the basis of what is known as the Fischer–Tropsch synthesis [234, 235]. In 1927, in the United States, Commercial Solvents Corporation used the high-pressure technology to produce methanol from CO_2/H_2 mixtures obtained as fermentation by-product gases [234–236]. At the same time, DuPont began the production of both methanol and ammonia in the same plant using syn-gas produced from coal. In the 1940s, steam reforming of natural gas began in the United States, based in part on the earlier developments of BASF in the 1930s. From then on, coal was slowly abandoned as a feedstock for syn-gas in favor of the cleaner, cheaper and plentiful natural gas. Over the years, other feedstocks, including heavy oil and naphtha, have also been used, albeit to a much lesser extent. The steam reforming process of methane, because of the very high purity of the syn-gas, opened the way to the technical realization of the low-pressure methanol process, introduced commercially in 1966 by Imperial Chemical Industries (ICI). This new process, using a more active Cu/ZnO catalyst and operating at 250–300 °C and 100 atm [237], put an end to the high-pressure methanol synthesis technology, which operated under much more severe conditions. The use of these highly active catalysts was made possible by the lower content of catalyst poisons such as sulfur or metal carbonyls in the syn-gas feed. Not much later, Lurgi launched its own process with even lower operating temperatures and pressures (230–250 °C, 40–50 atm). During the past 40 years, considerable further improvements have been made in methanol synthesis from carbon oxides (CO containing some CO_2) and hydrogen, making this technology a rather mature one. Using the low-pressure process, selectivity for methanol is now in excess of 99.8% with an energy efficiency of nearly 75%. Current research is aimed at developing new ways to synthesize methanol at even lower temperature and pressure from diverse origin carbon oxides/hydrogen feeds, as well as by direct oxidation of methane, which is a superior approach from an energetic viewpoint.

Today, worldwide, almost all methanol is produced from syn-gas. However, as discussed in Chapter 12, new ways for its production directly from methane (natural gas) without going through syn-gas, as well as by hydrogenative chemical recycling of carbon dioxide, are being developed.

Besides being present on Earth to a limited extent naturally (in fruits, grapes, etc.) and produced on a large industrial scale, methanol has also been found recently in outer space. Astronomers have observed an enormous cloud of methanol around a nascent star in deep space that measures across some 460 billion km [238]. Even if its concentration in the near vacuum of space is extremely low, the overall amount is mind boggling. The methanol in this vast cloud is believed to be formed on the surface of dust particles present in space, from carbon monoxide and hydrogen at temperatures only slightly above absolute zero [239]. When these particles are heated up by the energy released from the nascent stars being formed, the methanol present on the surfaces desorbs and forms a cloud, which has been observed by radio telescopes. There must be ongoing processes of methanol formation to allow its presence to be observed as clouds in view of its inevitable decomposition. Numerous other methanol clouds have also been spotted, always close to massive young stars. Methanol clouds are therefore good indicators for detecting and studying star forming regions of our universe. Besides being formed around massive stars, methanol has also been observed in comets. Likewise, DME (dimethyl ether) has also been found in regions of massive star formations [240].

11.2
Chemical Uses of Methanol

Today, methanol is mainly a feedstock for the chemical industry and used for the production of varied chemical products and materials. It is manufactured on a large scale [over 32 million tonnes per year in 2004 [241] and 40 million tonnes in 2007 (source: Methanol Institute, PCI Ockerbloom & Co., Inc.)] as an intermediate for the production of various chemicals (Figure 11.1). Worldwide, almost 65% of the methanol production is used to obtain formaldehyde (39%), methyl *tert*-butyl ether (MTBE, 14%) and acetic acid (11%). Methanol is also a starting material for chloromethanes, methylamines, methyl methacrylate (MMA), dimethyl

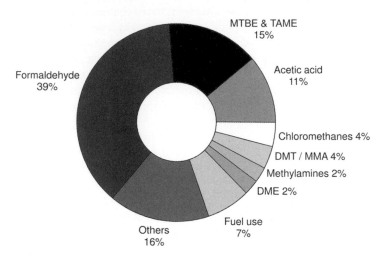

MTBE: Methyl *tert*-butyl ether
TAME: Tertiary-amyl methyl ether

Figure 11.1 World demand for methanol in 2007. (Based on data from the Methanol Institute, PCI Ockerbloom & Co., Inc.)

terephthalate, and so on, [232]. These chemical intermediates are then processed to manufacture many products of our daily life, including paints, resins, silicones, adhesives, antifreeze and plastics [242].

Formaldehyde, the largest chemical product derived from methanol, is mainly used to prepare phenol-, urea- and melamine-formaldehyde and polyacetal resins as well as butanediol and methylenebis(4-phenyl isocyanate) (MDI). MDI foam is, for example, used as insulation in refrigerators, doors and in motor car dashboards and fenders. The formaldehyde resins are then predominantly employed as adhesives in the wood industry in a wide variety of applications, including the manufacture of particle boards, plywood and other wood panels. The market for MTBE, an oxygenated gasoline additive and blending component, which became the second largest use of methanol globally, grew strongly during the 1990s. In the United States, it accounted in 2001 for 37% of methanol consumption. Because of its high octane rating it replaced the phased-out lead-based anti-knock compounds. At the same time this oxygenated compound, when added to gasoline, helped to reduce air pollution from motor cars. In recent years, however, MTBE has come under serious environmental attacks, especially in California, due to MTBE contamination discovered in groundwater, primarily as the result of leaking underground storage tanks from local filling stations. Increased maintenance, stricter control or replacement of these tanks would certainly have gone a long way in solving this problem. Nevertheless, as a consequence of the contamination, the use of MTBE has been phased out in most of the United States and will probably also be phased out in other countries. Proposed substitutes include ethanol and ethyl *tert*-butyl ether (ETBE). There is also the question of whether any oxygenated additives to mitigate air pollution are necessary in reformulated gasoline for today's internal combustion engines, which, for the most part, use direct fuel injection systems and oxygen sensors, allowing the effective control and substantial reduction of emissions, even without the addition of any oxygenated compounds. This improved efficiency, however, does not negate the fact that oxygenated additives provide superior performance and cleaner-burning fuels [243].

11.3
Methanol as a Transportation Fuel

Methanol and derived dimethyl ether (DME) have excellent combustion characteristics that make them ideal fuels for today's ICE-driven vehicles [243] and diesel engines, respectively. Establishing an infrastructure for methanol fuels would also greatly ease the introduction of advanced fuel cell-based vehicles using methanol as a fuel either by on-board reforming to hydrogen or directly with direct methanol fuel cells. The use of methanol and DME as transportation fuels is one of the main aspects of the Methanol Economy. Their increasing use in this sector will fundamentally change their need and production capabilities.

11.3.1
Development of Alcohols as Transportation Fuels

The use of alcohol (methanol or ethanol) as a fuel is as old as the ICE itself. Some of the earliest ICE models, developed at the end of the nineteenth century by Nicholas Otto, Karl Benz and others, were actually designed to run on alcohol. These alcohol-powered engines had started to replace steam engines for farm machinery and train locomotives. Also used in automobiles, alcohol engines were advertised as less polluting than their gasoline counterparts. Most European countries with few or no oil resources were especially eager to develop ethanol as a fuel because it could be readily distilled from fermentation of various domestic agricultural products. Germany, for example, went from a production of almost 40 million liters of ethanol in 1887 to about 110 million liters in 1902 [244]. During the first decade of the twentieth century, many races were held between alcohol- and gasoline-powered automobiles, and there were lively debates as to which fuel gave the best performances. On an economic basis, however, ethanol could hardly compete with gasoline, especially in the United States, which had plentiful petroleum resources at the time and a very powerful proponent of gasoline: the Standard Oil Trust, which was understandably hostile towards the introduction of any alternative fuel. World War I saw an important expansion of the ethanol industry not only for fuel uses but also for gunpowder and war gas manufacture. As mentioned, in the Soviet Union after the Bolshevik Revolution Lenin suggested wide use of agricultural grain-based alcohol for industrial uses (fuels and production of ethylene and its products). Owing to opposition to diverting the Russian's beloved Vodka to these purposes, the plans were soon abandoned by Lenin himself. In the United States, the prohibition, which became effective at the beginning of 1920, outlawed the consumption of any alcoholic beverage and greatly hindered the large-scale production of ethanol, even for fuel purposes. The 1930s saw the introduction of Agrol, a blend of alcohol and gasoline, which had some success in Midwestern States for a limited time but was vigorously opposed by oil companies. At the same time, European countries such as Germany, France and England – as well as Brazil and New Zealand, which were relying on imported oil – strongly encouraged the production of alcohol by offering subsidies, or even made the blending of alcohol with gasoline mandatory. Interestingly, the very same Standard Oil of New Jersey (one of the companies resulting from the dismantling of the Standard Oil Trust), which fought against the use of alcohol fuels in the United States, was promoting its own alcohol blend in England under the name Discol. Nowhere, however, was the effort of developing alternative fuels as ambitious as in Germany, where the Nazi government had the goal to achieve energy independence, mainly because of military considerations. Hitler was in fact convinced that Germany's failure in World War I was in a large part due to fuel shortages and that such a situation should never hinder the Nazi war machine. From the time Hitler came to power, the production of ethanol (mainly from potatoes) grew dramatically, reaching almost 1.5 million barrels in 1935 [244]. At the same time, methanol produced from coal via syn-gas using the process invented by BASF also expanded dramatically. Chemical production of ethanol from calcium carbide also became possible. Methyl and ethyl alcohol were blended with gasoline and sold under the name Kraftspirit. One of the difficulties with blending alcohols with gasoline, however, is the phase separation that can take place when moisture is present, leading to the stalling of the engine. Besides methanol synthesis, the large coal reserves, especially in the Ruhr valley, were also used to produce, via syn-gas, large amounts of synthetic gasoline following the Fischer–Tropsch process developed by I. G. Farben, the world's largest chemical company at the time. Shortly before World War II, all alternative fuels produced through the Fischer–Tropsch as well as the Bergius process accounted for more than 50% of Germany's total light motor-fuel consumption. During the war, a large part of the alcohol produced was diverted to other uses, including ammunition [245], medicines and synthetic rubber manufacture. Switzerland, however, while struggling to remain neutral during the conflict, turned to methanol when its petroleum supplies were cut off. After the war the interest in alcohol-based fuels declined rapidly in the United States and most other countries due to the ready availability of large quantities of cheap oil. Only with the oil crises of the 1970s and concerns about pollution did the interest in alcohol fuels grow again. The large-scale development of ethyl alcohol fuel was most successful in Brazil, which launched in 1975 its National Alcohol Program (PNA). More than 20 years later, the production of ethanol, mainly from sugarcane and its residues, amounted to about 15 million m^3 per year in 1997/1998 (equivalent to some 220 000 barrels of oil per day [246]), allowing millions of vehicles to run on alcohol. With production costs averaging about $ 35–45 per barrel of oil equivalent, the discovery of large oil reserves off the Brazilian shore and the low oil prices of the 1990s, however, the share of alcohol-fueled motor cars, in total sale of news cars, fell from 96% in 1985 to 0.07% in 1997! The PNA project was kept afloat by blending up to 24% of ethanol into gasoline [246]. With increasing oil prices and the introduction of flexible fuel vehicles (FFV), able to run on any mixture of gasoline and ethanol, interest in ethanol has been revived. The utilization of ethanol fuel in developed countries, where relatively cheap and efficient sources such as sugar cane are generally not available, is limited. The production of ethanol from corn, however, in recent years gained significance in the United States, although compared to the size of the automobile fuel market it still covers only a very small fraction of the demand. Ethanol is also increasingly in demand as an oxygenated additive to replace MTBE. Although Brazil's government and agricultural producers in the United States fought to establish ethanol as an alternative fuel, worldwide interest in alcohol fuels in the automotive sector also is increasingly focused on methanol. Methanol is a very flexible fuel for ICEs and can be made from a wide variety of both renewable and fossil fuel resources: natural gas, coal, wood, agricultural and municipal waste, and so on, at a cost generally lower than that for ethanol. The idea of using methanol as an automotive fuel was slowly revived beginning in the 1970s. Thomas Reed, a researcher at the Massachusetts Institute of Technology (MIT), was one of the first to advocate methanol as a fuel in the United States, publishing in 1973 a paper in *Science* that explained some of its advantages. In passing, Reed also talked about its use on a wider scale, mentioning the name of a Methanol Economy, without further elaborating on it or even ever using the name further [245]. He stated that adding 10% methanol to gasoline improved performance, gave better mileage and reduced pollution. Similar results were obtained in Germany by Volkswagen (VW), with the support of the West German government. In 1975, VW began an extensive test involving a fleet of 45 vehicles using a 15% blend of methanol in gasoline. After minimal modifications to existing engines, VW was able to operate these vehicles efficiently on methanol blends with only minor problems [244]. At the time, the methanol blends were described as already competitive with gasoline. Furthermore, methanol (like ethanol) acted as an octane booster, with the methanol/gasoline blend delivering more power than pure gasoline. Five vehicles running on pure methanol were also tested by VW. Cold-start problems due to the lower volatility of methanol were successfully solved by using small amounts of additives such as butane or pentane. The use of methanol significantly improved the cars' performance. Methanol, being considered also safer than gasoline, has been the fuel of choice at the Indianapolis 500 races since the mid-1960s [247]. In the United States, most oil companies were at best apathetic if not clearly opposed to the introduction of methanol as an alternative automotive fuel. The interest of American motor car manufacturers for methanol-fueled cars was also very limited. In California, however, which was trying to reduce its significant air pollution problems and reliance on imported fuel, a research program on methanol was started in 1978 at the University of Santa Clara, where a Ford vehicle running on pure methanol was extensively tested. This test was followed by several fleets of Ford and VW vehicles operated in various state and local agencies. The 84 vehicles [248], which accumulated a total mileage of over 2 million km, showed good fuel economy and engine durability that was comparable to that of gasoline vehicles. At about the same time in 1980, the Bank of America, based in San Francisco, decided to convert most of its vehicle fleet into methanol fuel in response to high oil prices. During their lifetime, the more than 200 methanol-fueled vehicles accumulated over 30 million km on the roads. The Bank of America concluded that, compared to gasoline engines, the use of neat methanol was found to be cheaper, increased the engine's lifespan and greatly decreased exhaust pollutants. Nevertheless, the bank's interest in methanol declined rapidly with the sharp drop in oil prices of the mid-1980s. The state of California, however, continued its efforts on promoting methanol, mostly in the form of a blend composed of 85% methanol and 15% gasoline, called M85. A small fleet of cars using this methanol blend, driven on a daily basis at the Argonne National Laboratory near Chicago, also showed that, even in the frigid climate of northern Illinois, the cars had no problems with cold starting [249]. At the end of the 1980s, automobile companies began to develop cars powered by alternative fuels because they were concerned about meeting the new air pollution standards. These vehicles were first introduced in California, which has one of the most stringent emission control standards of the US. General Motors, Ford, Chrysler, Volvo, Mercedes and others transformed existing models to run on methanol with an additional cost of, at most, a few hundred dollars. Because of methanol's higher octane rating, Ford found that acceleration from 0 to $100\,\mathrm{km\,h^{-1}}$ was up to one second faster with the methanol-fueled models compared to the gasoline versions [249]. Owing to the limited number of filling stations dispensing the methanol-blended fuel, most of the models were designed as flexible fuel vehicles (FFVs), being able to run on gasoline alone in case the M85 methanol blend was not available. This concept enabled the bypassing of any lack of availability that stands in the way of most alternative fuels in the early stages of their introduction. The number of methanol-fueled vehicles in use in the United States (mostly in California) reached a maximum in 1997 of 20 000 units [250]. During the 1990s, different technological advances were achieving wide acceptance in the automobile industry: direct fuel injection, three-way catalytic converters, reformulated gasoline, and so on, reducing dramatically the emission problems associated with gasoline-powered vehicles but decreasing at the same time the interest in methanol-based fuels. However, the recent dramatic increase in oil prices, combined with growing concerns about human-caused climate changes, has revived interest in alternative fuels, among which methanol can play an important role.

The flexibility of FFVs represents a powerful means to circumvent the fuel supply conundrum. In the short term, FFVs are a necessary key transition technology that will allow the build-up of demand for, and widespread acceptance of, methanol and help make the Methanol Economy a reality. It must, however, be kept in mind that FFVs are only a compromise, and do not offer the best performance achievable in either emissions or fuel economy. In the long run, the use of cars optimized to run on pure methanol (M100) would be preferable, and would also greatly facilitate the transition to methanol-powered fuel-cell vehicles.

Looking back at the history of methanol in the automotive sector, it must be observed that the fate of methanol fuel is extremely dependent on economic aspects, and especially oil prices. Resistance to the widespread introduction of methanol by the oil industry, special interest groups (some of which favor agricultural ethanol), energy security issues, government energy and emission policies and other political considerations also play an important role. With diminishing oil and gas reserves, a new realization for the need of finding alternative solutions is finally achieving a foothold, and the future of methanol as a transportation fuel is entering a new period. Methanol is also easily dehydrated to dimethyl ether, which is an effective fuel particularly in diesel engines due to its high cetane number and favorable properties. Haldor Topsoe first promoted its use as an efficient diesel fuel in the 1990s. Interest in DME is rapidly growing.

11.3.2
Methanol as Fuel in Internal Combustion Engines (ICE)

In contrast to gasoline, which is a complex mixture containing many different hydrocarbons and some additives, methanol is a simple chemical. It contains about half the energy density of gasoline, which means that 2 L of methanol contains the same energy as 1 L of gasoline. Even though methanol's energy content is lower, it has a higher octane rating of 100 [average of the research octane number (RON) of 107 and motor octane number (MON) of 92], which means that the fuel/air mixture can be compressed to a smaller volume before it is ignited by the sparkplug. This higher octane rating allows the engine to run at a higher compression ratio (10–11 to 1 against 8–9 to 1 for gasoline engines) and thus also more efficiently than a gasoline-powered engine. Efficiency is also increased by methanol's higher "flame speed," which enables a faster and more complete fuel combustion in the cylinders. These factors explain why, despite having half the energy density of gasoline, less than double the amount of methanol is necessary to achieve the same power output. This is true even in engines that are only modified gasoline engines and not specifically designed for methanol's properties. Methanol-specific engines, however, provide even better fuel economy [243]. Methanol also has a latent heat of vaporization that is about 3.7 times higher than gasoline, so that methanol can absorb a much larger amount of heat when passing from the liquid to gaseous state. This helps to remove heat away from the engine so that it may be possible to use air-cooled radiators instead of heavier, water-cooled systems. For similar performance to a gasoline-powered car, a smaller, lighter engine block, reduced cooling requirements, better acceleration and mileage are to be expected from methanol-optimized engines in the future [243]. In addition, methanol vehicles have low overall emissions of air pollutants such as hydrocarbons, NO_x, SO_2 and particulates.

Problems remain, however, which arise mainly from the chemical and physical properties of methanol, and these need to be addressed. Methanol, not unlike ethanol, is miscible with water in all proportions. It has a high dipole moment as well as a high dielectric constant, making it a good solvent for ionizable substances such as acids, bases, salts (contributing to corrosion problems) and some plastic materials. Gasoline, in contrast, as already mentioned, is a complex mixture of hydrocarbons, most of which have low dipole moments, low dielectric constants and are non-miscible in water. Gasoline is therefore a good solvent for non-polar, covalent materials.

Owing to the different chemical characteristics of gasoline and methanol, some of the materials used in gasoline distribution, storage, devices and connectors are, predictably, often incompatible with methanol. Methanol, consequently, can corrode some metals, including aluminum, zinc and magnesium, though it does not attack steel or cast iron [251]. Methanol can also react with some plastics, rubbers and gaskets, causing them to soften, swell or become brittle and fail, resulting in eventual leaks or system malfunctions. Therefore, systems built specifically for methanol use must be different from those used for gasoline, but are expected to be only marginally, if at all, more expensive. Specific lubricating engine oils and greases that are compatible with methanol already exist, but must be further developed.

With pure methanol, cold-start problems can occur because it lacks the highly volatile compounds (butane, isobutane, propane) generally found in gasoline, which provide ignitable vapors to the engine even under the most frigid conditions [232]. The addition of more-volatile components to methanol is usually the preferred solution. In FFVs using M85, for example, the 15% gasoline provides enough vapors to allow the motor to start even in the coldest climates. Another possibility is to add a device to vaporize or atomize methanol into very small droplets which are easier to ignite. Dimethyl ether, directly derived from methanol, has also been proposed as a starting aid for methanol at low temperature [252, 253].

Recently, the Lotus car company unveiled a "tri-flex-fuel" car, the Exige 270E able to run on any mixture of gasoline, ethanol and methanol (Figure 11.2) [254]. Lotus is actively pursuing and promoting the use of methanol, especially produced from CO_2 and hydrogen, as a renewable and sustainable fuel for the future (the basis of the Methanol Economy). According to Lotus:

> methanol possesses properties better suited to internal combustion engines than today's liquid fuels, giving improved performance and thermal efficiencies. It is ideal for pressure-charging (turbo-charging and supercharging) already being introduced by manufacturers to downsize engines to improve fuel consumption [254].

Figure 11.2 Lotus Exige 270E tri-fuel – able to run on any mixture of gasoline, ethanol and methanol. Courtesy of Lotus Engineering (Lotus Cars Ltd.)

Only small changes to the engine, fuel storage and delivery system are needed [254, 255]. Lotus adds that:

> Ultimately, emerging processes to recover atmospheric CO_2 will provide the required carbon that can entirely balance the CO_2 emissions at the tailpipe that result from the internal combustion of synthetic methanol. The result is that a car running on synthetic methanol, such as the Exige 270E Tri-fuel, would be environmentally neutral [256, 257].

As discussed, using methanol as a fuel in ICE cars, including blends with gasoline containing methanol in concentrations greater than 10% necessitates adjustments to the vehicles, albeit at a modest cost. Flex-fuel cars that are now mass produced contain most of these adjustments or can be readily adapted. When adding methanol in lower concentrations to gasoline, no adjustments are needed and this mix could be introduced without additional costs or changes in dispensing infrastructure and used in existing cars.

In the 1980s EPA already granted a waiver to the Clean Air Act to DuPont and other companies, allowing the commercialization of blends of gasoline containing 5% methanol and 2.5% of other alcohols (ethanol, propanol, tertiary butyl alcohol and/or butanols) to reduce vehicle emissions [248, 258]. However, the use of such mixed fuels never caught up. Recently, Olah and Prakash developed a ternary fuel mix containing methanol and bio-ethanol allowing to overcome some of the problems associated with phase separation and other disadvantages and making bio-ethanol readily adaptable to gasoline mixes [259].

Technical problems have to be expected during the development of any new technology. The technological difficulties facing methanol as a blending component or substitute for gasoline in ICE vehicles, however, are relatively easy to solve and, indeed, most of them have already found solutions.

11.3.3
Methanol as Fuel in Compression Ignition (Diesel) Engines

Particulate matter – whether carcinogenic compounds are absorbed onto them or not – has been identified as a significant health hazard, especially in large cities. Diesel fuel generally produces particles during combustion. When combusted, methanol does not produce smoke, soot or particulates. This, and the fact that methanol produces very low emissions of NO_x because it burns at lower temperatures, makes methanol attractive as a substitute for diesel fuel [251].

Compression ignition (CI, diesel) engines are quite different from gasoline engines. Instead of using sparkplugs to ignite the fuel/air mixture in the engine's cylinders, diesel motors rely on the self-ignition properties of the fuel to ignite under specific high-temperature and high-pressure conditions. While a typical gasoline engine has a compression ratio of about 8–9 to 1, a diesel engine has generally a compression ratio in excess of 17 to 1. In the past, such engines were used mainly in heavy-duty vehicles such as buses, tractors, trucks, locomotives and ships. However, in the early 1970s, due to their better fuel economy compared to gasoline engines, diesel engines were increasingly used to power personal automobiles. Today, in Western Europe for example, diesel-fueled cars represent about 50% of all cars in operation.

Like gasoline, diesel fuel is composed of many hydrocarbons that have a wide boiling range. The physical and chemical properties of diesel fuel are, however, quite different. Whereas gasoline contains predominantly branched alkanes with three to ten carbon atoms as well as aromatic compounds to ensure a high octane rating, diesel fuel is mainly composed of straight-chain alkanes with 10–20 carbon atoms. A fuel's propensity to self-ignite under high heat and pressure is measured by the "cetane" rating. While diesel fuel has cetane ratings that range from 40 to 55, methanol's cetane rating is only about 3 [217]. With methanol and diesel fuels being substantially immiscible, the possibility of using any blends of methanol and diesel fuel in diesel vehicles is made difficult [251, 260].. Further, to overcome the low cetane rating of methanol, diesel motors must be adapted. Additives can be included to increase the cetane rating of methanol to levels close to diesel fuels. These ignition improvers, added in the order of a small percentage to methanol, are typically composed of nitrogen-containing compounds such as octyl nitrate and tetrahydrofurfuryl nitrate [251], although many of them are toxic and/or carcinogenic. Non-toxic cetane enhancers based on peroxides and higher alkyl ethers have also been developed. If neat methanol is used, ignition through sparkplugs or glow plugs is necessary.

The energy content of methanol on a volume basis is about 2.2 times less than that of diesel fuel [251], which means that the fuel tank must be larger than a conventional diesel fuel tank to provide the same amount of energy. Methanol also has a significant vapor pressure compared to diesel fuel. The higher volatility allows heavy-duty engines to start easily in the coldest weathers, thereby avoiding the white smoke that is typical of cold starts with conventional diesel engines.

As with many other alternative fuels, transit buses have been the main test field for methanol-powered diesel engines. The Detroit Diesel Corporation (DDC) in particular developed a methanol version of its 6V-92TA diesel engine, which in the early 1990s was the lowest emission heavy-duty diesel engine certified by the Environmental Protection Agency (EPA) and the California Air Resource Board (CARB) [251]. Using methanol instead of diesel fuel also allowed a dramatic reduction in particulate as well as NO_x emissions. Methanol contains no sulfur; hence, SO_x emissions that lead to acid rain are nearly eliminated. Fleets of buses equipped with methanol-fueled diesel engines were tested in different cities around the United States, including Los Angeles, Miami, Denver (Figure 11.3) and New York [261]. The Metropolitan Transit Authority (MTA) of Los Angeles in particular operated a large fleet of some 330 methanol-powered transit buses for some years in the 1990s. Compared to conventional diesel buses, higher maintenance costs due to some technical problems in the fuel system and engine were experienced using methanol. Similar operating problems were also encountered when these buses were converted to run on ethanol. Most of the difficulties experienced with alcohol use in diesel fuels are connected with the fuel delivery system and the use of non-compatible materials. Nevertheless, these

Figure 11.3 Methanol-powered regional transit bus in Denver, Colorado. (Source: Gretz, Warren DOE/NREL.)

technical problems are not insurmountable and can be solved. Stable ethanol-diesel fuel blends, such as E-diesel and O_2Diesel™, containing generally between 7 and 15% ethanol have also been proposed [262, 263]. O_2Diesel is now being used commercially in the Mid-West and in California, for example at the port of Long Beach, to reduce smoke, particulate matter and NO_x emissions. Similar blends with methanol could be envisioned, although co-solvents or additives might be needed to increase its poor solubility in diesel fuel. Methanol, however, must also compete with other alternative fuels. When considering transit buses, for example, where the bulkiness of fuel tanks is not a major issue, compressed natural gas (CNG) has now become the preferred fuel because of its low emissions, price and widespread availability. Improved diesel engines enabling cleaner operation, particle filters as well as advanced diesel fuels containing less-polluting impurities are also being developed.

11.4
Dimethyl Ether as a Transportation Fuel

Another possibility for diesel engines is to use dimethyl ether (DME), a superior and more calorific fuel than methanol, which can be readily obtained by dehydration of methanol.

Dimethyl ether, the simplest of all ethers, is a colorless, non-toxic, non-corrosive, non-carcinogenic and environmentally friendly chemical that is mainly used today as an aerosol propellant in various spray cans, replacing banned CFC gases. It is also used as a fuel of choice by glass blowers to obtain a soot-less flame. DME has a boiling point of $-25\,°C$, and is a gas under ambient conditions. DME, however, is generally handled as a liquid and stored in pressurized tanks, much like LPG (liquefied petroleum gas), which contains primarily propane and butane, commonly used for cooking and heating purposes. Significantly, DME, contrary to diethyl ether and other higher ethers, does not form an explosive peroxide, thereby making it safe to use. The interest in DME as an alternative transportation fuel lies in its high cetane

Table 11.2 Properties of dimethyl ether (DME).

Chemical formula	CH_3OCH_3
Molecular weight	46.07
Appearance	Colorless gas
Odor	Slightly sweet smell
Chemical composition (%)	
Carbon	52
Hydrogen	13
Oxygen	35
Melting point	$-138.5\,°C$
Boiling point	$-24.9\,°C$
Density of liquid at $20\,°C$	$668\,kg\,m^{-3}$
Energy content	$6880\,kcal\,kg^{-1}$
	$317\,kcal\,mol^{-1}$
Cetane number	55–60
Flash point	$-41\,°C$
Autoignition temperature	$350\,°C$
Flammability limits in air	3.4–17%

rating of 55–60, compared with 40–55 for conventional diesel fuel and much higher than that of methanol. Therefore, DME can be effectively used in diesel engines, as has been introduced by Haldor Topsoe. Like methanol, DME is clean burning, producing no soot, black smoke or SO_2, and only very low amounts of NO_x and other emissions even without exhaust gas after-treatment (Tables 11.2 and 11.3).

Today, DME is produced almost exclusively by the dehydration of methanol. Being directly derived from methanol, DME can be produced from a large variety of feedstocks: coal, natural gas, biomass, and so on, or reductive CO_2 recycling (Chapter 12). A method to synthesize DME directly from syn-gas by combining the methanol synthesis and dehydration steps in a single process has also been developed [264]. The direct synthesis of DME from CO_2 and H_2 has also been studied [265]. Until 2004, the global demand for DME was only about 150 000 tonnes per year. Since then, however, the demand for DME has grown tremendously as ever larger quantities are consumed as a fuel. In China alone, DME production reached more than 2 million

Table 11.3 Comparison of the physical properties of DME and diesel fuel.

Property	DME	Diesel fuel
Boiling point (°C)	−24.9	180–360
Vapor pressure at $20\,°C$ (bar)	5.1	—
Liquid density at $20\,°C$ ($kg\,m^{-3}$)	668	840–890
Heating value ($kcal\,kg^{-1}$)	6880	10 150
Cetane number	55–60	40–55
Autoignition temperature (°C)	350	200–300
Flammability limits in air (vol.%)	3.4–17	0.6–6.5

Figure 11.4 DME-fueled Volvo bus developed in Denmark.
(Courtesy: Danish Road Safety and Transport Agency.)

tonnes in 2008 and is continuing to grow [266]. DME vehicles are being developed in many parts of the world. In Europe, according to Volvo, which is currently working on its third generation DME technology and running tests on DME-powered buses and trucks, DME is one of the most promising fuels for substituting conventional diesel oil (Figure 11.4) [267, 268]. DME is particularly attractive for Sweden, Volvo's home country, as it can be produced from black liquor, a residual product from paper pulp production. It can be used in an ordinary diesel engine equipped with a new fuel injection system, producing the same performance, whilst dramatically reducing emissions. Road tests with several DME-powered Japanese trucks and buses led to conclusions comparable to those obtained by Volvo. Isuzu, a well-known diesel engine manufacturer, is taking part in these tests and is confident that diesel oil can be supplanted in the future by improved technologies, including the use of DME. In China, at the Shanghai Jiao Tong University, researchers in collaboration with several companies have developed a transit bus powered by DME, while the Chinese Ministry of Science has supported the production of 30 such buses. The cost to adapt each bus to run on DME has been estimated to be around only $1200. The city of Shanghai plans to have a fleet of about 1000 DME-fueled buses in time for the World Expo in 2010.

Replacing diesel fuel with DME in compression ignition (CI) engines not only significantly reduces emissions, especially of particulate matter and NO_x, but also lowers the noise of the motor, while keeping the fuel consumption similar to diesel fuel on an energy basis. The engine noise was reduced to the level of a gasoline spark-ignited engine [269, 270]. To continue to improve on the energy efficiency of the motor and reduce emissions, new engine and fuel delivery designs have been investigated for DME, including the low compression ratio direct injection (LCDI) diesel engine, the homogeneous charge compression ignition (HCCI) engine and the compound charge compression ignition (CCCI) [270, 271]. Lowering the compression ratio in the LCDI motor, for example, allowed the DME compression ratio to be reduced from 18 : 1 to as low as 12 : 1. As a consequence, the pressure in the engine cylinders was lowered, meaning that lighter engines compared to current diesel engines could be used, thereby decreasing the cost and increasing fuel efficiency.

Besides the many advantages of DME as a clean-burning fuel, it is also important to recall that, due to the lower energy density of DME compared to diesel fuel, a DME fuel storage tank must be about twice the size of a conventional diesel fuel tank to achieve an equivalent driving range to a comparable diesel vehicle. DME has a lower viscosity, which can lead to leakage. Nevertheless, the viscosity of DME at 0.15 centipoise falls between that of propane and butane at 0.10 and 0.18 centipoise, respectively [269]. Given the millions of engines running on propane and butane (LPG), and the experience gathered, solving this problem through selection of adequate components is relatively simple. DME also has low lubricity, which can lead to premature wear and eventual failure of pumps and fuel injectors. Additives, such as Lubrizol and Hitec 560, have been used to increase the lubricity of DME, and the commonly used additives have been those developed for reformulated diesel. DME is not compatible with some types of plastics and rubbers, so that careful selection of materials is necessary to prevent deterioration of seals after prolonged exposure to DME. DME-filled vessels can be sealed with materials such as poly-tetrafluoroethylene (PTFE) and butyl-n (Buna-N) rubber. The most effective type of seals, also resistant to high temperature, are non-sparking metal-to-metal seals. Although research on improving DME wear, lubricity and material compatibility is still ongoing, most of the technical problems associated with the use of DME are already resolved.

Interestingly, it was also found that DME/LPG blends can be used as fuel substitutes for gasoline-powered ICE vehicles. The octane number of DME is relatively low (RON of 35 and MON of 13), but when mixed with adequate amounts of propane, a component of LPG with a higher octane rating (RON of 111 and MON of 100), fuels with desired octane ratings can be obtained. A mixture containing about 20 wt% of DME had an average octane rating of 89.5 (average of RON and MON), well in the range of gasoline used in current cars [272]. Vehicles designed or modified to run on LPG could thus be used for DME/LPG mixtures, given the materials, especially rubber and plastics in contact with the fuel in the fuel storage and delivery system, have been tested for DME service.

11.5
DME Fuel for Electricity Generation and as a Household Gas

Realizing that future LPG supplies may not meet demand, Japan, China and other countries are studying DME as an option not only for the transportation sector but also for electric power generation, as well as household and industrial uses (Table 11.4) [273]. Japan, which widely uses imported LPG for domestic applications, has an extensive LPG infrastructure that can be easily adapted to DME. DME is very competitive against LPG prices, and this utilization is one of the most important potential markets for DME. In China, DME started to be used as a substitute for LPG in 2003. Most of the DME currently produced in China is indeed used as a mixture with LPG (nearly 90% in 2007 [266]), mainly for heating and cooking purposes. Up to 20% DME can be blended with LPG with no or very limited modifications to existing

Table 11.4 Comparison of the physical properties of DME and liquefied petroleum gas (LPG).

Property	DME	Main components of LPG	
		Propane	Butane
Boiling point (°C)	−24.9	−42.1	−0.5
Vapor pressure at 20 °C (bar)	5.1	8.4	2.1
Liquid density at 20 °C (kg m^{-3})	668	501	610
Heating value (kcal kg^{-1})	6880	11 090	10 920
Flammability limits in air (vol.%)	3.4–17	2.1–9.4	1.9–8.4

equipment. New national standards are being drafted that will allow up to 20% DME to be mixed into LPG for civil use nationwide. Existing infrastructures for LPG, such as tanks and refrigerated tankers, could also be used for pure DME. DME is not corrosive towards conventional construction materials. LPG is used as an automotive fuel in many countries. Fuelling stations for DME can be built using the technologies developed for LPG – far simpler than those for CNG or hydrogen fuel.

General Electric showed DME to be an excellent fuel for gas turbines for power generation, with emissions and performances comparable to those of natural gas [274]. DME can be fired in existing turbines currently burning natural gas or liquid fuels such as naphtha or distillate oil, with some modifications, primarily to the fuel delivery system. Owing to similar combustion characteristics, cooking stoves designed for natural gas can use DME without any modification [275].

Two major Japanese consortia composed of leading companies are currently assessing the economical viability of DME. One consortium, led by the giant NKK Corp., created DME International Corp. and DME Development Company to investigate the economics and facilitate the introduction of DME as a fuel with commercial production in the 850 000–1 650 000 tonnes per year range. The other consortium, Japan DME, led by Mitsubishi Gas Chemical, is planning to build a large scale DME/methanol plant in Papua New Guinea. The Japanese government, through substantial financial support from the Ministry of Economy, Trade and Industry (METI), is also committed to the development of mass produced, low-cost DME.

Developing countries in Asia, such as China and India, are also very active in DME, given their rapidly increasing needs and growing demands for diesel and LPG, as well as their deteriorating air quality. China in particular, because of its enormous reserves of coal, is interested in coal-to-DME liquefaction technology, and has recently emerged as the most fervent promoter of DME as an alternative fuel. China's annual domestic production of DME was only about 50 000 tonnes in 2002, but has increased dramatically in recent years, from about 400 000 tonnes in 2006 to 2 million tonnes in 2007 and an expected 4 million tonnes in 2008. Following estimates, the DME production in China could reach about 14 million tonnes in 2010! A partnership including China National Group, Sinopec and Shenergy Group is currently constructing a 3 million tonnes-per-year DME plant in Inner Mongolia. Several other plants based on coal with capacities ranging from 1 to 3 million tonnes per year DME production capacity are also being constructed in that same coal-rich region. The produced DME will be transported with a new DME pipeline from Inner Mongolia to the Chinese coast, from where it could be shipped to important consuming centers. Chinese government regulations from the National Development and Reform Commission (NDRC) in 2006 have banned any coal-based DME projects with a design capacity lower than one million tonnes. To promote DME as an alternative fuel, the Chinese authorities have also lowered the value-added tax (VAT) on DME from 17% to 13% in July 2008. Of course, all of this coal-based development generates very large amounts of carbon dioxide that, ideally, should be captured and then sequestered or preferably chemically recycled.

Countries that have low-cost natural gas reserves, but are distant from important consuming centers, such as the Middle East, Australia, Trinidad and Tobago and other locations, are also interested in DME as a convenient way to transport energy to markets in highly populated areas. In 2007, a plant with a production capacity of 800 000 tonnes per year of DME for fuel uses was completed in Iran. In the Middle East [276], BP is also involved in a planned DME plant to produce 1.8 million tonnes per year [277].

Besides DME, dimethyl carbonate (DMC), which also has a high cetane rating, can be blended into diesel fuel at a concentration of up to 10%, thereby reducing the fuel viscosity and improving emissions. In China, plans to produce DMC (mainly from coal or some natural gas resources) on a commercial scale as a diesel additive are under way [277]. Owing to its melting point of 3 °C, however, neat DMC is not an ideal fuel because of freezing problems at lower operating temperatures. The commercial route to DMC has used the reaction of methanol with phosgene. However, phosgene, being highly toxic, has been replaced by the oxidative carbonylation of methanol, developed by EniChem and other companies [278].

Blends of diesel fuel and dimethoxymethane (DMM) have also been studied and tested [279, 280]. Fuels blends containing up to 30% DMM were found to exhibit satisfactory fuel efficiency and emission characteristics. DMM is a liquid with a cetane number of 30 that can be manufactured by oxidation of methanol or by reaction of formaldehyde with methanol. It can also be produced by catalytic oxidation of DME [281].

11.6
Biodiesel Fuel

Another way to use methanol in diesel engines and generators is through biodiesel fuels. These can be made from a large variety of vegetable oils and animal fats containing fatty acid esters, which are reacted with methanol in a transesterification process to produce compounds known as fatty acid methyl esters, the constituents of biodiesel. Biodiesel can be blended without major problems with regular diesel oil in any proportion. It is a renewable, domestically produced fuel that also reduces the emission of unburned hydrocarbons, carbon monoxide, particulate matter, sulfur compounds as well as CO_2. The use of biodiesel has grown significantly during the past few years, mainly in Europe and the United States. However, as pointed out in Chapter 6, the natural feedstocks for biodiesels are limited, and consequently biodiesel can cover only a relatively small portion of our fuel needs. Biodiesel alone will be unable to replace or even significantly impact diesel fuel obtained from fossil fuels in the quantities required for our transportation systems.

Another approach to biodiesel is the use of ethanol or methanol from agricultural sources to produce ethylene and propylene, which can be processed together with petroleum oil feeds to regular diesel fuels. Brazil, for example, has started to use this approach.

11.7
Advanced Methanol-Powered Vehicles

Methanol or its derivatives (DME, DMC, DMM, biodiesel) can already be used as substitutes for gasoline and diesel fuel in today's automobiles, with only minor modifications to existing engines and fuel systems. ICE is a much-proven and reliable technology that has been continuously improved and perfected since its invention over a hundred years ago. Fuel economy compared to generated power is now better, and emissions lower, than ever before. Hybrid cars, combining an ICE with an electric motor, are commercialized by a growing number of companies (Toyota, Honda, Ford, etc.) and can reduce even further fuel consumption and emissions. In these vehicles, too, gasoline and diesel fuel can be easily substituted by methanol or its derivatives. Their use on a large scale is realizable in the short term. In the foreseeable future, however, to further increase efficiency and lower emissions, fuel cell technology will be the best alternative to ICEs in the transportation field. Much effort and financial resources are currently being invested by major motor car manufacturers and governments to make fuel cell vehicles (FCVs) an affordable and viable option for consumers. FCVs promise to be much quieter, cleaner and to require less maintenance than ICEs because of fewer moving parts. Proton exchange membrane fuel cells (PEMFCs) are currently the favored type of fuel cell to power cars because of their relatively light weight, low operating temperature and high power output. As described in Chapter 9, these vehicles operate primarily on hydrogen, which can be stored in liquid, gaseous or solid metal hydride forms, or even reformed onboard from different liquid fuels, including gasoline and methanol.

11.8
Hydrogen for Fuel Cells Based on Methanol Reforming

In seeking to overcome the difficult problems associated with hydrogen storage and distribution, numerous approaches have set out to use liquids rich in hydrogen, such as gasoline or methanol, as a source of hydrogen via on-board reformers. In contrast to pure hydrogen-based systems, these liquid fuels are compact (containing on a volume basis more hydrogen than even liquid hydrogen) and are easy to store and handle without pressurization. The possibility of generating hydrogen with more than 80% efficiency by the on-board reforming of gasoline has been demonstrated. However, the process is expensive and challenging because it involves high temperature and needs considerable time to reach a steady operational state. The advantage is that the distribution network for gasoline already exists, though this would not solve the problems of diminishing oil resources and dependence on oil-producing countries. On the other hand, methanol and DME steam reformers operating at much lower temperature (250–350 °C) [13], albeit still expensive, are more adaptable to on-board applications. In methanol and DME, the absence of C—C bonds, which are difficult to break, greatly facilitates its transformation into high-purity hydrogen with 80–90% efficiency [282]. Methanol and DME furthermore contain no sulfur, a contaminant in fuel cells. With the reformer operating at low temperature, no nitrogen oxides are produced. The use of an on-board reformer enables the rapid and efficient delivery of hydrogen from a liquid fuel that can be easily distributed and stored in the vehicle. To date, methanol is the only liquid fuel that has been developed and demonstrated on a practical scale in fuel cells for transportation applications. The disadvantages of this system are, however, the added weight, complexity and cost, as well as trace emissions that may be produced from the reformer when it burns some of the methanol to provide the necessary heat for hydrogen production [247].

The potential for on-board methanol reformers to power FCVs has been demonstrated by several prototypes constructed and tested by various automobile companies. In 1997, Daimler presented the first methanol-fueled FCV, the Necar 3, a modified A-Class Mercedes-Benz compact vehicle equipped with a 50-kW PEM fuel cell and a driving range of 400 km. In 2000, an improved version with an 85-kW fuel cell, the Necar 5 was introduced (Figure 11.5) [283]. In this vehicle, which was described by the company as being fit for practical use [247], the entire fuel cell and reformer system has been accommodated in the underbody of the car, enabling five passengers and their luggage to be transported at a maximum speed of 150 km h^{-1}, with a driving range approaching 500 km [283]. In 2002, this FCV was the first to complete a coast-to-coast trip across America, from San Francisco to Washington DC, or a distance of more than 5000 km, by refueling with methanol every 500 km [284].

Figure 11.5 Daimler's methanol-fueled Necar 5 fuel cell vehicle (introduced in 2000). (Courtesy of Daimler.)

Figure 11.6 Methanol fuel cell buses developed at Georgetown University in front of the US Capitol (2002). (Source: Georgetown University.)

Daimler also presented, in 2000, a fuel cell/battery hybrid Jeep Commander SUV powered by methanol. Based on a Ford Focus, Ford constructed the TH!NK FC5 [285], a methanol-fueled FCV with a fuel cell/reformer system beneath the vehicle's floor and characteristics similar to the Necar 5. Other companies that have developed methanol-powered FCVs include General Motors, Honda, Mazda, Mitsubishi, Nissan and Toyota. However, most of these companies, including Daimler, have recently concentrated their efforts on vehicles with on-board storage of pure hydrogen.

Georgetown University of Washington DC has been at the forefront in the development of transportation fuel cells for some 20 years. Supported by the US Federal Transit Administration, the university has developed several fuel cell transit buses running on methanol [286]. In 1994 and 1995, Georgetown produced three buses that were the world's first FCVs able to operate on a liquid methanol fuel (Figure 11.6). These methanol buses, each powered by a 50-kW phosphoric acid fuel cell (PAFC) combined with a methanol steam reformer, are still operating today. In 1998, an improved second-generation bus using a more powerful 100-kW PAFC provided by UTC Fuel Cells was introduced, followed in 2001 by the first urban transit bus powered by a liquid-fueled 100-kW PEMFC system manufactured by Ballard Power System, a major fuel cell developer in Vancouver, Canada. Batteries provide surge power and a means to recover braking energy by regeneration. These two buses, each able to seat 40 passengers, meet all the requirements of the transit industry, are much quieter than their ICE-powered counterparts and have a driving range of some 560 km between refueling. The use of methanol allows the refueling to be as quick and easy as with diesel buses. PM and NO_x emissions are virtually eliminated, and other emissions are well below even the cleanest CNG buses on the road with the most stringent clean air standards [286].

Besides on-board methanol reforming, methanol is also seen as a convenient way to produce hydrogen in fueling stations to refuel hydrogen FCVs. Mitsubishi Gas Chemical has developed a process to produce high-purity hydrogen by steam reforming of methanol using a highly active catalyst that allows operation at relatively low temperature (240–290 °C) and enables rapid start-up and stopping, as well as flexible operation. These methanol-to-hydrogen (MTH) units, which range in production capacity from 50 to 4000 m³ H₂ per hour, are already used by various customers in the electronic, glass, ceramics and food processing industries [287, 288]. They provide excellent reliability, prolonged life service and minimal maintenance [282]. Based on the technology developed by Mitsubishi Gas Chemical, the first fueling station to supply hydrogen by methanol reforming was constructed in Kawasaki, Japan as part of the Japan Hydrogen & Fuel Cell Demonstration Project (JHFC), which is studying a large array of possible feedstocks for hydrogen generation. According to JHFC, methanol is the safest of all materials available for hydrogen production [289]. Operating at relatively low temperature, the MTH process has a clear advantage over the reforming of natural gas and other hydrocarbons, which must be carried out above 600 °C. A smaller amount of energy is necessary to heat methanol to the appropriate reaction temperature. At the Kawasaki station, methanol and water are evaporated and reacted over a catalyst. After purification and separation, the hydrogen produced is compressed and stored to provide FCVs with high-pressure hydrogen [289]. Although methanol reforming is a convenient and attractive way of producing hydrogen, it does not solve the problems associated with the costly and difficult on-board storage of hydrogen.

Significant investigations aimed at further improving methanol reforming to hydrogen, whether focused on on-board or stationary applications, are under way. Hydrogen obtained by methanol reforming in current processes always contains more than 100 ppm CO, a poison for PEM fuel-cell catalysts operating below 100 °C. At present, reformed gas has thus to be cleaned to remove CO, lowering the total efficiency of the process. At the Brookhaven National Laboratory, new catalysts have been designed that produce hydrogen with high yield and generate only marginal amounts of CO. Using a process known as oxidative steam reforming, which combines steam reforming and the partial oxidation of methanol, and different novel catalyst systems, the National Industrial Research Laboratory of Nagoya, Japan, has also achieved the production of high-purity hydrogen with either zero or only trace amounts of CO, at high methanol conversion and temperatures as low as 230 °C. Oxidative steam reforming of methanol also has the advantage of being – contrary to steam reforming – an exothermic reaction, thereby minimizing energy consumption. The exothermicity of the reaction, however, can also be a drawback since the generated heat and consequently the reactor's temperature may be difficult to control. In an ideal case, the reaction should therefore only produce enough energy to sustain itself. This is the principle of autothermal reforming. The autothermal reforming of methanol, which combines steam reforming and partial oxidation of methanol in a specific ratio, is an idea first developed in the 1980s by Johnson-Matthey. It is neither exothermic nor endothermic, and thus does not require any external heating once the reaction temperature has been reached. For a fast start-up, the methanol/oxygen feed ratio can be varied, as has been shown, for example, in Johnson-Matthey's "Hot-Spot" methanol reformer.

Steam and autothermal reforming of DME can be carried out with similar processes. The transformation of DME into hydrogen is generally a two-step reaction. In the first step DME is hydrolyzed to methanol over a mildly acidic catalyst. In the second step the formed methanol is reformed to hydrogen using any of the catalysts and processes described above. Both steps are generally carried out in the same reactor using a mixture of hydrolysis and reforming catalyst [290]. Autothermal reforming will most likely be the process implemented for on-board vehicle fuel processing as well as static hydrogen production.

11.9
Direct Methanol Fuel Cell (DMFC)

In contrast to hydrogen fuel cells, direct methanol fuel cells are not dependent upon hydrogen generation by processes such as natural gas, hydrocarbon, methanol or DME reforming. As mentioned earlier, the storage and distribution of hydrogen fuel will require an entirely new infrastructure or a complete overhaul of existing systems, all of which constitutes a major barrier for entry into the commercial market. Methanol, in contrast, is a clear liquid fuel (bp $= 64.6\,°C$, density $= 0.791\,g\,mL^{-1}$) that does not require special cooling at ambient temperature, and will fit into existing storage and dispensing units with only small modifications.

Methanol has a relatively high volumetric theoretical energy density compared to other systems such as conventional batteries and the H_2-PEM fuel cell (Figure 11.7). This is of basic importance for small portable applications, as battery technology may not be able to keep up with the demand for laptops and mobile phones that are lightweight and have extended operating time [291, 292].

There is more hydrogen in 1 L of liquid methanol than in 1 L of pure cryogenic hydrogen (98.8 g of hydrogen in 1 L of methanol at room temperature compared to 70.8 g in liquid hydrogen at $-253\,°C$). Therefore, it transpires that methanol is a safe carrier fuel for hydrogen.

In the past, methanol-based PEM cells have used a separate reformer to release the hydrogen from liquid methanol, after which the pure hydrogen is fed into the fuel cell stack. However, since 1990 researchers at the Jet Propulsion Laboratory, and our own group at the University of Southern California [293, 294], have developed a simple DMFC that consists of two electrodes separated by a PEM and connected via an external circuit that allows the conversion of free energy from the chemical reaction of methanol with air to be directly converted into electrical energy (Figure 11.8).

The anode is exposed to a methanol/water mixture fed by flow from an external container, where it is oxidized to produce protons, which travel through the PEM by ionic conduction, and electrons that travel through the external circuit by electronic

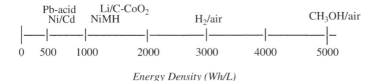

Figure 11.7 Theoretical energy density of batteries, H_2-PEM fuel cells and DMFCs.

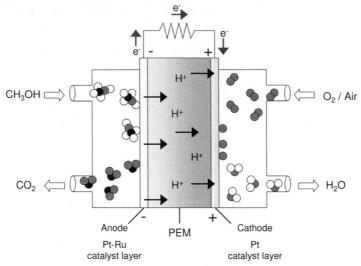

Figure 11.8 Direct methanol fuel cell (DMFC).

conduction. The cathode containing platinum as catalyst is exposed to oxygen or air, which may be either ambient or pressurized. The PEM is coated on both sides with layers of catalyst (1:1 Pt–Ru catalyst at the anode and Pt catalyst at the cathode), usually supported by a gas diffusion electrically conductive carbon (graphite) electrode at the cathode and a liquid feed-type carbon electrode structure at the anode that facilitates reduction of oxygen and oxidation of methanol, respectively (Equations 11.1 and 11.2).

Cathode reaction:

$$1.5O_2 + 6H^+ + 6e^- \rightarrow 3H_2O \tag{11.1}$$

Anode reaction:

$$CH_3OH + H_2O \rightarrow CO_2 + 6H^+ + 6e^- \tag{11.2}$$

Overall reaction:

$$CH_3OH + 1.5O_2 \rightarrow CO_2 + 2H_2O \tag{11.3}$$

At room temperature, the overall reaction (Equation 11.3) gives a theoretical open circuit voltage of 1.21 V with a theoretical efficiency close to 97%. Although fundamentally simple, DMFCs (direct methanol fuel cells) presently still perform well below their theoretical Nernstian potential, even under open-circuit conditions, because of sluggish redox kinetics and fuel crossover. Advances in both catalysts and membranes have – and are being – made and promise to conquer these issues in terms of both performance and cost.

PEMs intended for H_2-PEM fuel cells are not good candidates for DMFCs due to the issue of methanol crossover from anode to cathode. High crossover rates have a deleterious effect on DMFC performance. As oxygen is reduced to produce a cathodic current, oxidation of crossover methanol simultaneously produces an anodic current that results in a mixed potential and overall reduced cathode potential. Methanol may poison the cathode catalyst (Pt) and block cathode catalyst sites, further reducing its ability to efficiently reduce oxygen, and necessitating an increase in oxygen flow above and beyond stoichiometric requirements. Furthermore, the chemical oxidation of methanol also produces excessive water that hampers cathode performance due to flooding. Instead of meaningful electrical energy, waste heat is generated and fuel utilization efficiency is lowered. Methanol permeates through the PEM by two methods: (i) by simple diffusion due to a concentration gradient and (ii) by electro-osmotic drag from proton migration when the cell is under an applied current. New membranes based on hydrocarbon/hydrofluorocarbon materials with reduced cost and cross-over characteristics have been developed that allow room temperature efficiency of 34% [295, 296].

With the advances made in all aspects of DMFC development, many companies, such as Toshiba, are now actively developing low-power DMFCs for portable devices such as cellular phones and laptop computers [297, 298, 299]. The success of these initial devices is pivotal to transition away from rechargeable batteries, which can in theory deliver only $600\,W\,h\,kg^{-1}$ at best. Currently, rechargeable commercial lithium-ion batteries deliver a power density anywhere between 120 and $150\,W\,h\,kg^{-1}$. Consumers will soon enjoy the benefits of DMFC-powered devices, including longer cellphone talk time, extended usage time on laptop computers, rapid rechargeability and lighter weight contribution of the power source.

Many issues arise when single fuel cells are assembled into practical stack assemblies such as temperature and pressure control, resistance and water management. Various materials and designs have been employed to deal with these issues. The Jet Propulsion Laboratory [300] has developed a small, six-cell DMFC stack with a total active area of $\sim 48\,cm^2$ (anode and cathode catalyst loading of 4–$6\,mg\,cm^{-2}$) using ambient air with 1 M methanol at room temperature, having a power density from 6 to $10\,mW\,cm^{-2}$. The cell is a flat array, where the cells are externally connected in series and share a single membrane. The cathode catalyst, when applied to teflonized carbon supports, imparted excellent water removal properties to the system, although the design also provides increased ohmic resistance. Three two-flat-pack arrays are necessary to power a mobile phone, and a 10 h operating time is estimated before methanol replenishment.

Los Alamos National Labs, in conjunction with Motorola [302], have also developed a stack for mobile phones utilizing ceramic fuel plates with microfluidic channels for efficient delivery of methanol and water and removal of CO_2. Their four-cell stack with a total active area $\sim 60\,cm^2$ gives a power density between 12 and $27\,mW\,cm^{-2}$ with catalyst loading of 6–$10\,mg\,cm^{-2}$ at room temperature.

The Korea Institute of Energy Research (KIER) has developed a 10 W DMFC stack [301]. The cell has a bipolar plate design with six cells each of $52\,cm^2$ active area. Most notably, the stack was operated with 2.5 M methanol, and achieved 6.3 W at room temperature using ambient oxygen flow. In addition, the Korea Institute of Science and Technology (KIST) has developed and assembled a six-cell monopolar stack [303] with a total active area of $27\,cm^2$, and which produced a power density of $37\,mW\,cm^{-2}$ using 4 M methanol and ambient air.

Figure 11.9 Daimler DMFC go-cart. (Courtesy of Daimler.)

Toshiba has developed a promising DMFC prototype stack for laptop computers. The stack has an average output of 12 W and may be continuously used for 5 h with a 50 mL methanol storage cartridge. To minimize the size of the cartridge, the cell collects output water for recombination with methanol. Sensors are hooked up directly to the PC to tell users when the cartridge needs replacing. NEC has developed similar stacks and, within the next few years, anticipates the stack to have a 40 h operation time.

Both stationary and portable DMFC stacks are now available for consumer purchase. For instance, The Fuel Cell Store offers its SFC A25 Smart Fuel Cell capable of 25 W continuous output, and can cover four days worth of energy demand using only 2 kg of fuel. A larger 50 W model is also available.

In the transportation area, Daimler, working on DMFC for automotive purposes, has constructed a one-person go-cart prototype vehicle powered by a 3 kW DMFC (Figure 11.9) [304]. Recently, a similar vehicle, equipped with a 1.3 kW DMFC, known as "JuMOVe," has been developed in Germany by the Jülich Research Center [305]. The same center also unveiled in 2008 a DMFC powered forklift (Figure 11.10). A similar DMFC forklift was developed by Oorja in the US. In Japan, Yamaha presented in 2003 its FC06 prototype, the first two-wheeler motor cycle powered by a DMFC with an output of 500 W. Equipped with a 300 WAC outlet, this bike can also serve as

Figure 11.10 DMFC powered forklift. (Source: Forschungszentrum Jülich.)

(a)   (b)

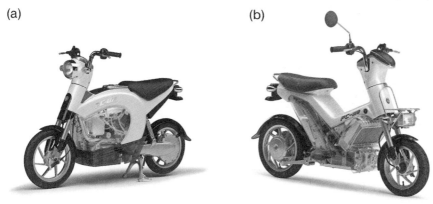

Figure 11.11 Yamaha FC-me (a) and FC-Dii (b) two wheelers
powered by a direct methanol fuel cell (DMFC). (Courtesy
©Yamaha Motor Co.)

an electric power source for outdoor activities or during emergencies [306]. An
advanced version of this motor bike, the FC-me, was in practical use on a lease basis
in Japan (Figure 11.11a). A further improved version, the FC-Dii (Figure 11.11b),
presented in 2007, is powered by a DMFC with an output around 1 kW. The high-
performance cell stack developed by Yamaha is lightweight, compact and offers the
highest power density for a 1 kW class DMFC system. The system efficiency of 30% is
also one of the highest for a DMFC [307]. In the United States, Vectrix has developed a
hybrid fuel cell scooter powered by a 800-W DMFC attached to a rechargeable battery
[308]. The fuel cell continuously recharges the battery, which powers the electric
motor. Regenerative braking technology also captures the energy usually dissipated
during braking to provide additional battery charging. With a maximum speed in
excess of $100\,km\,h^{-1}$ and a range of about 250 km at cruising speed, it has
characteristics comparable to conventional scooters.

Considerable development efforts are still needed to make larger DMFCs practical,
for example to be able to power motor cars, but ongoing progress is impressive.
DMFCs offer numerous benefits over other proposed technologies in the transportation sector. By eliminating the need for a methanol steam reformer, the vehicle's
weight, cost and the system's complexity can be significantly reduced, thereby
improving fuel economy. DMFC systems also come much closer to the simplicity
of a direct hydrogen-fueled fuel cell, without the cumbersome problem of either on-
board hydrogen storage or hydrogen-producing reformers. By emitting only water
and CO_2, other pollutant emissions (NO_x, PM, SO_2, etc.) are eliminated. As methanol
will eventually be made by recycling atmospheric carbon dioxide, CO_2 emissions will
not be of any concern and there will be no dependence on fossil fuels.

A direct DME fuel cell (DDMEFC), very similar to a DMFC, has also been proposed,
and studies are currently being conducted to determine the potential for this
technology. The use of DME could have some advantages. The energy losses due
to fuel crossover in the DDMEFC are expected to be smaller than that in a DMFC.
DME also has a theoretical energy density ($8.2\,kWh\,kg^{-1}$) higher than methanol

($6.1\,\mathrm{kWh\,kg^{-1}}$) and close to ethanol ($8\,\mathrm{kWh\,kg^{-1}}$) [309]. Owing to its low boiling point, DME can be used in its gaseous form. Because of its solubility in water ($76\,\mathrm{g\,L^{-1}}$, i.e., $1.65\,\mathrm{mol\,L^{-1}}$) it can also be used in a liquid form. This solubility corresponds to the concentration range generally used in a direct methanol fuel cell and direct ethanol fuel cell (DEFC) [310]. As with the electrooxidation of ethanol, DME produces 12 electrons.

Anode reaction:

$$CH_3OCH_3 + 3H_2O \rightarrow 2CO_2 + 12H^+ + 12e^-$$

Cathode reaction:

$$3O_2 + 12H^+ + 12e^- \rightarrow 6H_2O$$

Overall reaction:

$$CH_3OCH_3 + 3O_2 \rightarrow 2CO_2 + 3H_2O$$

The data available on DDMEFC is still relatively limited, and many more studies will have to be conducted to determine the potential of this new technology.

Methanol as a transportation fuel has various important advantages. In contrast to hydrogen, methanol does not need any energy-intensive procedures for pressurization or liquefaction. Because it is a liquid, it can be easily handled, stored, distributed and carried on board vehicles. Methanol is already used today in ICE vehicles. Through on-board methanol reformers it can act as an ideal hydrogen carrier for FCVs, and be used in the future directly in DMFC vehicles. Employing the same fuel from present ICEs to advanced vehicles equipped with DMFCs will allow a smooth transition between existing and new technologies. With the progressive phase-out of MTBE as a gasoline additive, substantial methanol production over capacity is immediately available to be used as a transportation fuel.

Over the next decade new innovations, such as novel proton-conducting materials, membrane-less fuel cells, and cheaper and more efficient catalysts, may lead DMFC technology away from the traditional cell structure and design. DMFC is poised to play a critical role in electricity production from methanol in the overall methanol economy infrastructure.

11.10
Fuel Cells Based on Other Methanol Derived Fuels and Biofuel Cells

Direct oxidation fuel cells based on other fuels, such as ethanol, formaldehyde, formic acid, dimethoxymethane and trimethoxymethane, have been studied in laboratories worldwide. However, none of these have shown the promise of either the H_2-PEM fuel cell or DMFC, although application of fuel mixes is feasible.

Biofuel cells use biocatalysts for the conversion of chemical energy into electrical energy. As most organic materials undergo combustion with the evolution of energy, the biocatalyzed oxidation of organic substances by oxygen or other oxidizers at two-electrode interfaces provides a means for the conversion of chemical into electrical energy. Abundant organic raw materials such as ethanol, hydrogen sulfide, organic acids or glucose can be used as substrates for these oxidation processes, while molecular oxygen or H_2O_2 can be reduced. Intermediate formation of hydrogen as a potential fuel is also possible. Biofuel cells can use biocatalysts, enzymes or even whole-cell organisms, including certain types of bacteria. The power produced in such devices is miniscule (microwatt to nanowatt range), although such devices have potential uses as chemical and biological sensors. Biofuel cells can also be used for waste water purification [311].

11.11
Regenerative Fuel Cell

A regenerative fuel cell concept based on methanol/formic acid fuel cells has also been proposed (Figure 11.12) [312]. The key to the success of such an approach is efficient capture of CO_2 and its electrochemical reduction to either HCOOH or CH_3OH in high current efficiencies. Intense research to achieve efficient electrochemical reduction of CO_2 is currently under way in many laboratories.

11.12
Methanol and DME as Marine Fuels

About 90% of world trade (measured in tonnes) is carried by ship. Shipping is one of the freight options with the lowest CO_2 emissions per tonne and kilometer. It only emits about 10–15 g of CO_2 per tonne-kilometer (tkm), compared to 19–41 tkm for rail, 51–91 tkm for trucking and 673–867 tkm for aviation [313]. Nevertheless, the world's fleet of ships was estimated to be the source of 3% of the global carbon dioxide emissions in 2007 [314] and is continuing to grow. Between 1985 and 2007 the global

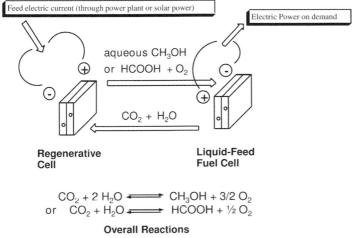

$$CO_2 + 2\,H_2O \rightleftharpoons CH_3OH + 3/2\,O_2$$
$$\text{or} \quad CO_2 + H_2O \rightleftharpoons HCOOH + \tfrac{1}{2}\,O_2$$

Overall Reactions

Figure 11.12 Regenerative fuel cell system based on CO_2.

Figure 11.13 Wärtislä 20-kW SOFC.

maritime trade doubled [313]. Marine fuels used in ships are relatively high in sulfur and, due to a lack of strict emission regulations (especially in international waters), maritime vessels using large diesel engines are responsible for 10–15% of global NO_x emissions and 4–6% of the SO_x emissions [315]. To reduce these emissions, the use of methanol as a clean and easy to handle fuel was proposed. In Iceland, the fishing fleet, a vital part of the national economy, runs mainly on imported diesel fuel. To reduce its dependence on foreign oil, the use of methanol as a fuel to power fishing vessels was suggested. The needed methanol would be produced from CO_2 from aluminum and ferrosilicon plant exhausts and H_2 obtained by electrolysis of water using geothermal and hydropower (Chapter 12) [316]. Alternatively, DME, because it is an excellent diesel substitute, would be a fuel of choice to run ships mostly powered by diesel motors.

The METHAPU project, funded under the Sixth Research Framework Programme of the European Union, aims at developing a methanol powered SOFC (solid oxide fuel cell) for use onboard commercial marine vessels as an auxiliary power source [317]. A 20-kW methanol-SOFC prototype built by Wärtislä is being installed in a cargo ship to assess the suitability of this technology for the shipping sector (Figures 11.13 and 11.14). Larger 250-kW modules are already planned, which could be clustered to produce, for example, a 1 MW auxiliary power source with four modules. Although these SOFCs are presently destined to be used in marine vessels only as an auxiliary power system, they might in the future provide the ship's main propulsion power.

11.13
Methanol and DME for Static Power and Heat Generation

Transportation and other mobile applications are not the only areas where methanol can be used as a fuel; indeed, it is also an attractive fuel for static applications. It can be

Figure 11.14 Type of cargo ship in which the SOFC for the generation of auxiliary power is being tested.

used directly as a fuel in gas turbines to generate electric power. Gas turbines use typically either natural gas or light petroleum distillate fractions as fuels. Compared to these fuels, tests conducted by many institutions, beginning in the 1970s, have shown that methanol can achieve higher power output and lower NO_x emissions due to lower flame temperatures. Since methanol does not contain sulfur, SO_2 emissions are also eliminated [318, 319]. Operation using methanol offers the same flexibility as using natural gas and distillate fuels, including the ability to start, stop, accelerate and decelerate rapidly, following the electric power needs. Existing turbines, designed originally for natural gas and other fossil fuels, can be relatively easily and inexpensively modified to run on methanol. For this application, fuel-grade methanol with lower production costs than higher purity chemical-grade methanol can be used. Considering increasing natural gas prices, methanol produced at low cost from remote natural gas resources in the Middle East or other regions, and shipped much more easily and less expensively than LNG, also offers an alternative for power generation in large consuming centers such as North America, Europe and Japan. Similarly to methanol, in test runs conducted by General Electric Power Systems, DME was also found to be an excellent gas-turbine fuel with emission properties comparable to natural gas [274]. India is considering DME-fired turbines to supply power to its southern regions [320].

For static uses, the size and weight of fuel cells are of lesser importance compared to mobile applications. Besides PEM fuel cells and DMFCs, phosphoric acid, molten carbonate and solid oxide fuel cells (PAFC, MCFC and SOFC), all of which are ill-suited for automobiles, can also be used for static power and heat generation. As described in Chapter 9, these fuel cells are already being used in the production of electricity in facilities sensitive to power outages such as airports, hospitals, military complexes and banks. Whereas the present cost of these installations is still high, their price is expected to decrease with further development and the number of units produced. Fuel cell units intended for the production of electricity for the residential home market are also being developed [321]. For such static applications, liquid methanol, as well as DME, which are easy to handle, deliver and store, would be the fuels of choice [322, 323].

Methanol and DME can also play a significant role as cooking and household fuels. In developing countries, methanol has been proposed as a substitute cooking fuel in place of wood and expensive and inconvenient kerosene. The consumption of large quantities of wood for cooking purposes by more than 2.5 billion people is in fact one of the major causes of deforestation and all the ecological (desertification, excessive erosion, land-slides, etc.) and socio-economical problems associated with it in the developing areas of the world. Wood-burning stoves used in these countries are generally also very inefficient, producing much smoke, fumes and soot, all of which are serious health hazards. In eliminating these drawbacks, stoves designed specifically for methanol have been developed [324, 325]. The use of DME as a component or replacement for household gas (LPG but also natural gas) has already been discussed and is of growing significance.

11.14
Methanol and DME Storage and Distribution

In parallel to the development of methanol-fueled vehicles, a widespread distribution network for methanol will have to be established to make it as easily available for the consumer as are petroleum-based fuels today. While the passage from ICE to fuel cell-powered methanol vehicles represents a radical technological change, the development of a related fueling infrastructure, including mixed fuels and household gas, is not. Refueling stations dispensing methanol will be almost identical to today's fueling stations, reflecting very little change to consumer habits. Rather than gasoline or diesel fuel, they will simply fill their tanks at the local service station with a different liquid fuel. Existing household gas distribution networks can also be easily adapted to deliver DME.

For the transportation sector, the installation of methanol storage tanks and distributing pumps in existing facilities or specifically designed stations is quite straightforward, and is in any case no more difficult than the installation of their gasoline counterparts. Starting in the late 1980s, a network of almost one hundred methanol refueling stations was built in California to fuel the private and state-owned 15 000 or so methanol-powered vehicles. Most of these were flexible fuel vehicles (FFVs), able to run on any mixture of methanol and gasoline, and usually fueled with M85 (85% methanol and 15% gasoline), though others were designed specifically to run on pure methanol (M100). Other methanol pumps were also installed across the United States and Canada [326].

For retail stations, the conversion costs are minimal. Converting existing double-walled underground gasoline or diesel fuel storage tanks and installing new piping and dispenser pumps compatible with methanol is quite trivial. For some $20 000, an existing 40 000 L tank can be cleaned and the remainder of the system equipped with methanol-compatible elements [326]. The complete operation takes only about one week. The cost of adding a new double-walled underground methanol storage tank with a 40 000 L capacity and methanol-compatible piping, dispensers, valves, and so on to an existing service station is around $60 000–65 000. In rural areas, or where space is available and local codes allow, an above-ground storage tank can be installed and the overall cost reduced to about $55 000 [326]. This means that in the United States, an investment of about $1 billion would enable 10% of the 180 000 service stations to distribute methanol, and for less than $3 billion methanol pumps could be added to one-fourth of the service stations [247, 304]. This cost amounts to a fraction of the more than $12 billion that has been spent by the oil industry to introduce reformulated gasoline to United States service stations [304].

Methanol fueling stations are also much less capital intensive than an infrastructure based on hydrogen, which would need special equipment and materials to handle high pressures or very low temperatures. General Motors has estimated that to build 11 700 new hydrogen fueling stations, $10–15 billion would have to be invested [327], which represents about $1 million per station. Besides the high cost, the technology to dispense hydrogen is presently still immature and has not yet reached the degree of convenience and safety that consumers have come to expect with conventional liquid fuels. Numerous regulations also stand in the way of hydrogen. In the United States, the National Fire Protection Association (NFPA) currently prohibits the placing of hydrogen fueling equipment within 25 m of gasoline pumps [304]. This makes hydrogen pumps difficult – or even impossible – to install in most existing fueling stations, especially in cities. Further higher costs are thus expected if hydrogen has to be dispensed in hydrogen-only stations. Ease of delivering liquid methanol from production centers to local stations avoids all the difficulties encountered for hydrogen transportation, whether under high pressure or in cryogenic form.

Today, already, methanol is a widely available commodity with extensive distribution and storage capacity in place. More than 500 000 tons of methanol are presently transported each month to diverse and scattered users in the United States alone [282], by rail, boat and truck. Overland, transport by railway – where methanol is moved in rail cars each holding about 100 tons – is the preferred option for the long-distance transportation of bulk quantities. The railroad system in the United States, Europe, Japan and other major consuming countries is generally very comprehensive, enabling methanol shipments to be made to all major markets. For smaller volumes and distribution to local markets, tanker trucks with capacities of up to 30 tons are generally used. Where inland waterborne shipment through rivers and canals is possible, methanol can be transported by barges, which typically contain some 1250 tons (10 000 barrels) of methanol. This is the largest inland transportation method, and is especially adapted to deliver methanol to large consumers and inland methanol hubs for regional redistribution [328]. Another means of transporting large quantities of liquids, and one that is also used extensively for oil, natural gas and their products, is via pipelines. At present, methanol pipelines are only viable in regions where major methanol producers and users are concentrated in close proximity, such as on the Texas Gulf Coast between Houston and Beaumont. For long-distance transportation, the volumes of methanol to be shipped are generally insufficient to justify the high investments needed to build a pipeline. In the future, however, if methanol has to be increasingly used as a fuel, the much larger amounts of methanol to be moved overland will improve the economics and make transportation through pipelines not only viable but also indispensable. Technically, transporting methanol through pipelines does not pose any problems, as has been demonstrated successfully in two test runs conducted in Canada. One demonstration used the Trans Mountain crude oil pipeline running from Edmonton, Alberta to Barnaby, British Columbia over a distance of 1146 km; the other involved the Cochin pipeline, primarily used for LPG, over a distance of nearly 3000 km [329]. In both cases the quantity of methanol shipped was the same (4000 tons) and the quality of the delivered product was well suited for fuel applications. When methanol is produced in remote locations where cheap natural gas is available, it is shipped throughout the world by dedicated methanol ocean tankers that range in size from 15 000 to almost 100 000 dead weight tons (DWT) in the case of the latest super tanker used by Methanex (Figure 11.15) [282, 330], one of the world leaders in methanol production. When transported in such large vessels, the costs of shipping methanol will become similar and ultimately equal to that of crude oil. Once delivered, methanol can be easily stored in large quantities, much like petroleum and its products, in tanks with capacities exceeding 12 000 tons. Such tanks can be constructed from various materials, including carbon steel and stainless steel, which are compatible with methanol.

Dimethyl ether, having physical properties similar to LPG fuels, can use the existing land-based and ocean-based LPG infrastructures. For ocean transport of dimethyl ether, conventional LPG tankers can be used. DME can be offloaded and stored at a receiving station using the same methods and equipment as those used for LPG with only minor modifications to the pumps, seals and gaskets. On land, transport by trains, trucks or pipelines, as well as storage and distribution of DME, would require similar modifications. Since there are numerous refilling stations for LPG, a transitioning to dimethyl ether using the same technologies could be less costly than building a completely new infrastructure. Additional refueling stations would be built as the demand for dimethyl ether increases.

Figure 11.15 The methanol tanker *Millennium Explorer.* (Courtesy Mitsui O.S.K. Lines.)

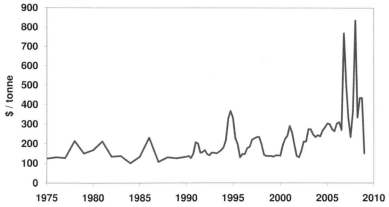

Figure 11.16 Historic methanol prices. (Source: Methanol Institute, Methanex and *Chemical Week*.)

11.15
Price of Methanol and DME

Since 1975, the average wholesale price for methanol has been around $175–200 t^{-1}, but has fluctuated roughly between $100 and $400 t^{-1} (Figure 11.16). As with any other commodity, methanol is subject to fluctuations, depending on supply and demand. The high prices experienced in 1994–1995, which reached more than $350 t^{-1}, were due to an increase in demand for major methanol derivatives such as MTBE, formaldehyde and acetic acid, coupled with production problems at methanol plants. However, with increased capacity and competition, as well as concerns about the use of MTBE as a gasoline additive, prices have decreased rapidly. In recent years, high natural gas prices – especially in North America – have driven the price of methanol (which is still mainly produced from natural gas) to higher levels. As methanol production in North America is gradually phased out, new methanol production facilities are being constructed in regions rich in natural gas but far from main consuming centers, such as the Middle East. Recently, methanol prices have experienced high peaks of more than $500 t^{-1} at the end of 2007 and end of 2008 for short periods of time, before dropping back to more reasonable levels around $150 t^{-1}, in line with historical levels. These peaks were mainly the results of plants shutting down for scheduled maintenance or repairs, although increasing demand from the growing economies in Asia also contributed. The subsequent economic slow-down resulted in the lowering of methanol prices. The construction of extremely efficient mega-methanol plants with very low production costs in regions rich in natural gas such as the Middle East will allow the price of methanol to remain at a relatively low level as long as sufficient natural gas reserves are available. The production cost for methanol in mega-methanol plants has been estimated to be well below $100 t^{-1} (equal to less than 8.5¢ per liter, or 30¢ per gallon) [286]. Even considering its relatively lower energy content (half that of gasoline), methanol will then be quite competitive with gasoline and diesel fuels. At oil prices of $30 to $150 a barrel, as experienced recently, a liter of crude oil costs between 20¢ and 95¢ (from

$0.8 to $3.6 a gallon), without including costs of further processing in refineries to create a suitable fuel from the raw material. Methanol production from feedstocks other than natural gas (in particular coal) is generally higher because of the added cost of generating and purifying the syn-gas necessary for the methanol synthesis. In regions rich in coal, such as the United States and China, the production of methanol from coal on a large scale would, however, bring the costs down, providing an alternative domestic route to methanol.

DME is currently derived directly from methanol and thus has a slightly higher production cost than methanol. NKK Corp. estimated that at a natural gas price of $4 Gcal^{-1} ($1 per MMBtu) the production cost for DME was about $100 t^{-1} ($14.7 Gcal^{-1}) in a plant with a daily capacity of 10 000 tonnes [331]. As recently as 2008, a gallon of diesel fuel sold for some $4 a gallon in the US, including taxes, refining and distribution. At this retail price, the cost of the crude oil to produce a gallon of diesel represents about 60% of the price, or $2.4 ($100 per barrel) [332]. In energy terms, this amounts to about $73 Gcal^{-1}, not including refining, which is much more costly for diesel fuel than for gasoline. With an estimated cost of only $14.7 Gcal^{-1}, DME, requiring no further refining, is therefore already competitive with diesel fuel. Even considering natural gas prices as high as in the US ($8–$9 per MMBtu on average in the last 4 years), the production of DME would probably still be competitive with diesel at oil prices above $100 per barrel.

The cost of producing methanol and DME from sources other than fossil fuels and, most importantly, from the recycling of CO_2 with hydrogen is still difficult to estimate closely at this stage. The energy cost to produce hydrogen is, however, a determining factor. Nevertheless, it was shown that methanol and DME even made from atmospheric CO_2 compares favorably with liquid hydrogen as a transportation fuel, in terms of well-to-wheel efficiency as well as cost [333, 334]. The lower distribution, transportation and storage costs for methanol compared to hydrogen offset the energy needed for separating CO_2. As with any other synthetic material or fuel, the price of methanol and DME from alternative routes is nevertheless expected to be more costly than existing pathways via fossil fuels. It is more difficult and energy intensive to manufacture a convenient fuel that is not directly derived from natural sources. At the same time, chemical recycling of CO_2 to methanol has a significant environmental value that should be taken into account. Fossil fuels are a gift from nature, which have allowed humankind to reach unprecedented levels of development. They served us well, but now – due to their finite nature – we should start to replace them with more sustainable sources of energy.

11.16
Safety of Methanol and DME

Methanol, as mentioned earlier, is a colorless liquid with a mild alcoholic odor. It is a widely used chemical intermediate and solvent in industry and is present in various consumer products. This includes, for example, the windshield washer fluid that most car owners are familiar with, which is composed in large part of methanol.

Methanol is also a deicing fluid, antifreeze and camping cooking fuel. This implies that almost every household has and uses methanol. Even if caution is always required, no significant problems have been associated with its use by the general public. With its increasing usage as an automotive fuel, exposure to methanol will increase. As shown by several studies, the risks for the consumer will, however, remain minimal and not greater than those associated with the use of gasoline or diesel fuel.

Methanol is, like all other motor fuels, toxic to humans and should be handled with the same care as gasoline or diesel fuel with regard to its adverse effects on human health. It is readily absorbed by ingestion, inhalation and more slowly by skin exposure. The ingestion of 25–90 mL of methanol [335] may be fatal if not treated in time (compared to 120–300 mL of gasoline). A tragic, deadly methanol poisoning was, for example, documented to have occurred in 2000 at a South Pole research station to a scientist having ingested more than 100 mL of methanol. The poisoning was not recognized by the station's doctor and remained therefore untreated. The same occurred in instances during World War II of soldiers drinking alcohol that was in fact methanol. Shortly after exposure, methanol causes a temporary effect on the brain, of similar nature, but of lesser strength, to that of ethanol. Methanol is converted in the body by metabolism (i.e., enzymatic conversion) in the liver into formaldehyde, and then to formic acid, which can be excreted in the urine or further metabolized to CO_2 (Figure 11.17). As methanol is metabolized, the most severe effects are delayed for up to 30 h, being caused mainly by the formic acid produced, which humans metabolize very slowly. Higher concentrations of formic acid lead to

Metabolism of methanol in the human body

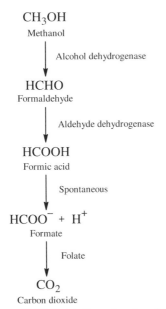

Figure 11.17 Metabolism of methanol in the human body.

increased acidity of the blood. Symptoms may include weakness, dizziness, headache, nausea and vomiting, followed by abdominal pain and difficulties in breathing. In severe cases methanol poisoning may progress to coma and death. Another well-known symptom associated with methanol poisoning is that of visual impairment, which ranges from blurring to total loss of vision, and is caused by formic acid affecting the optic nerve.

Several treatments can be applied to combat methanol poisoning, and these generally lead to complete recovery if administered in a timely manner. Early treatment with sodium carbonate counters the higher blood acidity and prevents or reverses vision impairment. Dialysis is effective in removing both methanol and formate from the bloodstream. In addition, 4-methylpyrazole (Antizolr, fomepizol) [335, 336], an antidote approved by the US Food and Drug Administration (FDA), and acting in the same manner for ethanol ingestion, though without its side effects (it is also effective against ethylene glycol poisoning), can be administered either intravenously or orally.

Although overexposure to methanol can be dangerous to human health, it is also important to realize that both methanol and formate are naturally present in our bodies from our diet, and also as a result of metabolic processes. Methanol is ingested when consuming fresh fruits, vegetables or fermented foods and beverages. Aspartame, a widely used artificial sweetener included in many diet foods and soft drinks, is also partially converted into methanol during its digestion process. According to the FDA, a daily intake of up to 500 mg of methanol is safe in an adult's diet [337]. Methanol and formate are naturally present in the blood in concentrations of approximately 1–3 and 10 mg L^{-1}, respectively; moreover, formate is an essential building block for many biomolecules, including components of DNA [335]. Methanol itself is not considered to be either a carcinogenic or mutagenic hazard; in contrast to gasoline, which contains several compounds that are considered to be hazardous, including (amongst others) benzene, toluene, xylene, ethylbenzene and *n*-hexane, some of which are known carcinogens and mutagens.

Refueling a motor car with methanol at a service station equipped with current refueling systems is only expected to result in low-dose exposures (23–38 ppm during the refueling process [335]) to the general public. By inhalation, a small oral intake of 2–3 mg of methanol is thus expected during a typical refueling. For comparison, this is much less than drinking a single 0.35 L can of diet soda containing 200 mg of aspartame, which will produce some 20 mg of methanol via the body's digestive system. Using vapor recovery systems, exposure to methanol during refueling can be further reduced to the 3–4 ppm level, adding insignificantly to the methanol balance of the body. Even considering a worst-case scenario involving a malfunctioning vehicle in an enclosed garage where the methanol concentration is estimated to reach 150 ppm, a 15-minute exposure would only add some 40 mg to the body's intake, or the equivalent of drinking 0.7 L of diet soda. Exposure to methanol can be readily minimized through the correct design of fueling systems and fuel containers.

To avoid spills, spill-free nozzles have been developed that make it virtually impossible for the consumer to come into contact with methanol during refueling. Nevertheless, in case of skin contact, the affected area should be washed thoroughly with water and soap. To avoid accidental ingestion of methanol, the addition of agents of distinct taste and odor should be considered. A dye may also be added to give methanol-containing fuel a distinctive color. As shown, however, by some 35 000 annual cases of gasoline ingestion in the United States alone (mostly through mouth siphoning when fuel is transferred from one tank to another), unpleasant taste and smell may not be enough to prevent accidents. Thus, refueling systems should also be designed to make siphoning impossible and to allow only the vehicles themselves to be refueled with methanol. Use of containers not meant specifically for methanol, as their improper labeling could lead to mistakes or misuse, should be prohibited. To prevent the accidental ingestion of methanol by heavy alcohol drinkers or the ill-informed public, the name "methyl alcohol" should also be avoided for methanol to minimize possible confusions with ethyl alcohol (i.e., ethanol). Above all, common-sense caution in handling methanol should make its use safe. After all, gasoline and diesel fuel are also not intended for human consumption and the public is not known to misuse them.

Fire and explosion are major hazards associated with the use of transportation fuels, and these are also of concern for the safety of methanol. Compared to gasoline, methanol's physical and chemical properties significantly reduce the risk of fire. Combined with its lower volatility, methanol vapor in air must be four times more concentrated than gasoline for ignition to occur. If it does ignite, methanol burns about four times slower than gasoline and releases heat at only one-eighth the rate of gasoline fires. Because of the low radiant heat output, methanol fires are less likely to spread to surrounding ignitable materials. In tests conducted by the EPA and the Southwest Research Institute [232], cars – one fueled by methanol and the other by gasoline – were allowed to leak fuel on the ground adjacent to an open flame. Whilst the gasoline ignited rapidly, resulting in a fire that consumed the entire vehicle within minutes, methanol took three times longer to ignite and the resulting fire damage affected only the rear of the car. The EPA has estimated that switching fuels from gasoline to methanol would reduce the incidence of fuel-related fires by 90%, saving annually in the United States more than 700 lives, preventing some 4000 serious injuries and eliminating property losses extending to many millions of dollars [338]. Methanol has been the fuel of choice for Indianapolis-type race cars since the mid-1960s because, in addition to achieving superior performances, it is one of the safest fuels available. Unlike gasoline fires, methanol fires can be quickly and easily extinguished even by simply pouring water on them. Methanol burns with little or no smoke, reducing the risks of injuries associated with smoke inhalation and allowing a better visibility around the fire, enabling easier fire fighting. Methanol's combustion generates a light blue flame that is visible in most situations, but may not be easily seen in bright sunlight. In most fires, however, the burning of materials other than fuel, such as upholstery, engine oil and paint, would impart color to the flames, making them visible in any situation. In confined areas such as fuel tanks and reservoirs, an ignitable methanol/air mixture can form at ambient temperature. This property of methanol is, however, unlikely to lead to fires or explosions even in the event of a collision, and has been addressed by simple fuel tank modifications or addition of a volatile compound that makes the vapor space in the tank too rich to ignite [248].

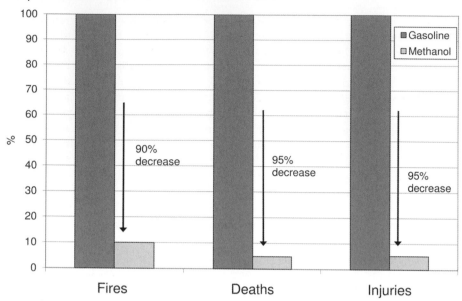

Figure 11.18 Comparative fuel-related fires, deaths and injuries. (Source: US Environmental Protection Agency, EPA 400-F-92-010.)

In summary, compared to gasoline, methanol fires are far less likely to occur, and are much less damaging when they do (Figure 11.18).

DME is non-toxic, non-carcinogenic and has no known teratogenic or mutagenic effects [339, 340]. At low concentration in air the gas hardly has any odor and causes no known negative health effects. Even at higher concentrations there is no long-term effect on human health except some temporary narcotic effects after long exposure and it may be recognized by a slightly etherated odor [341]. Under controlled laboratory conditions, short exposures of up to 10% DME produced mild yet reversible effects on the central nervous system. Human exposure to concentrations higher than 15% resulted in unconsciousness after about half an hour [340]. As a comparison, CO_2 concentrations greater than 10% cause difficulty in breathing, impaired hearing, nausea, vomiting, a strangling sensation, sweating, stupor within several minutes and loss of consciousness within 15 minutes. Several deaths have been attributed to exposure to CO_2 concentrations greater than 20%. DME displays a visible blue flame when burning over a wide range of air–fuel ratios, similar to natural gas, which is a significant safety warning characteristic. Nevertheless, the operation of DME combustion systems needs the adoption of rigorous procedures for safe operation due to its wide flammability limits (3.4–17% in air). Because DME has a very mild odor, an odorant should also be added for safety.

Most ethers, such as ethyl ether and tetrahydrofuran (THF), form peroxides upon exposure to air, light and contaminants. These peroxides, when exposed to friction, impact or heating, may cause explosions. Studies have found that compared to ethyl ether and diisopropyl ether, DME formed no peroxides. Extremely harsh conditions with simultaneous exposure to UV light, presence of contaminants and exposure to air are needed to produce any demonstrable amount of peroxide. The risk of peroxide formation from DME under normal conditions is extremely small. In addition, many references have stated that no cases of peroxide formation from storage and use of DME have ever been reported. In any event, a small amount of free radical inhibitor could be added to DME to avoid even the remotest possibility of peroxide formation [339].

11.17
Emissions from Methanol- and DME-Powered Vehicles

Transportation-associated air pollution is a major problem in large metropolitan areas. CO, NO_x, volatile organic compounds (VOCs), SO_2 and particulate matter (PM) emitted by automobiles, trucks and buses can have serious effects on the population's health, especially in children, the elderly and other sensitive persons. As discussed earlier, the use of clean-burning methanol in ICEs would greatly help to reduce these emissions [342]. One of the VOCs of concern is formaldehyde, an air-toxic and ozone precursor (it is present naturally in low concentrations in the atmosphere) that is produced in small quantities by the incomplete combustion of not only gasoline and diesel but also methanol, and is classified as a possible carcinogen. The issue of formaldehyde formation in methanol-powered ICEs has been successfully addressed by the development and use of effective catalytic mufflers that remove the relatively reactive formaldehyde by catalytic oxidation. One should bear in mind that, although methanol is an inherently cleaner fuel, gasoline- and diesel-fueled ICE vehicles themselves have made – and will continue to make through improved technology – considerable progress in emission control, and thus effectively compete with alternative fuels. In ICE cars, however, due to the sophistication of the systems needed to keep emissions low, a lack of proper maintenance and regular inspection can easily lead to higher emissions as the vehicles age.

In the long term the use of methanol-powered FCVs will almost eliminate all current air pollutants from vehicles. Furthermore, unlike ICE vehicles, the emission profile of methanol-powered vehicles will remain almost unchanged as they age. Emissions of methanol FCVs equipped with an on-board methanol reformer are expected to be even much lower than the already stringent limits set by the State of California for super ultra low emission vehicles (SULEV). Direct methanol FCVs are expected to be virtually zero emission vehicles (ZEVs) [247]. Tests conducted with Georgetown University's reformed methanol fuel cell bus have shown that it is almost a ZEV, releasing only negligible amounts of carbon monoxide and hydrocarbons, and no NO_x or PM (Figure 11.19) [343].

Diesel vehicles, especially trucks, are responsible for almost all the particulate matter emission and most of the NO_x emissions generated by the transportation sector. They have therefore a significant impact on air quality. The emission of PM,

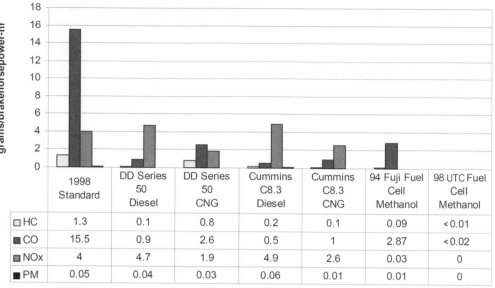

	1998 Standard	DD Series 50 Diesel	DD Series 50 CNG	Cummins C8.3 Diesel	Cummins C8.3 CNG	94 Fuji Fuel Cell Methanol	98 UTC Fuel Cell Methanol
☐ HC	1.3	0.1	0.8	0.2	0.1	0.09	< 0.01
■ CO	15.5	0.9	2.6	0.5	1	2.87	< 0.02
▨ NOx	4	4.7	1.9	4.9	2.6	0.03	0
▨ PM	0.05	0.04	0.03	0.06	0.01	0.01	0

Figure 11.19 Pollutant emission from methanol-powered fuel cell buses. (Source: Georgetown University.)

NO_x and other pollutants has serious health effects and has been linked to problems such as difficulty in breathing, asthma, increased occurrence of heart attacks and cancers. The use of DME as a fuel for cars and trucks could dramatically reduce emissions compared to diesel fuel, even with the existing CI engine technologies. Containing no sulfur, DME generates no SO_2 emissions. Owing to the absence of C–C bonds it also emits virtually no particulate matter (PM). At the same time, emission of NO_x is also reduced. It was already shown early on in the development of DME as an alternative fuel, at the beginning of the 2000s, that the emission of PM could be reduced by about 75% and the emission of acrolein, propionaldehyde, ethylene and propylene could be reduced by 99% or more compared to diesel fuel. Similarly, acetone and acetaldehyde emissions were reduced by about 80% [344]. Some formaldehyde was also formed, which can, however, be easily lowered to negligible levels with an oxidation catalyst. Over the years the emission characteristics of DME-fueled CI engines have been continuously improved. DME trucks developed and tested in Japan are satisfying the most stringent emission regulations of Japan, the EU and the US (Figures 11.20–11.23) [345, 346]. The tested trucks emit only 0.11 g NO_x and 0.001 g of particulate matter per kWh. The emission of NMHC (non-methane hydrocarbons) and CO is also extremely low. The PM emission is mainly caused by minute amounts of lubricating oil used in the motor. Owing to the intrinsic clean burning characteristics of DME, the exhaust treatment system will be simpler than in existing diesel vehicles with emphasis put on further reduction of NO_x, CO and hydrocarbon emissions. In fuel cell vehicles the use of DME as a fuel should bring benefits comparable to methanol fuel: virtually no emission apart from CO_2 and water.

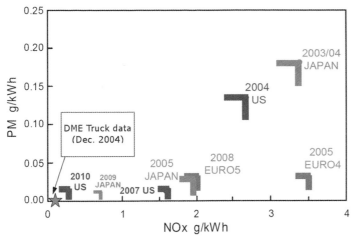

Figure 11.20 Emission regulation for diesel engines in US, Japan and Europe compared to emission from DME truck.

11.18
Environmental Effects of Methanol and DME

Methanol is emitted into the atmosphere from several natural sources, including volcanoes, vegetation, microbes, insects, animals and decomposing organic matter [335]. The anthropogenic (human-caused) release of methanol into the environment

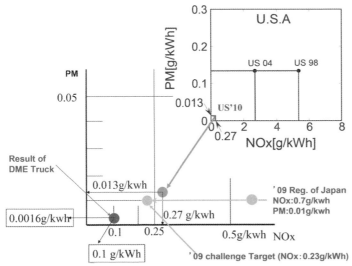

Figure 11.21 DME truck NO_x and particulate matter (PM) emission compared to emission regulations for diesel engines in the US and Japan.

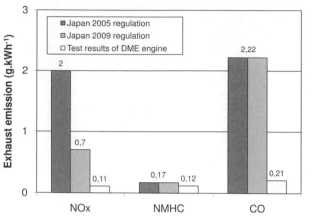
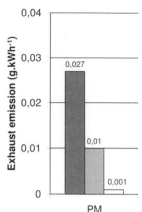

Figure 11.22 NO_x, non-methane hydrocarbons (NMHC), CO and PM emissions generated by DME engines compared to emission regulations in Japan.

is presently mainly due to its use as a solvent through evaporation. Releases to the water and ground are less significant. In the environment, methanol is readily degraded by photooxidation and biodegradation processes. This degradation, along with other factors, is why methanol is widely used, for example, in windshield washer fluids. Methanol is rapidly degraded under both aerobic (in the presence of air) and anaerobic (in the absence of air) conditions in fresh and salt water, groundwater, sediments and soils with no evidence of bioaccumulation. It is a regular growth substrate (source of carbon and energy) for many microorganisms, which are found in the ground and are able to degrade methanol to CO_2 and water. Methanol can even promote plant growth serving as a carbon source instead of CO_2 [232]. Methanol is of low toxicity to aquatic and terrestrial organisms, and effects due to environmental exposure to methanol are unlikely to be of consequence under normal conditions [347]. In fact, methanol is used for the denitrification of wastewater in sewage treatment plants. The addition of methanol in this process accelerates the conversion of nitrates into harmless nitrogen gas by anaerobic bacteria. If discharged in large quantities, nitrates can accumulate in rivers, lakes and oceans and have devastating effects on the water ecosystem. Excess nitrogen from nitrates causes an overgrowth of algae, and this prevents oxygen and sunlight from penetrating deeply

Figure 11.23 DME-powered truck in Japan.

into the water; the result is often suffocation of the fish and all aquatic life below. Currently, more than 100 wastewater treatment plants across the United States use methanol in their denitrification process. The Blue Plains Wastewater Treatment Facility, which serves the Washington DC area, is one of the largest such treatment plants in the country, and by using methanol denitrification avoids the release each day of some 10 tonnes of nitrogen into the Potomac River [348].

The accidental release of methanol into the environment during its production, transportation and storage, although possible, would cause much less damage than a corresponding crude oil or gasoline spill, due to the above-mentioned favorable chemical and physical properties of methanol. A large accidental methanol spill into surface water would have some immediate impact on the ecosystem in close proximity to the spill. However, since the methanol is totally miscible with water, it would rapidly be diluted and dissipated into the environment by wave action, winds or tides to non-toxic levels. It has been calculated that the release of 10 000 tonnes of methanol into the open sea would result in a concentration of only 0.36% within the first hour of the spill, and this would be much less during the following hours [349]. At this point, biodegradation through the action of microorganisms would take care of the dilute methanol in a matter of days. A similar accident involving petroleum oil, which is not miscible with water, covers large surfaces and does not easily dissipate into the environment, would most likely lead to an ecological disaster. Methanol, in leaving no residues, also avoids the long and fastidious cleaning of beaches, shorelines, birds and wild life needed after a crude oil spill. On land, the accidental release from a tank, truck or rail car transporting methanol, or from an underground storage tank, are possible scenarios. Depending on the size and location of the spill, different situations might be encountered, but in most cases the miscibility of methanol with water should allow it to be diluted and the effects to be dispersed rapidly, also allowing rapid biodegradation by microorganisms present in the ground. Notably, the behavior of methanol in the environment is very different from that of MTBE, which is not easily degraded, thus contributing to its ban as an oxygenated additive for transportation fuels [247]. Methanol leaks are also less hazardous than gasoline leaks because the latter contains many toxic and carcinogenic compounds (e.g., benzene) that biodegrade slowly and persist for a longer time in the environment. To further minimize the risk of leaks in underground storage, the use of double-walled storage tanks and leak detectors are preferable for all fuels. Compared to gasoline or diesel fuel, methanol is clearly environmentally much safer and less toxic.

Dimethyl ether is a volatile organic compound, but is non-toxic and environmentally benign. DME is soluble in water, with a solubility of 5.7% by weight at 20 °C. Possible contamination of groundwater from accidental leakage from underground storage must therefore be monitored. Like methanol, however, DME can be degraded by microorganisms. Accidental release of DME into surface water bodies is also of relatively low concern. Owing to its low boiling point, DME would simply volatilize. It has been determined that the volatilization half-life of DME (the time it takes for half of the DME to be volatilized) is about 3 hours from a river and 30 hours from a pond [339].

Table 11.5 Global warming potentials of DME and other compounds.

Gas	Time horizon (years)		
	20	100	500
DME	1.2	0.3	0.1
CO_2	1	1	1
Methane	62	23	7
N_2O	275	296	156

Based on data from the IPCC and Reference [350].

In the atmosphere, DME degrades very quickly. In the troposphere, the lowest part of our atmosphere, the lifetime of DME is only 5.1 days [350]. Little dimethyl ether would therefore be transported into the stratosphere and the impact on the stratosphere's ozone layer would be very limited if reactions from its degradation products could affect ozone. When released into the atmosphere, DME has a lower ozone-forming potential than LPG and the hydrocarbons present in gasoline. The global warming potential of DME was determined to be 1.2 at a 20 year time horizon, 0.3 at a 100 year time horizon and 0.1 on a 500 year time horizon (Table 11.5). Based on all these facts, dimethyl ether's effect on the environment therefore appears to be benign.

11.19
Beneficial Effect of Chemical CO_2 Recycling to Methanol on Climate Change

Today, methanol is still manufactured almost exclusively from syn-gas produced by catalytic reforming of natural gas or coal (i.e., from fossil fuel sources). In contrast to natural gas, our coal reserves are extensive. However, because of coal's deficiency in hydrogen, its conversion into methanol produces the most CO_2 compared to all other fossil fuels. Currently, on a large scale, the cleanest, most efficient and most economical way to produce methanol is from natural gas-generated syn-gas. In the area of ongoing research and development, however, much progress has been made in the direct oxidative conversion of methane into methanol without passing first through the generation of syn-gas. It should also be recognized that (as discussed in Chapter 4) there are large resources of coalbed methane as well as methane hydrates tied up in vast areas of the subarctic tundra and under the seas in the areas of the continental shelves. All of these resources will eventually be utilized, but they will only extend the inevitable exhaustion of our natural hydrocarbon resources.

Using methanol produced from natural gas in traditional ICE vehicles and FCVs will only moderately reduce the CO_2 emissions compared to their gasoline and diesel equivalents. To further reduce CO_2 emissions and to provide a sustainable alternative source of energy for the long term, methanol will have to be produced in a renewable or regenerative way from feedstocks other than fossil fuels. Production from biomass is one possibility, but this could provide for only a minor portion of our growing energy needs. Methanol, however, can also be obtained from the chemical recycling of CO$_2$ by catalytic reduction with hydrogen or by electrochemical reduction in water (Chapter 12). The flue gases of coal- and other fossil fuel-burning power plants, as well as from facilities such as cement and steel factories, contain high concentrations of CO$_2$, and these will increasingly be captured and disposed of. Instead of sequestration, the chemical recycling of CO$_2$ into methanol and DME, besides providing useful fuels and a source for synthetic hydrocarbons, will also mitigate human-caused climate changes. Eventually, the CO$_2$ content of the atmosphere itself will be similarly recycled, freeing humankind from its dependence on fossil fuels, with the required energy produced from non-fossil fuel sources (renewable energy sources and safe atomic energy) for the production of hydrogen and conversion into methanol.

The chemical recycling of CO$_2$ to methanol will make carbon fuels environmentally CO$_2$ neutral and renewable on the human timescale. We believe this recycling represents a new and feasible solution to one of the major challenges of our time.

12
Production of Methanol: From Fossil Fuels and Bio-Sources to Chemical Carbon Dioxide Recycling

In 2004, the worldwide production of methanol was 32 million tonnes, while it stood at 40 million tonnes in 2007 and is continuing to increase steadily. Although virtually any hydrocarbon source (coal, petroleum oil, naphtha, coke, etc.) can be converted into methanol via derived syn-gas, natural gas accounted for a majority of the feedstock used for methanol manufacture. Most existing plants have production capacities ranging from 100 000 to 800 000 t year^{-1}. In the past, plants were generally constructed close to large methanol-consuming centers in the United States and Europe. However, with decreasing local reserves and increasing prices of natural gas in these areas, production has shifted to natural-gas producing countries that still have large reserves but lie far from centers of major consumption, such as Chile, Trinidad and Tobago, Qatar, and Saudi Arabia (Figure 12.1). In the United States, for example, the annual methanol production capacity peaked at some 7.4 million tonnes in 1998, supplying 70% of the domestic requirements. In 2006, after the progressive phase-out of methyl *tert*-butyl ether (MTBE, a reformulated gasoline additive), together with the closure of aging plants and high natural gas prices, the annual production capacity fell to less than 1 million tonnes, covering merely 10–15% of the United States domestic demand [351, 352]. Plans were announced to close down altogether methanol production in North America. In China, in contrast, production of methanol, mainly from coal, has rapidly increased in the last decade – from two million tonnes per year in 2000 to more than 11 million tonnes in 2008. Not only the geographical location, but also the size of new methanol plants has changed. Facilities with capacities of 1 million tonnes per year or more – termed "mega-methanol plants" – are becoming the norm. Two such mega-methanol plants, each able to produce about 1 million tonnes of methanol per year, are currently operating in Qatar and Saudi-Arabia [353]. A plant with a 1.7 million tonne per year capacity was recently commissioned in Saudi Arabia and two plants with similar production capacities have been operating since 2005 in Trinidad and Tobago [352, 354]. Further plants are either planned or already under construction in China, Qatar, Saudi-Arabia, Iran, Chile, Malaysia and other countries. Designs for plants with capacities up to 3.5 million tonnes per year are now offered by major methanol technology developers (Johnson Matthey, Lurgi, Mitsubishi Gas Chemical and Haldor-Topsoe).

Beyond Oil and Gas: The Methanol Economy, Second updated and enlarged edition
George A. Olah, Alain Goeppert, and G. K. Surya Prakash
Copyright © 2009 WILEY-VCH Verlag GmbH & Co. KGaA, Weinheim
ISBN: 978-3-527-32422-4

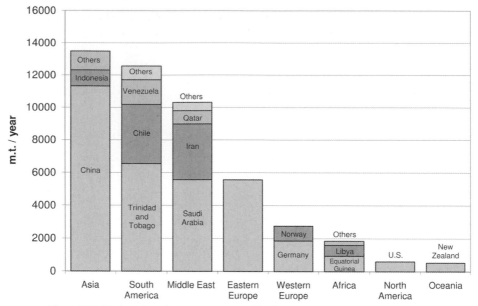

Figure 12.1 Methanol production capacity worldwide in 2006.
(Source: Adapted from data in *Chemical Week*, May 23, 2007.)

The economy of scale brought by the construction of ever larger and more efficient syn-gas-based plants is expected to considerably lower the cost of methanol production, and is starting to open up the market for methanol as an alternative fuel and petrochemical feedstock (Figure 12.2). New methods for the direct oxidative

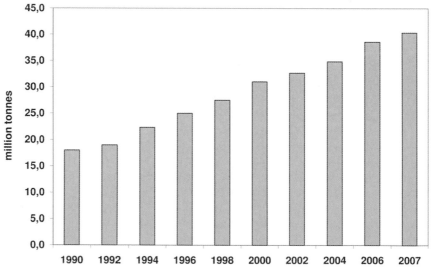

Figure 12.2 Recent global demand for methanol.
(Source: Methanex and Methanol Institute.)

conversion of existing natural gas (methane) sources into methanol without going through syn-gas, as well as the hydrogenative recycling of CO_2 (*vide infra*) are essential new technologies for the proposed "Methanol Economy."

There are today more than 600 million personal automobiles and some 200 million light and heavy trucks registered worldwide. Together, they represent 80% of the energy consumed in the transportation field. Air, maritime, rail and pipeline transport account for the remaining 20%. In 2006, the transportation sector alone consumed some 2.1 billion tonnes of oil [42], yet with increasing demand – especially from developing countries – this figure is expected to grow substantially to reach more than 3.2 billion tonnes in 2030, with 1.3 billion vehicles on the road [355]. Evidently, given these massive numbers, any alternatives to oil-based fuels will have to be economically produced on a very large scale. Methanol, as discussed earlier, has for the same volume only about half the energy content of gasoline or diesel fuel. Replacing only 10% of our current energy needs for transportation would require the production of some 420 million tonnes of methanol per year (more than ten times today's world production). However, as methanol fuel cell vehicles (FCVs) are expected to be twice as energy efficient as present ICE vehicles, this would decrease the amount of fuel necessary by about half. With the widespread use of fuel-cell technology, the amount of methanol required for the transportation sector would thus become comparable to that of oil-based fuels. For methanol to become a major fuel of the future, a significant growth in production capacity and new technologies, to some degree already developed, are needed.

Although conventional natural gas reserves are currently the preferred feedstock for the production of syn-gas-based methanol, unconventional gas resources such as coalbed methane, tight sand gas and eventually vast methane hydrate resources will also be used, as will any other available fossil fuel source. Coal in particular, with its vast reserves, widespread availability and low cost is already experiencing a revival, especially in China. Coal, at present, is thus the best alternative to natural gas (methane)-based syn-gas. Large domestic coal supplies in major energy-consuming countries such as the United States and China offer price stability and enhanced independence from geopolitically unstable oil- and gas-producing regions. Coal, however, must be mined and has the drawback of releasing upon its combustion large amounts of pollutants in addition to carbon dioxide into the atmosphere. Considering effects on climate change, it also produces more CO_2 per unit of energy than natural gas and petroleum oil. To minimize CO_2 releases, biomass represents another possible source for methanol but, due to the enormous needs of our modern society, biomass will be able to cover only a fraction of the demand. To lessen our dependence on fossil fuels and also to address the CO_2 emission problem, new and improved methods of methanol production are needed. Besides directly converting methane (natural gas) into methanol without going through syn-gas, the reaction of hydrogen, generated from water with use of renewable or nuclear energy sources, and CO_2 from emissions from fossil fuel-powered electric plants, other industrial sources and eventually air itself will allow independence from diminishing fossil fuel resources.

12.1
Methanol from Fossil Fuels

12.1.1
Production via Syn-Gas

Today, methanol is almost exclusively produced from syn-gas, which is a mixture of hydrogen, carbon monoxide and some CO_2, over a heterogeneous catalyst according to Equations (12.1–12.3):

$$CO + 2H_2 \rightleftharpoons CH_3OH \qquad \Delta H_{298K} = -21.7 \, \text{kcal mol}^{-1} \qquad (12.1)$$

$$CO_2 + 3H_2 \rightleftharpoons CH_3OH + H_2O \qquad \Delta H_{298K} = -11.9 \, \text{kcal mol}^{-1} \qquad (12.2)$$

$$CO_2 + H_2 \rightleftharpoons CO + H_2O \qquad \Delta H_{298K} = 9.8 \, \text{kcal mol}^{-1} \qquad (12.3)$$

The first two reactions are exothermic, with heats of reaction equal to -21.7 and $-11.9 \, \text{kcal mol}^{-1}$, respectively. They both result in a decrease of volume as the reaction proceeds. According to Le Chatelier's principle, the conversion into methanol is therefore favored by increasing pressure and decreasing temperature. Equation 12.3 describes the endothermic reverse water-gas shift reaction (RWGSR), which also occurs during methanol synthesis, producing carbon monoxide that can then further react with hydrogen to produce methanol. In fact, (12.2) is simply the sum of the reactions in (12.1) and (12.3). Each of these reactions is reversible, and thus limited by thermodynamic equilibrium depending on the reaction conditions, including temperature, pressure and composition of the syn-gas.

Synthesis gas for methanol production can be obtained by reforming or partial oxidation of any available carbonaceous material such as coal, coke, natural gas, petroleum, heavy oils and asphalt. Economic considerations dictate the choice of the raw material. However, the long-term availability of raw materials, energy consumption and environmental aspects also play important roles.

The composition of syn-gas is generally characterized by the stoichiometric number S. Ideally, S should be equal to or slightly above 2. Values above 2 indicate an excess of hydrogen, whereas values below 2 mean a hydrogen deficiency relative to the ideal stoichiometry for methanol formation:

$$S = \frac{(\text{moles } H_2 - \text{moles } CO_2)}{(\text{moles } CO + \text{moles } CO_2)}$$

Synthesis gas from coal has less than the optimum hydrogen to carbon content. Treatment of the gas before methanol synthesis or addition of hydrogen is therefore needed to avoid the formation of undesired by-products. Reforming of feeds with a higher H/C ratio, such as propane, butane or naphthas, leads to S values in the vicinity of 2, which is ideal for conversion into methanol. Steam reforming of methane, in contrast, yields syn-gas with a stoichiometric number of 2.8–3.0. In this case, the addition of CO_2 can lower S close to 2. Excess hydrogen can also be used in an adjacent plant to co-produce ammonia.

The production of methanol from syn-gas on an industrial scale using high pressures (250–350 atm) and temperatures (300–400 °C) was first introduced by BASF in Germany during the 1920s. From then until the end of World War II, most methanol was produced from coal-derived syn-gas and off-gases from industrial facilities such as coke ovens and steel factories. The use of such feedstocks, containing high levels of impurities, was made possible by the design of a catalyst system consisting of zinc oxide and chromium oxide, which is highly stable to sulfur and chlorine compounds. After World War II, the feedstock for methanol synthesis shifted rapidly to natural gas, which became widely available at low cost, particularly in the United States. Whereas 71% of the methanol in the United States was still derived from coal in 1946, by 1948 almost 77% was obtained from natural gas. Natural gas became the preferred feedstock for methanol production because it offers, in addition to a high hydrogen content, the lowest energy consumption, capital investment and operating costs. Furthermore, natural gas contains fewer impurities, such as sulfur and halogenated compounds, which can poison the needed catalysts. When present, these impurities (mostly sulfur in the form of H_2S, COS or mercaptans) can, however, be removed relatively easily. Lower levels of impurities in the syn-gas allowed the use of more active catalysts, operating under milder conditions. This led, during the 1960s, to the development by the former ICI Synetix (now Johnson Matthey) in England of a process using a copper-zinc-based catalyst, allowing the conversion of syn-gas into methanol at pressures of 50–100 atm and temperatures of 200–300 °C. This low-pressure route is the basis for most current processes for methanol production. The formation of by-products (dimethyl ether, higher alcohols, methane, etc.) associated with the old high-pressure technology was also drastically reduced or even eliminated [233]. Production using the high-pressure process is no longer economical, and the last plant based on this technology closed in the 1980s. Worldwide, the production capacity for methanol is dominated by few processes, with Johnson Matthey (formerly ICI Synetix) accounting for some 60% of the installed capacity and Lurgi for 27% (Figure 12.3).

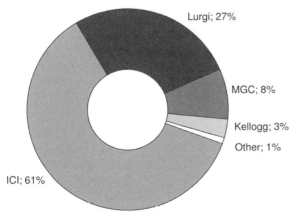

Figure 12.3 Worldwide methanol production capacity by process. (Source: ICI Synetix (now Johnson Matthey).)

All present processes use copper-based catalysts that are extremely active and selective, and are almost exclusively used in gas-phase processes. They differ only in the type of reactor design and catalyst arrangement (fixed bed, tube, suspension, etc.). Control of the temperature is important, and overheating of the catalyst should be avoided, as it will rapidly decrease its activity and shorten its lifetime [232]. In a single pass over the catalyst, only a part of the syn-gas is converted into methanol. After separation of methanol and water by condensation at the reactor's outlet, the remaining syn-gas is recycled to the reactor. Quite recently, liquid-phase processes for methanol production have also been introduced. Air Products, in particular, has developed the liquid-phase methanol process LPMEOH, in which a powdered catalyst is suspended in an inert oil, providing an efficient means to remove the heat of reaction and control the temperature (Figure 12.4). The syn-gas is simply bubbled into the liquid. This process promotes a higher syn-gas to methanol conversion, so that a single pass through the reactor is generally sufficient [356].

Generally, modern methanol plants have a selectivity for methanol higher than 99%, with energy efficiencies above 70% (Figure 12.5). Crude methanol leaving the reactor contains, however, water and small amounts of other impurities, the composition of which will depend on the gas feed, reaction conditions and type and lifetime of the catalyst. Impurities that may be present in methanol include dissolved gases (methane, CO, CO_2), dimethyl ether, methyl formate, acetone, higher alcohols (ethanol, propanol and butanol) and long-chain hydrocarbons.

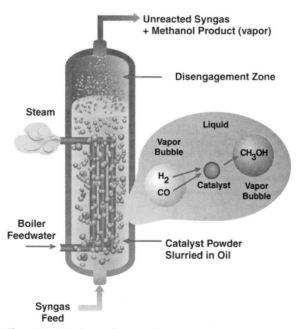

Figure 12.4 Synthesis of methanol in the LPMEOH™ slurry reactor developed by Air Products. (Source: US Department of Energy.)

Figure 12.5 Atlas mega-methanol plant in Trinidad. (Courtesy of Methanex.)

Commercially, methanol is available in three grades of purity: fuel grade, "A" grade (generally used as a solvent) and "AA" or chemical grade. The latter has the highest purity with a methanol content exceeding 99.85% and is the standard generally observed by the industry for methanol production. Depending on the amount of impurities and purity desired, methanol is purified by distillation using one or more distillation columns.

Aside from the methanol synthesis step, the most crucial part of present methanol plants is the syn-gas generation and purification system, which depend on the nature and purity of the feedstock used. Although natural gas is generally the preferred feedstock due to the simplicity of obtaining an adequate syn-gas with low levels of impurities, other routes are also used under given circumstances. Regions rich in coal or heavy oils, but having limited natural gas sources, in view of the high natural gas prices can turn to these resources to produce methanol despite higher cost for the needed syn-gas purification system.

12.1.2
Syn-Gas from Natural Gas

12.1.2.1 Steam Reforming of Methane

In methane steam reforming, methane is reacted in a highly endothermic reaction with steam over a catalyst, typically based on nickel, at high temperatures and under pressure (800–1000 °C, 20–30 atm) [357] to form CO and H_2. A part of the CO formed reacts consequently with steam in the water gas shift (WGS) reaction to yield more H_2 and also CO_2. The gas obtained is thus a mixture of H_2, CO and CO_2:

$$CH_4 + H_2O \rightleftharpoons CO + 3H_2 \qquad \Delta H_{298K} = 49.1 \, kcal \, mol^{-1}$$

$$CO + H_2O \rightleftharpoons CO_2 + H_2 \qquad \Delta H_{298K} = -9.8 \, kcal \, mol^{-1}$$

The ratio of the components depends on the reaction conditions: temperature, pressure and H_2O/CH_4 ratio. The efficiency of syn-gas generation from methane, however, increases with increasing temperature and decreasing pressure. With increasing temperatures, the WGS reaction becomes also less dominant and the main products are CO and H_2 [357]. Since the overall methane steam reforming process is highly endothermic, heat must be supplied to the system, generally by burning a part of the natural gas used as feedstock. Steam reforming is the most widely used technology to produce not only syn-gas for methanol synthesis but also hydrogen for the synthesis of ammonia and other products. The stoichiometric number S obtained by steam reforming of methane is close to 3, which is far above the desired value of 2. This can generally be corrected by addition of CO_2 to the steam reformer's exit gas or use of excess hydrogen in above-mentioned other process such as ammonia synthesis.

Methane reforming can also be affected by the thermal "coking" process. It involves the formation of carbon, which may deposit as soot or coke on the catalyst (greatly reducing its activity) as well as all the internal parts of the reformer and downstream equipment, resulting in possible clogging. Carbon may be formed by CH_4 decomposition or CO disproportionation (Boudouard reaction):

$$CH_4 \rightleftharpoons C + 2H_2 \qquad \Delta H_{298K} = 18.1 \, \text{kcal mol}^{-1}$$

$$2CO \rightleftharpoons C + CO_2 \qquad \Delta H_{298K} = -40.8 \, \text{kcal mol}^{-1}$$

In practice, the undesired carbonation, or soot formation, is largely prevented by the use of excess steam and short residence times in the reactor. It can, however, be more problematic for the partial oxidation process operating at higher temperature than methane reforming.

12.1.2.2 Partial Oxidation of Methane

Partial oxidation is the reaction of methane with insufficient oxygen, which can be performed with or without a catalyst [358]. This reaction, which is exothermic and operated at higher temperatures (800–1500 °C) [358], generally yields syn-gas with a H_2/CO ratio of 2, which is ideal for methanol synthesis. The problem with partial oxidation is that the products, CO and H_2, can be further oxidized to form undesired CO_2 and water in highly exothermic reactions, raising safety concerns and leading to S values typically below 2:

$$CH_4 + \frac{1}{2}O_2 \rightleftharpoons CO + 2H_2 \qquad \Delta H_{298K} = -8.6 \, \text{kcal mol}^{-1}$$

$$CO + \frac{1}{2}O_2 \rightleftharpoons CO_2 \qquad \Delta H_{298K} = -67.6 \, \text{kcal mol}^{-1}$$

$$H_2 + \frac{1}{2}O_2 \rightleftharpoons H_2O \qquad \Delta H_{298K} = -57.7 \, \text{kcal mol}^{-1}$$

The production of excess energy in the form of heat is not desirable and wasteful if no immediate use for it can be found for other processes.

12.1.2.3 Autothermal Reforming and Combination of Steam Reforming with Partial Oxidation

To produce syn-gas without either consuming or producing much excess heat, modern plants usually combine exothermic partial oxidation with endothermic steam reforming to have an overall thermodynamically neutral reaction while obtaining syn-gas with a composition suited for methanol synthesis (*S* close to 2). This process is called "autothermal reforming." Partial oxidation and steam reforming can be conducted simultaneously in the same reactor by reacting methane with a mixture of steam and oxygen. Having only one reactor lowers the cost and complexity of the system. However, because the two reactions are optimized for different temperature and pressure conditions, they are generally conducted in two separate steps. After the steam reforming step, the effluent from the reformer's outlet is fed to the partial oxidation reactor where all the residual methane is consumed [359]. The oxygen required for the oxidation step means that an air separation plant is needed, but to avoid the construction of such a unit the use of air rather than oxygen is also possible. The produced syn-gas in this case will, however, contain a large amount of nitrogen and require special processing before conversion into methanol. Thus, most modern methanol plants use pure oxygen.

12.1.2.4 Syn-Gas from CO_2 Reforming of Methane

Syn-gas can also be produced by the reaction of CO_2 with methane or natural gas, generally termed CO_2 or "dry" reforming, because it does not involve any steam. With a reaction enthalpy of $\Delta H = 59.1\,\text{kcal mol}^{-1}$ this reaction is more endothermic than steam reforming ($49.1\,\text{kcal mol}^{-1}$) [360]:

$$CO_2 + CH_4 \rightleftharpoons 2CO + 2H_2 \qquad \Delta H_{298K} = 59.1\,\text{kcal mol}^{-1}$$

The reaction is carried out commercially at temperatures around 800–1000 °C using generally catalysts based on nickel (Ni/MgO, Ni/MgAl$_2$O$_4$). The syn-gas produced has a H_2/CO ratio of 1, which is much lower than the values of around 3 obtained with steam reforming. While this lower ratio is a disadvantage for methanol synthesis, it makes a suitable feed gas for other processes, especially iron ore reduction and Fischer–Tropsch hydrocarbon synthesis. Several industrial processes, particularly iron production, have taken advantage of the high CO and low water contents of synthesis gas produced by CO_2 reforming. For methanol production, however, hydrogen generated from other sources would have to be added to the obtained syn-gas. Combination of steam and CO_2 reforming, which can also be performed using the same catalysts, to reach adequate syn-gas composition is also possible (bi-reforming, *vide infra*).

12.1.3
Syn-Gas from Petroleum Oil and Higher Hydrocarbons

Natural gas is not the only hydrocarbon source used for syn-gas generation for methanol production. Although on a smaller scale, liquefied petroleum gas and different fractions obtained from oil refining (especially naphtha) are also employed to produce syn-gas for the manufacture of ammonia, methanol and higher alcohols (propanol, butanol, etc.). Crude oil, heavy oil, tar and asphalt can all be transformed into syn-gas. The methods used are similar to those for natural gas: steam reforming and partial oxidation, or a combination of both:

$$C_nH_m + nH_2O \rightleftharpoons nCO + \left(n + \frac{m}{2}\right)H_2$$

$$C_nH_m + \frac{n}{2}O_2 \rightleftharpoons nCO + \frac{m}{2}H_2$$

The problem with higher carbon-rich feedstocks is generally their higher content in impurities, especially sulfur compounds, which can very rapidly poison the catalysts used for steam reforming and for subsequent methanol synthesis. Therefore, considerably more capital must be invested in the purification steps, and catalysts that are more resistant to poisoning may be employed. Heavy oils, tar sands and other hydrocarbon sources containing large and complex aromatic structures are also relatively low in hydrogen content, resulting in syn-gas rich in CO and CO_2 but deficient in hydrogen.

12.1.4
Syn-Gas from Coal

Coal was the original feedstock used in the industrial production of syn-gas. It is still widely used in China and South Africa for the production of methanol and ammonia, and was expected for years to become the preferred route for large-scale syn-gas generation in the United States, because of existing huge domestic coal reserves. Syn-gas is produced from coal by gasification, a process combining partial oxidation and steam treatment, according to the following reactions:

$$C + \frac{1}{2}O_2 \rightleftharpoons CO \qquad \Delta H_{298K} = -29.4\,\text{kcal mol}^{-1}$$

$$C + H_2O \rightleftharpoons CO + H_2 \qquad \Delta H_{298K} = 31.3\,\text{kcal mol}^{-1}$$

$$CO + H_2O \rightleftharpoons CO_2 + H_2 \qquad \Delta H_{298K} = -9.8\,\text{kcal mol}^{-1}$$

$$CO_2 + C \rightleftharpoons 2CO \qquad \Delta H_{298K} = 40.8\,\text{kcal mol}^{-1}$$

Different coal gasification processes have been developed and commercialized over the years. The selection of a particular design depends greatly on the characteristics of the coal used: lignite, sub-bituminous, hard coal and graphite have differing water contents, ash contents, levels of impurities and so on. Owing to the low H/C ratio of coal, the obtained syn-gas is rich in carbon oxides (CO and CO_2) and deficient in hydrogen. Before being sent to the methanol unit, the syn-gas must thus be subject to the WGS reaction to enhance the amount of hydrogen formed. Some of the CO_2 produced must also be separated and any H_2S removed to avoid poisoning of the very sensitive methanol synthesis catalyst.

12.1.5
Economics of Syn-Gas Generation

The investment in the syn-gas generation unit accounts generally for more than half of the total investment for natural gas-based plants producing methanol. For plants using coal as a feedstock, it represents even more, generally 70–80% [359], with the remaining balance accounting for the capital costs involved in the actual production of methanol. The price of methanol from coal is thus more dependent on the technology than the cost of the feedstock itself, which remains much cheaper and less fluctuating than natural gas, due to still large and easily accessible coal deposits to be found all around the globe. Despite the higher initial investment for the construction of a methanol plant, increasing natural gas prices created a new interest in the production of methanol from coal. A growing number of large plants in China, for example, are already using their still abundant and inexpensive coal sources to produce methanol. In the United States, only one plant – operated by the Eastman Corporation in Kingsport, Tennessee – presently produces methanol by coal gasification. The co-production of methanol and electricity in the same plant is also an attractive option.

12.2
Methanol through Methyl Formate

To reduce the pressure and temperature needed for the current methanol production process, and also to improve its thermodynamic efficiency, alternative routes to convert CO/H_2 mixtures into methanol under milder conditions have been developed. Among these, the most notable is the synthesis of methanol via methyl formate, first proposed in 1919 by Christiansen [361–363]:

$$CH_3OH + CO \rightarrow HCOOCH_3$$
$$\underline{HCOOCH_3 + 2H_2 \rightarrow 2CH_3OH}$$
$$CO + 2H_2 \rightarrow CH_3OH$$

This methanol synthesis route consists of two steps. Methanol is first carbonylated to methyl formate, which is subsequently reacted with hydrogen to produce two moles of methanol. The carbonylation reaction is carried out in the liquid phase using sodium or potassium methoxide ($NaOCH_3$ or $KOCH_3$) as a homogeneous catalyst. It is a proven and commercially available technology used in the production of formic acid from CO through methyl formate. High activities for methanol carbonylation have also been recently shown with Amberlyst and Amberlite resins used as heterogeneous catalysts. The subsequent reaction of methyl formate with hydrogen (hydrogenolysis) to produce methanol can be conducted either in the liquid or gas phase using a copper-based catalyst (copper chromite, copper supported on silica, alumina, magnesium oxide, etc.). Carbonylation and hydrogenolysis can be carried out in two separate reactors, but are preferably combined in a single reactor. To run the carbonylation and hydrogenolysis simultaneously in a single reactor, different combinations of catalysts have been investigated, in particular CH_3ONa/Cu and CH_3ONa/Ni. Nickel-based systems are very active and selective, but due to the volatility and high toxicity of the $Ni(CO)_4$ that may be formed during the reaction this process is considered difficult and hazardous to use in industrial settings. Copper systems are thus preferred because they offer similar activities and selectivities, without the toxicity problems associated with the nickel systems.

Overall, by the methyl formate route, methanol is produced from syn-gas, but at lower temperature and pressure than that employed in conventional methanol production processes. Patents from Mitsui Petrochemicals, Brookhaven National Laboratories and Shell, using the methyl formate route, claim to produce methanol at 80–120 °C and pressures of 10–50 atm. Another report states that continuous operation in a bubble reactor at 110 °C and a pressure of only 5 atm is also possible.

The problem with this process is the presence of CO_2 and water in the syn-gas, which will react with sodium methoxide, deactivating the catalyst and forming undesirable by-products. To minimize this deactivation, CO_2 and water must therefore be removed from the gas feed. Catalysts more tolerant to them (such as the recently reported $KOCH_3$/copper chromite systems) are also being developed [363]. Although further improvements are still needed, the synthesis of methanol via methyl formate at modest temperature and pressure could lead to an attractive alternative to current processes operating at 200–300 °C and 50–100 atm. At the same time, methyl formate can also be produced from the hydrogenative conversion of CO_2 into formic acid and methanol as well as by the dimerization of formaldehyde; the methyl formate to methanol route can also play a role in the secondary treatment of the oxidative conversion of methane into methanol, without going through syn-gas (*vide infra*).

12.3
Methanol from Methane without Producing Syn-Gas

As long as methane, the main component of natural gas, is still quite abundant, it will inevitably be used to produce methanol, and through it synthetic hydrocarbons and their products. It should also be recognized that (as discussed in Chapter 4) there are other large methane resources such as unconventional natural gas sources as well as methane hydrates tied up in vast areas of subarctic tundra and under the seas in the areas of the continental shelves. All of these resources will eventually be utilized to produce methanol as our traditional natural gas resources are increasingly depleted. New ways to convert methane into methanol more efficiently, and without going through the currently used syn-gas-based processes, are much needed [364–367]. Extensive research has been conducted towards this goal, and much progress has been made in recent years, especially in the oxidative direct conversion of methane into methanol.

12.3.1
Direct Oxidation of Methane to Methanol

A major disadvantage of the present technology of producing methanol through syn-gas is the large energy requirement of the initial highly endothermic step of methane steam reforming. The process is also inefficient in the sense that it first transforms methane in an oxidative reaction into carbon monoxide (and some CO_2) which, in turn, must be reduced again to methanol with hydrogen. The direct selective oxidative transformation of methane into methanol is therefore a highly desirable goal, but this is difficult to accomplish in a practical way (with high conversion and selectivity). It would, however, eliminate the need for producing syn-gas, increase the amount of methanol obtained and save on capital costs in commercial plants.

The main problem associated with the direct oxidation of methane to methanol is the higher reactivity of the oxidation products themselves (methanol, formaldehyde and formic acid) compared to methane, giving eventually CO_2 and water – that is, the thermodynamically favored complete combustion of methane:

$$CH_4 \begin{cases} \xrightarrow{0.5\,O_2} CH_3OH & \Delta H = -30.4 \text{ kcal.mol}^{-1} \\ \xrightarrow{O_2} CH_2{=}O + H_2O & \Delta H = -66.0 \text{ kcal.mol}^{-1} \\ \xrightarrow{1.5\,O_2} CO + 2H_2O & \Delta H = -124.1 \text{ kcal.mol}^{-1} \\ \xrightarrow{2\,O_2} CO_2 + 2H_2O & \Delta H = -191.9 \text{ kcal.mol}^{-1} \end{cases}$$

Reaction conditions to achieve high conversion of methane into methanol without complete oxidation producing CO_2 were consequently explored. At present, however, no process has yet succeeded in achieving the combination of high yield, selectivity and catalyst stability that would make the direct oxidative conversion competitive with conventional syn-gas-based methanol production methods.

Several ways exist for the oxidation of methane. These include homogeneous gas-phase oxidation, heterogeneous catalytic oxidation and photochemical and electrophilic oxidations [8].

12.3.2
Catalytic Gas-Phase Oxidation of Methane

In homogeneous gas-phase oxidation, methane is generally reacted with oxygen at high pressures (30–200 atm) and high temperatures (200–500 °C). Optimum conditions for the selective oxidation to methanol have been extensively investigated. The selectivity to methanol increases with decreasing oxygen concentration in the system. The best result (75–80% selectivity in methanol formation at 8–10% conversion) was achieved under cold flame conditions (450 °C, 65 atm, less than 5% O_2 content), using a glass-lined reactor, which seemed to suppress secondary reactions. Most other studies obtained under their best reaction conditions (temperatures of 450–500 °C and pressures of 30–60 atm) selectivities of 30–40% methanol with 5–10% methane conversion [366]. At these pressures, gas-phase radical reactions prevail, limiting any potentially favorable effect of solid catalysts. Selectivity to methanol can only be moderately influenced by controlling factors affecting the radical reactions such as reactor design, shape and residence time of the reactants.

At about ambient pressure (1 atm), catalysts can play a crucial role in the partial oxidation of methane with O_2. A large number, primarily metal oxides and mixed oxides, have been studied. Metals were also tested, but they tended to favor complete oxidation. In most cases the reactions were performed at temperatures ranging from 600 to 800 °C; formaldehyde (HCHO) was obtained as the predominant (and often only) partial oxidation product. Silica itself exhibits a unique activity in the oxidation of methane to formaldehyde. Higher methane conversions were, however, obtained with silica-supported molybdenum (MoO_3) and vanadium (V_2O_5) oxide catalysts. The yield of formaldehyde produced nevertheless remained in the range of 1–5%. An iron molybdate catalyst, $Fe_2O_3(MoO_3)_{2.25}$, was found to be the most active catalyst for partial oxidation of methane, with a reported formaldehyde yield of 23% [366]. Reaction over a silica-supported PCl_3-$MoCl_5$-R_4Sn catalyst gave 16% yield [368]. A high selectivity of 90% to oxygenated compounds (CH_3OH and $HCHO$) was obtained at methane conversions of 20–25% in an excess of steam, using a silica-supported MoO_3 catalyst [369]. In most cases, however, such high yields were difficult to reproduce by other research groups, and yields of oxygenates generally did not exceed 2–5%. At the high temperatures used for partial oxidation of methane, methanol formed on the catalyst's surface is rapidly decomposed or oxidized to formaldehyde and/or carbon oxides, which explains its absence from the obtained product mixture [366].

As methanol is formed together with formaldehyde and formic acid in the oxidation of methane, it has recently been found by Olah and Prakash that this mixture can be further processed in a second treatment step without the need of prior separation, resulting in a substantial increase in methanol content and making the overall process more selective and practical for the production of methanol [370].

The initial oxidation of methane to a mixture of methanol, formaldehyde and formic acid can utilize any known catalytic oxidation procedure. The formation of CO_2, however, must be kept to a minimum. To minimize overoxidation, the overall amounts of oxygenated products are kept between 20 and 30%, with unreacted methane being recycled.

The secondary treatment of formaldehyde and formic acid produced in the oxidation of methane, without any separation, can be carried out in different ways. One is by dimerizing formaldehyde over TiO_2 or ZrO_2 [366] to give methyl formate. Formaldehyde can also undergo conversion over solid-base catalysts such as CaO and MgO, a variation of the Cannizzaro reaction, giving methanol and formic acid, which again readily react with each other to form methyl formate:

$$2HCHO \xrightarrow{\text{TiO}_2 \text{ or ZrO}_2 \text{ catalyst}} HCOOCH_3$$

$$2HCHO \xrightarrow{\text{H}_2\text{O}} CH_3OH + HCO_2H$$

$$CH_3OH + HCO_2H \xrightarrow{-\text{H}_2\text{O}} HCOOCH_3$$

The methyl formate obtained can then be catalytically hydrogenated or electro-chemically reduced in water using suitable electrodes (made from copper, tin, lead, etc.), giving two molecules of methanol with no other by-product:

$$HCOOCH_3 + 2H_2 \rightarrow 2CH_3OH$$

Formic acid formed during the oxidation can itself serve as a hydrogen source to be reacted with formaldehyde in aqueous solutions at 250 °C or over suitable catalysts to produce methanol and CO_2:

$$HCHO + HCO_2H \rightarrow CH_3OH + CO_2$$

Suitable combinations of these reactions allow the oxidative conversion of methane into methanol with overall high selectivity and yield. Any CO_2 formed as by-product can be recycled to methanol (*vide infra*). As methanol is readily dehydrated to dimethyl ether, the oxidative conversion of methane into methanol is also well suited to produce dimethyl ether for fuel or chemical applications.

Avoiding any additional steps, and directly producing methanol from methane, remains a most desirable goal. Catalysts able to activate methane at lower temperatures are much sought after for the direct and selective synthesis of methanol from methane and air (oxygen).

An additional new development for the oxidation of methane is the use of O_2–H_2 gas mixtures. With $FePO_4$ as a catalyst, methanol is formed as the main product in the presence of O_2–H_2 at temperatures below 400 °C [366]. Nitrous oxide (N_2O) was also reported to be effective in producing both methanol and formaldehyde in the presence of silica-supported MoO_3 and V_2O_5 catalysts. Varied iron-containing catalysts exhibited interesting behavior in the partial oxidation of methane. Using the $FePO_4$ catalyst, partial oxidation of methane with N_2O gave methanol with a remarkable selectivity close to 100% in the presence of H_2 at 300 °C. The yield is, however, low, with only 3% methanol obtained at 450 °C. The O^- ion formed by the decomposition of N_2O at the catalyst's surface was postulated to be the active species, initiating the activation process of methane. Even if N_2O turns out to be a suitable oxidizing agent, its application on an industrial scale for methanol production is unlikely due to the costs of its production and its high activity as a greenhouse gas. Oxygen from the air is, and will remain, the most affordable, ubiquitous – and therefore preferred – oxidizing agent.

12.3.3
Liquid-Phase Oxidation of Methane to Methanol

To minimize the formation of side products and increase the selectivity to methanol, the use of lower reaction temperatures (<250 °C) is preferable. Lower temperatures and moderate pressures would also considerably decrease the cost of methanol plants. In current plants the higher temperatures (900–1000 °C) needed for the generation of syn-gas require the use of costly specialty materials for the reactor construction. Presently used catalysts for methane oxidation are, however, not sufficiently active at lower temperatures. The development of new generations of catalysts that could produce methanol directly from methane in sufficiently high yields and selectivity at lower temperatures is therefore highly desirable. As discussed, gas-phase reactions have met with limited success and require generally higher temperatures (>400 °C). This has prompted the investigation of liquid-phase oxidation reactions operating at more moderate temperatures.

During the 1970s it was first observed by Olah and coworkers that methanol could be obtained through electrophilic oxygenation (i.e., oxy-functionalization) of methane in superacids (acids many millions of times stronger than concentrated sulfuric acid) at room temperature. With hydrogen peroxide (H_2O_2), methane produces methanol in high yield and selectivity. From the results obtained, it was concluded that the reaction proceeds through the insertion of protonated hydrogen peroxide, $H_3O_2^+$, into the methane C–H bond. The formed methanol in the strong acid system is in its protonated form, $CH_3OH_2^+$, and is thus protected from further oxidation; this accounts for the observed high selectivity. The use of H_2O_2 is, however, not well suited for the large-scale production of methanol, and neither is the high cost of liquid superacids [8]. Studies were, however, continuing with varied peroxides and strong acid systems.

The concept of chemically protecting methanol formed in the oxidation of methane was successfully further developed by Periana and coworkers using predominantly metal salts and complexes as catalysts in sulfuric acid or oleum [371]. Several of these homogeneous catalyst systems, able to activate the otherwise very unreactive C–H bonds of methane at modest temperatures and with surprisingly high selectivity, have been developed [372]. At temperatures around 200 °C, the conversion of methane into methanol in concentrated sulfuric acid using a $HgSO_4$ catalyst was found to be an efficient reaction. The oxidation of methane produces methyl hydrogen sulfate (CH_3OSO_3H), which can be hydrolyzed to methanol in a separate step. At a conversion of 50%, an 85% selectivity to methyl hydrogen sulfate was achieved [373]. To complete the catalytic cycle, Hg^+ is reoxidized to Hg^{2+} by H_2SO_4. Overall, the process uses one mole of sulfuric acid for each mole of methanol produced. SO_2 generated during the process can be easily oxidized to SO_3, which, upon reaction with H_2O, will give H_2SO_4 that can be recycled:

$$CH_4 + 2H_2SO_4 \xrightarrow{Hg^{2+}} CH_3OSO_3H + 2H_2O + SO_2$$
$$CH_3OSO_3H + H_2O \rightarrow CH_3OH + H_2SO_4$$
$$\underline{SO_2 + 1/2O_2 + H_2O \rightarrow H_2SO_4}$$
$$CH_4 + 1/2O_2 \rightarrow CH_3OH \qquad \text{(Overall reaction)}$$

The cleavage of methyl hydrogen sulfate to methanol and its separation from sulfuric acid media is, however, an energy-consuming process. Furthermore, the use of poisonous mercury makes the process somewhat unattractive. Interest has therefore shifted towards other less toxic metals and organometallic complexes. Systems incorporating platinum, iridium, rhodium, palladium, ruthenium and others have been tested for homogeneous methane oxidation. Even gold was found to catalyze the oxidation of methane to methanol [374, 375]. The best results were obtained with a platinum complex in H_2SO_4. With this system, methane was converted into methyl hydrogen sulfate, and, subsequently, methanol, with yields in excess of 70% and selectivities over 90% [371].

Platinum complex

Water formed during the reaction accumulates progressively in sulfuric acid, decreasing its acidity. This makes the process problematic, as the activity, not only of the platinum but also of mercury and most other systems tested, rapidly decreases at lower acidities and can be inhibited by high water content. For this and other reasons, the conversion is advantageously conducted in oleum (a mixture of H_2SO_4 and SO_3). SO_3 reacts with the water formed to give H_2SO_4, avoiding a decrease in acidity. The use of a catalyst system exhibiting high activity, even at lower acidities, would be, however, preferable. A gold-based catalyst system using selenic acid (H_2SeO_4) as the oxidant has also been developed and shows some promise [371]. To replace the high-cost platinum and other noble metal-based catalysts, the development of a less expensive, but still effective and selective, catalyst would also be desirable.

Although our understanding of the reaction mechanism for the direct oxidation of methane in homogeneous systems has greatly improved over the years, many key problems have yet to be solved. There is also a need to develop more active, selective and stable catalysts before a commercial process based on this technology becomes a reality. Today, however, this goal appears increasingly achievable.

12.3.4
Methane into Methanol Conversion through Monohalogenated Methanes

Another possible alternative for the selective conversion of methane into methanol involves the intermediate catalytic formation of methyl chloride (CH_3Cl) or bromide (CH_3Br), which are then hydrolyzed to methanol (or dimethyl ether) with the by-product HCl or HBr being reoxidized [376, 377]:

$$CH_4 \xrightarrow{X_2} HX + CH_3X \xrightarrow{H_2O} CH_3OH (\text{or } CH_3OCH_3) + HX$$
$$2HX \xrightarrow{1/2O_2} X_2 + H_2O$$
$$X = Cl, Br$$

The chlorination of paraffins, discovered by Dumas in 1840, is the oldest known substitution reaction and is practiced on a large scale in industry. It is usually a free radical process initiated either thermally or photochemically (by heat or light). The major problem with free radical reactions, also experienced in oxidative methane conversion, is the lack of selectivity. With methane chlorination, all four chloro-methanes (CH_3Cl, CH_2Cl_2, $CHCl_3$ and CCl_4) are generally obtained:

$$CH_4 \xrightarrow{Cl_2} CH_3Cl \xrightarrow{Cl_2} CH_2Cl_2 \xrightarrow{Cl_2} CHCl_3 \xrightarrow{Cl_2} CCl_4$$

As methyl chloride is chlorinated more rapidly than methane itself, a high ratio of methane over chlorine of at least 10:1 is required to obtain methyl chloride as the main product. Chlorination of methane has also been reported over catalysts such as active carbon, kieselguhr, pumice, alumina, kaolin, silica gel and bauxite, but showed limited selectivity due to the free radical nature of the reaction. In industry, the strongly exothermic chlorination reaction is generally conducted in the absence of a catalyst at 400–450 °C without external heating under slightly elevated pressures.

In the 1970s, Olah and coworkers observed that chlorination of methane with superacidic SbF_5/Cl_2 at low temperature gives methyl chloride in high selectivity with only small amounts of methylene chloride and higher chlorinated compounds [378, 379]. The high selectivity obtained in this electrophilic reaction contrasted strongly with the low selectivity for radical chlorinations. Since the conversions were low, no adaptation of the reaction for practical use was made.

In extending the electrophilic halogenation of methane to catalytic heterogeneous gas-phase reactions during the 1980s, it was shown that methyl chloride could be formed with high selectivity and acceptable yields [376]. Furthermore, the catalytic mono-chlorination or -bromination of methane was achieved over solid superacids such as TaF_5/Nafion-H, SbF_5/graphite, supported metals such as Pt/Al_2O_3 and $Pd/BaSO_4$, or supported oxychlorides and oxyfluorides such as $ZrOF_2/Al_2O_3$ and GaO_xCl_y/Al_2O_3. The reactions were carried out at between 180 and 250 °C, giving 10–60% conversion with selectivities of methyl chloride (bromide) generally exceeding 90%. Methylene halides (CH_2Cl_2, CH_2Br_2) were the only higher halogenated methanes formed and no haloforms ($CHCl_3$, $CHBr_3$) or carbon tetrahalides (CCl_4, CBr_4) were observed.

In the halogenation of methane, hydrogen halides (HCl or HBr) are formed as equimolar by-products. Similarly hydrolysis of methyl halides also gives hydrogen halides. Their recycling is essential to utilize halogens only as a catalytic agent in the overall conversion of methane into methanol. The oxidation of hydrogen chloride to chlorine is industrially possible (Deacon process, Kellogg's improved Kel-Chlor process, Mitsui Toatsu Chemicals MT Chlor process), but it remains technologically difficult. An improved way to effect the reaction was also found by Benson and coworkers [380, 381]. In contrast, HBr is readily oxidized to bromine even by air. Its reoxidation and continuous recycling is thus easier.

Combination of the selective monohalogenation (preferentially bromination) of methane with subsequent catalytic hydrolysis to methanol/dimethyl ether was shown by Olah and coworkers to be an attractive alternative route for the preparation of methyl alcohol without going through syn-gas [376, 377]. Methyl halides obtained directly from methane also offer a convenient way to convert methane via zeolite or bifunctional acid–base-catalyzed condensation into ethylene and propylene, which are starting materials for synthetic hydrocarbons and their products (Chapter 13):

$$2CH_3X \xrightarrow[-2\,HX]{Cat} CH_2 = CH_2 \xrightarrow[-HX]{CH_3X,Cat} H_3C - CH = CH_2$$

Related processes have been developed by Dow Chemical [382] and U.C. Santa Barbara [382, 383]. The company Gas Reaction Technologies, created in 1999, has patented some of its processes and is continuing to develop the commercial development of this technology [384–386].

To produce methanol from methane in a single step, combining halogenation, hydrolysis and reoxidation of the hydrogen halide by-product with a $Br_2/H_2O/O_2$ or $Cl_2/H_2O/O_2$ mixture can also be achieved, although to date the conversions remain modest.

Iodine was found by Periana and coworkers to be a suitable catalyst for the selective conversion of methane into methanol in oleum [387]. At about 200 °C, methyl hydrogen sulfate was formed with up to 45% yield and 90% selectivity, and subsequently hydrolyzed to methanol. The intermediate formation of methyl iodide, followed by conversion into methyl hydrogen sulfate and reoxidation of HI to I_2 can be assumed, not unlike the discussed chlorine- and bromine-catalyzed conversions.

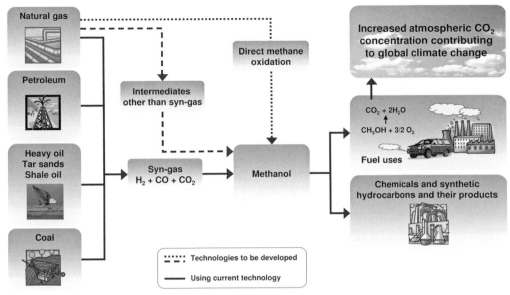

Figure 12.6 Methanol from fossil fuel resources.

Corrosion and related problems due to the use of a highly acidic media and iodine could, however, be quite significant:

$$CH_4 + 2SO_3 \xrightarrow{I_2} CH_3OSO_3H + SO_2$$

Methanol production from fossil fuels, as long as they are readily available, will remain a major source of methanol (Figure 12.6). Improved methods of direct methane conversion without going through syn-gas are being developed. Besides natural gas, methane hydrates and other sources will also be utilized. Biological production of methanol and most significantly chemical reductive recycling of CO_2 into methanol (*vide infra*) will, however, increasingly become important.

12.3.5
Microbial or Photochemical Conversion of Methane into Methanol

Certain bacteria, known as methanotrophs, can obtain all the energy and carbon they need from methane [364, 388]. The key step in their consumption of methane is its selective conversion into methanol using oxygen. In subsequent biological processes, methanol is further oxidized to formaldehyde, which in turn can be either incorporated into biomass or oxidized to CO_2, thereby providing the energy needed by the bacteria. Nature's catalyst for the conversion of methane into methanol is an enzyme called methane monooxygenase (MMO), which can operate in aqueous solution at ambient temperature and pressure [389, 390]. In such a system, composed of several imbricated proteins, the reactivity is controlled by molecular recognition and the regulation characteristics of the enzyme, allowing a virtually complete selectivity to methanol. The conversion is achieved by reductive activation of O_2 with a reductant known as NADH:

$$CH_4 + NADH + H^+ + O_2 \rightarrow CH_3OH + NAD^+ + H_2O$$

Two varieties of MMO systems have been found in methanotrophic organisms. The first, which contains copper, is a membrane-bound enzyme that has proven difficult to isolate and has eluded full characterization. Most attention has thus been focused on the soluble, iron-containing MMO, especially from the species *Methylosinus trichosporium* and *Methylococcus capsulatus*. The active site in this enzyme that is directly responsible for methane oxidation contains a pair of iron atoms. Given the ability of MMO to activate methane at room temperature and ambient pressure, much research has been devoted to reproducing such activity for the large-scale production of methanol. The understanding of the reaction mechanisms taking place in MMO has greatly improved over the years, and simpler catalyst systems that try to model the behavior of MMO have been developed and tested. However, because of their complexity, the direct use of MMO enzymes has proven difficult to apply for the practical production of methanol. Besides MMO systems and their ability to oxidize specifically the simplest alkane, methane, enzymes of the cytochrome P-450 family, with a less-complicated structure than MMO, have been found to catalyze the oxidation of many types of hydrocarbons to alcohols [391]. At the heart of the cytochrome P-450 is a porphyrin system containing at its center a metal atom, most commonly iron, that catalyzes the oxidation reaction. Although to date no P-450 system has been shown to be capable of effectively oxidizing methane, significant advances have been made in that direction through the genetic engineering of enzymes based on P-450, specifically designed for high activity towards small alkanes. As a result, modified P-450 systems able to convert ethane into ethanol have been disclosed [392, 393]. In attempting to approach Nature's ability to activate methane under mild conditions, methanol has also been obtained by reacting methane with oxygen and H_2O_2 in water at temperatures below 75 °C using a vanadium-based catalyst [394]. The finding of a practical, highly efficient and selective catalyst based on Nature's example is, however, still a long-term goal. Notably, MMO is not a perfect system. When O_2 is used as the oxidant, it requires the presence of a reductant, NADH. Alternatively, a reduced form of O_2, for example H_2O_2, is generally used in model experiments. In practical applications, however, this would most probably imply the use of hydrogen as a reducing agent, making the process less attractive. Ideally, methane should be oxidized by only O_2 to form methanol.

12.4
Methanol from Biomass, Including Cellulosic Sources

Although reserves of methane from natural gas are still very large, they are nevertheless finite and diminishing. The exploitation of unconventional natural gas resources and methane hydrates could significantly increase the amounts of methane available to humankind. Innovative processes continue to be explored for the transformation of methane into easy-to-handle liquid fuels, primarily methanol, but this will not solve the problem of increasing CO_2 concentrations in the atmosphere and its detrimental effect on the global climate. Even if methane does release less CO_2 than coal or petroleum, its increased utilization to fill the world's growing energy needs will still produce large quantities of CO_2. Thus, new ways are needed to fulfill humankind's ever increasing appetite for energy, without adversely affecting our environment. In the long term therefore, methanol will have to be produced from sources that release only minimal quantities of CO_2, if not none at all. In this respect, the use of biomass is one possibility, although natural resources will be at best able to cover only a part of our needs.

Biomass is referred to as any type of plant or animal material – that is, materials produced by life forms. This includes wood and wood wastes, agricultural crops and their waste by-products, municipal solid waste, animal waste and aquatic plants and algae. Methanol as mentioned was originally made from thermal destructive distillation of wood (hence the name "wood alcohol"), but due to its inefficiency and the advent of synthesis via syn-gas this route was rapidly abandoned in the first half of the twentieth century. Increased oil and natural gas prices, dependence on foreign countries for energy supply, growing concerns about CO_2-induced climate changes as well as the depletion of easily accessible fossil fuel resources have prompted recent reconsideration of a wider and more extensive use of biomass to fulfill our energy demands. Since biomass itself is both bulky and heterogeneous, one way to achieve this goal would be to convert it into a convenient liquid fuel, namely methanol. The modern production of methanol from biomass is very different, however, and is much more efficient than the methods used a century ago. Basically, not only wood but also any organic (i.e., carbon-containing) material obtained from living systems can be used in the process. Presently, technologies to transform biomass into methanol are generally similar to those used to produce methanol from coal. They imply the gasification of biomass to syn-gas, followed by the synthesis of methanol with the same processes employed in processes based on fossil fuels. Enzymatic conversion of biomass into methanol is also a possible, and attractive, alternative, raising interest and stimulating research. Most efforts in this area are, however, currently focused on the production of ethanol.

In the gasification process, the biomass feedstock is usually first dried and pulverized to yield particles of a uniform size and with moisture content no higher than 15–20% for optimal results. Gasification is a thermochemical process that converts biomass at high temperature into a gas mixture containing hydrogen, carbon monoxide, carbon dioxide and water vapor. The pretreated biomass is sent to the gasifier where it is mixed, generally under pressure, with oxygen and water. There, a part of the biomass is burned with oxygen to generate the heat necessary for gasification. The combustion gases (CO_2 and water) react with the rest of the biomass to produce carbon monoxide and hydrogen. With biomass being used as the heating fuel, no external heat source is necessary. The production of syn-gas from biomass in a single-step operation by partial oxidation is very attractive, but it has experienced technical problems. Gasification of biomass feedstocks is therefore generally a two-step process. In the first stage, called "pyrolysis" or "destructive distillation," the dried biomass is heated to temperatures ranging from 400 to 600 °C in an atmosphere too deficient in oxygen to allow complete combustion. The pyrolysis gas obtained consists of carbon monoxide, hydrogen, methane, volatile tars, carbon dioxide and water. The residue, which is about 10–25% of the original fuel mass, is charcoal. In the second stage of gasification, called "char conversion," charcoal residue from the pyrolysis step reacts with oxygen at 1300 to 1500 °C, producing carbon monoxide. Before being sent to the methanol production unit, the syn-gas obtained must be purified. Compared to most coals, the advantage of biomass is its much lower sulfur content (0.05–0.20 wt%), while heavy metals impurities (mercury, arsenic, etc.) are present only in insignificant quantities. Biomass gasification has, however, its own issues. The formation of tars in particular is most cumbersome and problematic for the commercialization of any biomass gasification technology. These tars, composed mainly of oxygenated compounds and higher molecular weight hydrocarbons, are problematic because they condense in pipes, boilers, transfer lines and particulate filters, leading to blockage, clogging and other operational difficulties. The nature and amounts of tars depends on the gasification technologies, operating conditions and biomass composition. Their level can be reduced significantly by choosing proper gasification conditions and technology [395].

In conventional systems, part of the biomass feedstock has to be burned to provide the necessary energy for gasification. To improve the utilization of biomass and reduce CO_2 emissions, other heat sources such as solar or nuclear could be used instead. Gasification of biomass dispersed in a molten salt medium heated at 800–1000 °C by concentrated solar energy has been proposed. The storage of energy in the form of heat in molten salt could allow the production of syn-gas without interruption from solar energy [396]. Solar energy can also be used to dry wet biomass prior to gasification, reducing the required energy input.

While the production of methanol from biomass is possible on a relatively small scale, large-scale methanol plants are preferable because they induce substantial economies of scale, thereby lowering the cost of production and increasing efficiency. The quantities of biomass needed to feed a world-scale 2500 tonnes-per-day methanol plant are, however, very large (on the order of 1.5 million tonnes per year). To deliver such amounts, biomass would have to be collected over vast areas of land [397]. The transportation of bulky biomass products with low energy density over long distances is not economical, and consequently its transformation into an easy-to-handle and store liquid intermediate through fast pyrolysis has been proposed. In fast pyrolysis, small biomass particles are heated very rapidly to 400–600 °C at atmospheric pressure to yield oxygenated hydrocarbon gases. The generated gases are then rapidly quenched to avoid their decomposition by cracking. The black liquid obtained, called "biocrude" [398, 399] (because of its resemblance to crude oil, Figure 12.7), has a wide range of possible applications. It can be processed into a replacement for fuel oil and used directly in furnaces for heat or electricity production. By altering its chemical composition, biocrude could also be combined and processed alongside petroleum crude in modified refineries. Although further development is required, biocrude holds promise as a domestically available synthetic alternative to petroleum. Besides biocrude, which is obtained in 70–80% yield,

Figure 12.7 "Biocrude."

fast pyrolysis also produces some combustible gases and char, a fraction of which is used to drive the process. A fraction of the char could also be finely ground and added to the biocrude to form a slurry; this, like biocrude, might be pumped, stored and transported via road, rail, pipelines and tankers in much the same way as crude oil today, facilitating greatly the handling of biomass feedstock. Fast pyrolysis is a relatively simple, low-pressure and moderate-temperature process that is applicable on both large and small scales. Since fast pyrolysis has been recently introduced, further development is required for its commercial application. In contrast, the production of syn-gas from liquid feedstocks from diverse fossil origins is a well-established process that is used in numerous plants worldwide. The production of methanol in large-scale plants (>2500 tonnes per day) through syn-gas obtained from biocrude generated in delocalized plants is therefore a feasible option depending on favorable specific local and economic conditions.

All biomass materials can be gasified for methanol production (Figure 12.8). The efficiency of the process will, however, depend upon the nature and quality of the feedstocks. Low-moisture materials such as wood and its by-products and herbaceous plants and crop residues are the most suitable. Using wood and forest residues, efficiencies as high as 50–55% have been recorded. Several demonstration projects are under way in different parts of the world. In Japan, for example, Mitsubishi Heavy Industries is operating a pilot plant for the production of methanol from cellulosic biomass. Various feedstocks such as Italian ryegrass, rice straw, rice bran and sorghum have been successfully tested, with the best yields having been obtained with sawdust and rice bran. In the United States, the Hynol process,

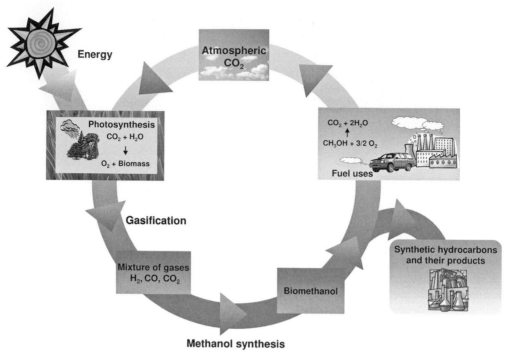

Figure 12.8 Carbon neutral cycle of biomethanol production and uses.

originally developed at the Brookhaven National Laboratory, has been tested on a pilot scale [400], and was successful in converting materials such as woodchips into methanol. In Europe, the paper industry has turned its attention to black liquor gasification as a possible source for methanol [401]. Black liquor is a pulp rich slurry obtained as by-product of the Kraft pulping operation used for paper production, and contains about half of the organic material that was originally in the wood. It has the great advantage of being already partially processed and in a pumpable, liquid form. With the gasification process, 65–75% of the biomass energy contained in black liquor can be converted into methanol. Worldwide, the pulp and paper industry currently produces about 170 million tonnes of black liquor each year. The United States alone could potentially produce 28 million tonnes of methanol per year from black liquor. Clearly, since the United States consumption of gasoline and diesel fuels totals the equivalent of more than 1000 million tonnes of methanol, this would replace only a small fraction of the petroleum-derived fuels. In smaller countries with an important paper industry and limited population (e.g., Finland and Sweden), methanol from black liquor could replace a substantial part of the motor fuel demand (50% and 28%, respectively). Accordingly, the European project BioDME, currently under way in Sweden, will produce in a demonstration plant about 4 tonnes per day of DME from syn-gas obtained through black liquor gasification. The produced DME will then be distributed in four filling stations to Volvo DME trucks. DME from biomass is one of the most energy efficient renewable fuels when the entire chain from production to vehicle operation, the so-called well-to-wheel efficiency, is taken into account [402–405]. Compared to synthetic diesel obtained through Fischer–Tropsch chemistry, DME is a much simpler compound that is easier and cheaper to produce. The production cost of DME from black liquor is estimated to be equivalent to $65 per barrel of oil without any subsidies [406]. Municipal solid waste (about 250 million tonnes are produced each year in the United States) is another possible feedstock for methanol – implying that landfills could actually become energy fields of the future.

Although waste products from wood processing, agricultural residues and by-products, as well as solid municipal waste, represent suitable possible feedstocks for methanol production in the short term, the quantities that can be generated from these resources are limited. In the long term, the growing demand for bio-methanol will necessitate a larger and reliable source of raw biomass material. Thus, crops selected specifically for energy purposes would have to be cultivated on a large scale if a significant amount of methanol were to be produced from biomass resources. Suitable "energy crops" are being identified, and the most promising – essentially fast-growing grasses and trees – are being field tested. Depending upon the climate, the species giving the best results are logically different. The potential impact of the large-scale culture of energy-devoted plants on ecosystem, soil erosion, water quality and wildlife is also being assessed. With regard to wood, fast-growing species such as poplar, sycamore, willow, eucalyptus and silver maple have been proposed [407] for cultivation in "short rotation" tree farms in the United States and Europe. In conventional silviculture for plywood or paper production, trees are grown for 20–50 years or more. The term "short rotation" means that fast-growing trees planted at close spacing are harvested after only 4–10 years to maximize productivity since, after harvesting, new trees can be planted. Most of the trees under consideration, however, have the ability to resprout from cut stumps in a process called "coppicing," thus eliminating the need to replant after each harvest. Such intensive tree culture has shown to produce generally between 4 and 10 tonnes of oven-dry wood per hectare per year under temperate climates. Higher rates can be obtained in tropical climates with appropriate species, such as eucalyptus in Florida, Hawaii and South America. Besides wood, energy crops such as switchgrass, sorghum or sugarcane have been identified as having a high efficiency in converting solar energy into biomass. To contain the cost of planting, growing and harvesting energy crops, highest productivity is expected. In most cases, this means that prime agricultural land would have to be used for their production. In large countries with a low population density such as the United States, Australia or Brazil, a significant portion of idle food-crop lands, pasture range and forest land could be used for energy crops. However, in more densely populated areas, such as Western Europe, Japan and China, most of the arable land is already in use for food production. Unless surplus land is available energy crops will directly and adversely compete with food and fiber production.

As with any form of plant life, high and sustainable productivity for energy crops over the years will only be maintained if the necessary nutrients for the plant's growth are present in the soil. In intensive agriculture, as practiced in most developed countries, this implies the massive use of fertilizers. Nitrogen fertilizers in particular, obtained from NH_3 derived from natural gas, if used for energy crops, could partially offset the benefits expected from the production of fuels from biomass. With nitrogen fertilizers on cultivated soils and animal wastes being responsible for 65–80% of the total emissions of N_2O (a powerful greenhouse gas, 296-fold more damaging than CO_2 [408]), a judicious choice of energy crops requiring minimum input of fertilizers is therefore necessary. Crop rotation, which is a widely accepted practice in agriculture to fight soil impoverishment, should also be applied. No-till techniques reducing soil erosion and improving soil health have also been proposed and are increasingly applied [409]. Although environmental risks must still be carefully assessed, biotechnology and genetic modifications are also expected to provide additional gains in crop yields.

From an industrial viewpoint, the production of energy crops in vast monocultures, including only one species, would be preferable. Such types of culture are, however, generally more prone to epidemics of pests and diseases, requiring frequent applications of pesticides and fungicides to keep them free from potentially disastrous infestations that might wipe out the entire plantation. To avoid this situation, several types or species of energy crops could be planted instead of only one; this would also allow a higher degree of biodiversity. At the same time, the introduction of nitrogen-fixing trees such alder or acacia in fast-growing poplar and eucalyptus plantations could also help to reduce fertilizer requirements [244]. It has been shown that high-diversity native grassland perennials had up to 240% higher bioenergy yield than monocultures after a decade, with greater greenhouse gas reduction and less agrochemical pollution. Furthermore, these perennial grasses are less susceptible to diseases and insects and sustainable even with low input of fertilizers. While limited phosphorus application might be needed, legumes, as part of the perennial grass mix, can supply the needed nitrogen. These low-input high-diversity (LIHD) mixtures of native grassland perennials are especially interesting for degraded lands where they can give much higher yields than monocultures of species such as switchgrass [410].

In avoiding the use of prime farmland, energy crops capable of being grown on marginal or degraded lands have attracted much attention as a means of adding value to vast areas unsuitable for most other agricultural purposes. The mesquite tree and its sugar-rich pods, for example, already thrives in the United States on tens of millions of hectares of semi-desert land that, without proper irrigation, would otherwise be of little value for the production of food crops or as pasture land. Owing to its extensive root system, which also helps to maintain the soil in place and to protect the land from erosion, the mesquite tree is able to reach far into the ground to capture water. Also present in the root systems are bacterial nodules that fix the nitrogen contained in air and provide not only the mesquite tree but also its neighboring plants with the nutrients necessary for growth. Although the productivity of plants such as the mesquite trees is less than for fast-growing poplar or eucalyptus, they have the advantage of growing, without irrigation needs, on vast areas of land that would otherwise be of limited value. In Australia, the National Science Agency devised a computer model to show that 30 million hectares of fast-growing trees planted over the next 50 years could produce enough methanol to replace 90% of the oil requirements of the country's 25 million people by 2050. It would, by the same time, create numerous jobs in rural areas and help to reduce the impact of salt that otherwise would be brought to the surface by rising water tables.

The European Union, with a population of over 450 million after its expansion to 25 member states, has much less potential for energy crops cultivation. Of the almost 400 million hectares total land area, about 170 million hectares are currently used for food production, while another 160 million hectares are covered by forests. Most of the remaining land is classified as unsuitable for the production of biomass (mountains, deserts, degraded lands, urban areas, etc.). To remain self-sufficient, only a small fraction of the land currently used for food production could be converted into energy crops. Forests are another potential source for large-scale methanol production from biomass in Europe. They could be used directly for wood production or be cleared to enable agriculture, but due to their importance for the environmental balance and biodiversity, transforming a large proportion of the forests into vast, highly standardized and mechanized tree plantations would run contrary to the sustainable use of natural resources. When considering possible feedstocks, biomass is thus not expected to represent more than 10–15% of Europe's total energy demand in the future.

In those developing countries already struggling to produce the necessary food for their growing population, the production capacity for energy crops is clearly limited.

12.4.1
Methanol from Biogas

Most mammals (including humans), as well as termites and other bio-organisms, produce a flammable gas called "biogas" when they digest their food. Biogas is also generated in wetlands, swamps and bogs where large amounts of rotting vegetation may accumulate. Biogas is formed when anaerobic bacteria break down organic material in the absence of oxygen. Anaerobic bacteria are some of the oldest forms of life on Earth, having evolved before the photosynthetic processes of green plants were able to release large quantities of oxygen into the primitive atmosphere.

Biogas is in fact a waste product of those microorganisms, composed mainly of methane and CO_2 in variable proportions, and trace levels of other compounds such as hydrogen sulfide (H_2S, gas of rotten egg smell), hydrogen or carbon monoxide. The biological process for biogas generation by anaerobic methanogenic bacteria, termed "methanogenesis," was discovered in 1776 by Alessandro Volta, and has been used since the nineteenth century in so-called "anaerobic digestion reactors." The process was used quite extensively in Europe when energy supplies were reduced during and after World War II. Interest in biogas was then revived after the oil crisis of the 1970s. Most common feedstocks are animal dung, sewage sludge and wastewater or municipal organic waste. The use of anaerobic digesters to treat industrial waste water has also rapidly gained popularity during the past decade. Wastewaters treated include those from food, paper and pulp, fiber, meat, milk, brewing and pharmaceutical plants. In Brazil, anaerobic digesters are used to treat the waste liquid (called vinasse) co-produced during ethanol manufacture from sugarcane. Anaerobic digestion is thus an important method of reducing the impact of waste disposal on the environment [411].

Anaerobic digesters can be constructed from concrete, steel, bricks or plastic, they may be shaped like silos, basins or ponds, and they may be placed either underground or on the surface. In large-scale operations, organic materials are constantly fed into a digestion chamber, and the gas produced is allowed to exit through a gas outlet placed on the top of the reactor. The sludge left behind in the digester, rich in nutrients, can be used as a soil fertilizer or conditioner.

The gas produced in anaerobic digesters, depending on the feedstock and effectiveness of the process, consists of 50–70% methane with the remainder being mostly CO_2. At present, the gas is mainly used to produce electricity or heat. Compressed methane obtained from biogas (basically the same as CNG) is also used as an alternative transportation fuel in light- and heavy-duty vehicles, notably in Brazil. Today, Sweden runs some 800 buses on biogas and a small biogas train powered by two biogas bus engines has been in use since 2005. After the removal of impurities (especially hydrogen sulfide), biogas could also be used directly for the production of methanol in the same way as natural gas. A fraction of the CO_2 already present in the gas could react with the excess hydrogen generated during methane steam reforming to form more methanol. Using this technology, the manure of 250 000 pigs is currently planned to be used in Utah to produce some 30 000 L of methanol per day.

Another source for biogas production is landfills which, worldwide, are the dominant method for municipal solid waste disposal. As in anaerobic digesters, the organic material present in the landfills is decomposed in the absence of oxygen by anaerobic bacteria. The resultant gas is composed of 50–60% methane, 40–50% carbon monoxide and traces of H_2S as well as other volatile organic compounds. The landfill gas, which for safety reasons is otherwise vented to the atmosphere or is flared, can also be collected through elaborate underground piping for use as a fuel or feedstock for methanol production.

In the United States, 250 million tonnes of municipal solid waste were produced in 2006. Of this, 32% was recycled or composted, 13% incinerated and 55% landfilled. From the more than 1700 landfills currently in operation, close to 450 have operational landfill gas utilization units in place [412]. Most of the gas produced is used to generate electricity or produce heat. The production of methanol from landfill gas is also possible. A molten carbonate fuel cell (MCFC) that can operate on methanol from municipal organic waste located in the city of Berlin, Germany, has been in use since 2004 (Figure 12.9).

Notably, however, in 2004 the combined production of electricity from all landfill gas energy projects presently installed in the United States was only 9 billion kWh [412] – similar to the output of a single large-scale nuclear reactor, or 0.2% of the national electricity production! While landfill gas utilization provides a good means of controlling methane greenhouse gas emissions from municipal solid waste, it has a very limited capacity for energy or methanol supply.

Figure 12.9 HotModule from MTU: The first bi-fuel molten carbonate fuel cell (MCFC) went into service at Vattenfall Europe AG in Berlin in September 2004. It is fuelled with natural gas or methanol or any mixture of these two. The methanol used is derived from waste generated locally. (Courtesy: MTU Friedrichshafen GmbH 2004.)

12.4.2
Aquaculture

Water, in the form of oceans, lakes, rivers and wetlands, covers more than 70% of our planet. Thus, although much less attention has been given to aquatic plants and organisms for the production of energy, those biomass sources could have considerable potential.

12.4.2.1 Water Plants
In freshwater, plants growing partially submerged, such as water hyacinth and cattail, have been identified as promising feedstock for energy applications. Water hyacinth, which is native to South America, is a free-floating herb that forms dense mats on the surface of the water. Introduced intentionally or by accident in most tropical and subtropical parts of the world, it has become a floating nightmare for the affected countries. In Africa, it infests every major river and almost every lake. In the United States, it flourishes in countless bodies of water throughout the South, from California to Virginia (Figure 12.10). Dense and rapidly expanding growth of water hyacinth can clog canals and water intakes, restrict navigation along rivers and lakes, and exclude native vegetation. In some areas, it must be continually cleared from vital waterways. It has a very high annual productivity of 30–80 tonnes of dry biomass per hectare.

Cattail (Figure 12.11), which thrives in marshland and lakes throughout the planet, is also viewed as a good candidate for energy production, because it can yield annually up to 40 tonnes of biomass per hectare. Once collected, the water hyacinth, cattail or

Figure 12.10 Water hyacinth infesting waterways in southern United States.

any other suitable water plant could be processed by anaerobic conversion to form methane and subsequently methanol.

12.4.2.2 Algae

Found in both marine and freshwater environment, algae have also been studied as potential sources of energy. Macroalgae – more commonly known as seaweed or kelp – are fast-growing plants that can reach a considerable size (up to 60 m in length) [413]. A few experimental seaweed farms were built along the coast of Southern California in the 1970s but, due to difficult weather conditions and rough waters, the open-sea project was rapidly abandoned. More protected from the elements, near-shore farming of *Macrocystis* algae gave better results, and yearly biomass production of more than 30 tonnes per hectare were reported (Figure 12.12). Growing algae in the open seas and oceans, such as the Sea of Japan, is also being explored. Seeding areas with iron to promote algae growth has also been proposed.

Microalgae – another form of algae – are microscopic and the most primitive form of photosynthetic organisms (Figure 12.13). Whilst their mechanism of photosynthesis is similar to that of higher plants, the algae are generally more efficient in converting solar energy into biomass because of their simple cellular structure and the fact that they grow in aqueous suspension surrounded by water, CO_2 and other nutrients necessary for their growth. Support from water also means that algae do not

Figure 12.11 Cattail.

Figure 12.12 Kelp forest. (Courtesy: Ocean Brite Systems.)

need to form cellulose and lignin, which are essential structural components of terrestrial plants such as grass and trees, but they are difficult to process economically into liquid fuels. Some species of microalgae are very rich in oil, and this can account for more than 50% of their mass, with most of the rest being starch and sugars. It has been claimed that microalgae can produce 30 times more oil per hectare than terrestrial oilseed crops. This prompted the National Renewable Energy Laboratory (NREL) from the US Department of Energy (DOE) to study these organisms for the manufacture of biodiesel in a program that ran from 1978 to 1996 [413]. Experimental production of microalgae was conducted in open, shallow ponds located in regions with high sun exposure such as California, Hawaii and New Mexico. Unlike higher, more complex organisms, microalgae reproduce very rapidly and can be harvested continuously as they are produced. To maintain high growth rates, the ponds were constantly fed with CO_2, water and other nutrients needed by the algae. For large-scale production in "algae farms," CO_2 could be obtained from several sources,

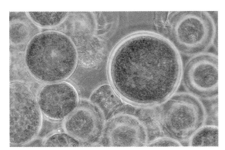

Figure 12.13 Microalgae.

especially fossil fuel-burning power plants. The carbon contained in fossil fuel could thus be used twice, reducing global emissions of CO_2. This would provide a practical approach for the recycling of CO_2 into a useable fuel. Ultimately it will, however, only postpone the release of CO_2 into the atmosphere. To be sustainable in the long term, algae should use the CO_2 contained naturally in air and dissolved in water. This will probably considerably lower the yield of biomass production, requiring vast areas of land to produce significant amounts of energy. According to the NREL's results, however, the production of biomass from aquatic plants has so far proven too expensive to compete with fossil fuels, or even plants grown on land. Japan, which is extremely dependent on energy imports, is nevertheless actively pursuing this approach to extend its domestic energy production. In recent years, this technology has also regained interest in the USA, where venture capital investments in algal biofuel accelerated, reaching more than \$180 million in 2008 compared to \$32 million in 2007, thus absorbing roughly 20% of all venture capital invested in biofuels in 2008 [160]. Numerous companies such as Solazyme, Sapphire Energy and Solix Biofuels are building experimental plants. Green Star Products, for example, after completing a microalgae growth study in a 40 000 L demonstration facility, announced in 2008 that it would build a \$140 million commercial algae-to-biodiesel plant paired with a biorefinery in Missouri.

Besides extracting their oil, one of simplest ways to use the biomass generated from macro- or microalgae is to produce methane. Because of the high water content of algae, gasification requiring prior drying of the material might not be attractive, and anaerobic digestion – which is anyway conducted in water – is therefore more adaptable. The methane generated can then be used in the same way as natural gas to produce methanol. More research is needed, however, to provide a better understanding of the biology of aquatic organisms and to identify the best species and growing conditions for optimal conversion of solar energy into biomass. Genetic engineering could also be used to improve the characteristics of such plants and algae.

12.5
Chemical Recycling of Carbon Dioxide to Methanol

We have discussed thus far the production of methanol from fossil fuels and bio-sources. When fossil fuels or for this reason any carbon-containing natural or synthetic product, including methanol, are combusted, they produce CO_2 and water. The great challenge, which is of increasing significance, is to reverse this process so as to produce – efficiently and economically by chemical regenerative conversion, that is, recycling – fuels, synthetic hydrocarbon and products and materials from CO_2 and water. Nature, via photosynthesis, captures CO_2 through trees, green plants, crops, and so on from the air and converts it with water, using the sun's energy and chlorophyll as the catalyst, into new plant life. Thus, plant life replenishes itself by recycling atmospheric CO_2. The difficulty is that the natural conversion of plant life into fossil fuels is a very long process, taking many millions of years. Of course, as discussed, plant life reforms itself in annual or short cycles. In the form of biomass it can be collected and processed to synthetic hydrocarbon fuels and products. In any event, the scope of the utilization of this natural carbon dioxide cycle compared with humankind's needs has limits. As a solution, we must therefore develop our own technological carbon dioxide recycling processes to achieve this goal within the required extremely short time scale. A promising new approach is to convert CO_2 chemically by catalytic or electrochemical hydrogenative methods into methanol or DME, and subsequently to synthetic hydrocarbons and products using hydrogen obtained from water electrolysis or other cleavage methods. Chemists have known since the early twentieth century on how to convert CO_2 and H_2 into methanol:

$$CO_2 + 3H_2 \xrightarrow{\text{catalyst}} CH_3OH + H_2O$$

In fact, some of the earliest methanol plants operating in the 1920s and 1930s in the US commonly used CO_2, obtained as by-products of fermentation processes, for methanol production. Efficient catalysts based on metals and their oxides (notably copper and zinc) have been developed for this process. These catalysts are very similar to those used presently in the industry for methanol production via fossil fuel based syn-gas. In view of our present understanding of the mechanism of methanol synthesis from syn-gas, this is not unexpected. It is now established that methanol is most probably almost exclusively formed by hydrogenation of CO_2 contained in syn-gas on the catalyst's surface. To be converted into methanol, the CO in the syn-gas first undergoes a water gas shift reaction to form CO_2 and H_2. The so-formed CO_2 then reacts with hydrogen to yield methanol [359, 414, 415]. The reaction on a commercial methanol catalyst of a CO/H_2 mixture, carefully purified from CO_2 and water, produces no or very little methanol.

Lurgi AG, a leader in methanol synthesis process technology, has developed and thoroughly tested a high-activity catalyst for methanol production from CO_2 and H_2 [416]. Operating at around 260 °C, slightly higher than for conventional methanol synthesis catalysts, the selectivity in methanol is excellent. The activity of this catalyst decreased at about the same rate as the activity of commercial catalysts currently used in methanol synthesis plants. The synthesis of methanol from CO_2 and H_2 has also been demonstrated on a laboratory pilot scale in Japan, where a 50 kg CH_3OH per day production with a selectivity for methanol of 99.8% was achieved [417]. A liquid-phase methanol synthesis process has also been developed, which allows a $CO_2 + H_2$ conversion into methanol of about 95% with very high selectivity in a single pass [356]. We ourselves have developed improved catalysts for the CO_2 into methanol conversion.

The first contemporary commercial CO_2 into methanol recycling plant using locally available, cheap geothermal energy is presently being developed after a successful pilot plant scale operation in Iceland by the Carbon Recycling International (CRI) company. The planned plant, with an initial annual capacity of 3500 tonnes of methanol, is based on the conversion of CO_2 accompanying the readily available local geothermal energy (hot water and steam) sources. The H_2

needed will be produced by water electrolysis using cheap geothermal- or hydropower-produced electricity [418]. Iceland embarked on this development as a means to exploit and possibly export its cheap and clean renewable electrical energy. In Japan, Mitsui chemicals has also announced the construction of a 100 tonne-per-year demonstration plant producing methanol from CO_2, obtained as an industrial by-product, and hydrogen generated by photochemical splitting of water using solar energy [419]. There is also significant industrial interest in CO_2 to methanol conversion in China, Australia, the European Union and other countries, including widespread research and development interest.

Varied improvements for producing methanol from CO_2 and H_2 have been reported [416, 417, 420, 421]. The capital investment for a methanol plant using CO_2 and H_2 is estimated to be about the same as that for a conventional syn-gas based plant [416]. The limiting factor for large scale-up of such processes is the availability and price of CO_2 and H_2 and, first of all, the necessary energy.

Carbon dioxide can be captured and purified relatively easily in large amounts from various sources, including exhausts of fossil fuel-burning power plants and various industrial plants such as cement factories, aluminum smelters and fermentation plants. There are also large natural sources accompanying, for example, natural gas and geothermal energy production. Eventually, even the CO_2 contained in air can be separated and chemically recycled to methanol and desired synthetic hydrocarbons and their products. Hydrogen is presently produced mainly from non-renewable fossil fuel-based syn-gas. In view of our diminishing fossil fuel resources, this is not the way of the future. In addition, the generation of syn-gas from fossil fuels also releases CO_2, which contributes further to the greenhouse effect. Cleavage of water by electrolysis, thermally or by any other method, as a hydrogen source is clearly the path to follow for the future, although in the meantime natural gas and other hydrocarbon sources will still be used. The Methanol Economy approach, beside its other advantages, offers the flexibility to utilize any energy source (conventional, renewable or atomic) to generate the needed hydrogen for the chemical recycling of carbon dioxide.

12.5.1
Carbon Dioxide into Methanol Conversion with Methane

To avoid excessive CO_2 emissions into the atmosphere, processes to produce methanol and its derivatives based on fossil fuels, while concomitantly generating less or even no net CO_2, have been developed and should be increasingly used. In a process called "Carnol," which was developed at the Brookhaven National Laboratory, hydrogen and solid carbon are produced by the thermal decomposition of methane [422]. The generated hydrogen is then reacted with CO_2 recovered from fossil fuel-burning power plants, industrial flue gases or the atmosphere to produce methanol. Overall, the net emission of CO_2 from this process is close to zero, because CO_2 released when methanol is combusted as a fuel is recycled from existing emission sources. The solid carbon formed as a by-product can be handled and stored much more easily than gaseous CO_2, and be disposed of or used as a commodity material, for example in tire production, soil conditioning or as a filler for road construction. The process ultimately depends on the availability of methane:

Methane thermal decomposition $\quad CH_4 \xrightarrow{>800\,°C} C + 2H_2 \quad \Delta H_{298K} = 17.9\,kcal\,mol^{-1}$

Methanol synthesis $\qquad CO_2 + 3H_2 \rightarrow CH_3OH + H_2O$

Overall Carnol process $\qquad 3CH_4 + 2CO_2 \rightarrow 2CH_3OH + 2H_2O + 3C$

Thermal decomposition of methane occurs when methane is heated at high temperature in the absence of air. To obtain reasonable conversion rates under industrial conditions, temperatures above 800 °C are required [422]. This process has been used for many years, not for the production of hydrogen but of carbon black used in tires and as a pigment for inks and paints. For the primary generation of hydrogen, different reactor designs have been proposed. Attention has recently been focused on reactors operating with a molten metal bath, such as molten tin heated to about 900 °C, into which methane gas is introduced. Being an endothermic reaction, the thermal decomposition of methane requires about $18\,kcal\,mol^{-1}$ to produce 2 moles of hydrogen from methane, or 9 kcal of energy for the production of 1 mol (i.e., 2 g) of hydrogen. As a comparison, this is still somewhat less than the highly endothermic methane steam reforming reaction, which requires about $39.3\,kcal\,mol^{-1}$ (combined energy of steam reforming and WGS reaction) to produce 4 moles of hydrogen, or $9.8\,kcal\,mol^{-1}$ hydrogen. It should also be clear that, whereas methane steam reforming produces four molecules of hydrogen for every molecule of methane used, methane decomposition, like partial oxidation of methane, yields only two. However, methane's thermal decomposition by-product (i.e., carbon) can be easily handled, stored and used without much further treatment. The suppression of CO_2 emissions generated by methane steam reforming or partial oxidation is more difficult and energy-intensive. The CO_2 must first be captured, purified, concentrated and finally transported to a suitable sequestration site; this may be distant from the production site and require an extensive pipeline network. The methane thermal decomposition process is still in its development stage, and will require considerable further investigation to become a mature and efficient technology for commercial application. In a carbon-constrained world that is aiming to reduce greenhouse gas emissions, methane thermal decomposition could be a useful alternative for the generation of hydrogen and methanol, avoiding the problems associated with CO_2 emission and sequestration. It would allow mankind to extensively use the world's remaining natural gas resources, including unconventional sources such as the huge methane hydrates deposits under the ocean floors and arctic tundra (given that effective means of processing them is found) with limited effects on the earth's atmosphere.

Another way to produce methanol from CO_2 by sequestering some of the CO_2 in the form of carbon is the combination of CH_4 decomposition and dry reforming. The result is the production of methanol and carbon. For 2 moles of CH_4 used, 1 mole of carbon is formed:

$$CH_4 + CO_2 \rightleftharpoons 2CO + 2H_2$$

$$\underline{CH_4 \rightleftharpoons C + 2H_2}$$

$$2CH_4 + CO_2 \rightleftharpoons \underbrace{2CO + 4H_2 + C}$$

$$\downarrow$$

$$2CH_3OH$$

The environmental benefit is not as high as with the Carnol process, but the economic cost may be lower.

12.5.2
CO$_2$ Conversion into Methanol with Bi-reforming of Methane

A further way to utilize more efficiently our still available natural gas resources to convert CO_2 into methanol is to react CO_2 with natural gas or other hydrocarbon sources to produce syn-gas. In *dry reforming*, which does not involve steam, CO_2 is reacted with natural gas to produce syn-gas with a H_2/CO of 1, which is, however, not suited for the synthesis of methanol requiring a H_2/CO ratio of close to 2. Hydrogen generated from other sources would thus have to be added to obtain the proper H_2/CO ratio.

To overcome this problem and produce a H_2/CO mixture with a ratio close to 2, suited for methanol synthesis, we have developed a specific combination of steam and dry reforming of methane, which we call "bi-reforming" (US Registered Trademark.) [18]. In its use for methanol synthesis, all the hydrogen ends up in methanol with no loss to by-product formation. The catalysts used for bi-reforming are those used for separate dry and steam reforming, combining the two streams thereafter. Dry and steam reforming can also be combined in a single step. The useful temperature range of the reactions is between 800 and 1000 °C. The energy needed can come from any external source, including renewable energy sources as well as atomic energy. As it does not require the combustion of a portion of the natural gas (methane) to provide the heat needed for the process, no excess harmful carbon dioxide is generated, yielding only methanol:

Steam reforming $\quad 2CH_4 + 2H_2O \rightarrow 2CO + 6H_2$
Dry reforming $\quad\quad \underline{CH_4 + CO_2 \rightarrow 2CO + 2H_2}$
Bi−reforming $\quad\quad 3CH_4 + 2H_2O + CO_2 \rightarrow 4CO + 8H_2 \rightarrow 4CH_3OH$

In practical use, natural gas is the major source of methane. Besides methane, natural gas also contains higher saturated hydrocarbons in various concentrations, which can also undergo bi-reforming according to the overall conversion:

$$3C_nH_{(2n+2)} + (3n-1)H_2O + CO_2 \rightarrow (3n+1)CO + (6n+2)H_2$$

The combination of steam and dry reforming in bi-reforming (in separate steps or in a single combined operation) can also be used for the chemical recycling of

CO_2 emissions. The process is also suited for converting CO_2, frequently accompanying natural gas wells and geothermal hot water and steam sources. Otherwise, this CO_2 would be vented into the atmosphere or, after its separation, sequestered underground or at the bottom of the seas. Some natural gas sources, such as some fields in Algeria, contain CO_2 in concentrations of up to 40%. The natural gas of the North Sea produced at the Sleipner platform in Norway contains 9% CO_2. This CO_2 is currently separated and sequestered beneath the North Sea in a deep saline aquifer.

Notably, building on the experience with autothermal reforming, the concept of "tri-reforming" of natural gas to syn-gas was developed based on the synergetic combination of dry reforming, steam reforming and partial oxidation of methane in a single step [423]. The exothermic oxidation of methane with oxygen produces the heat needed for the endothermic steam and dry reforming reactions, allowing a syn-gas mixture with a H_2/CO ratio of close to 2, suitable for methanol production, to be reached. As part of the natural gas is, however, burned to produce the needed heat, the process generates excess carbon dioxide that can be recycled into the process. This recycling would, however, require additional hydrogen from other sources and/or significantly reduce the amount of CO_2 that could be recycled. This is not the case in the discussed environmentally cleaner bi-reforming. The bi-reforming process of methane or natural gas is designed to provide utilization of all the hydrogen content in methane or natural gas for the complete conversion of CO_2 into methanol.

12.5.3
Dimethyl Ether Production from Syn-Gas or Carbon Dioxide

DME, a rapidly growing substitute for diesel and household fuels, is presently made mainly by dehydration of methanol. The conventional bimolecular dehydration of methanol to DME is readily carried out catalytically over varied solid acids such as alumina or phosphoric acid-modified γ-Al_2O_3:

$$2CH_3OH \rightarrow CH_3OCH_3 + H_2O$$

DME is already a significant industrial product in countries such as China, Japan and Korea with an annual production of millions of tonnes. The needed methanol is currently based on syn-gas obtained from either natural gas or coal, although it can also come from any other discussed source. In recent years, direct synthesis of DME from syn-gas, combining methanol synthesis and dehydration in a single step, has been studied extensively. The process essentially combines the methanol synthesis catalyst, based on $Cu/ZnO/Al_2O_3$, with a methanol dehydration catalyst that is operating preferably between 240 and 280 °C at pressures between 30 and 70 atm. Interestingly, the equilibrium conversion in the DME synthesis is significantly higher than that in the methanol synthesis even at low pressure [424]. Using this technology, a demonstration plant producing 100 t-DME per day based on a slurry reactor has been built and tested by JFE (Japan Steel Engineering) in Japan.

There are two main routes to produce DME directly from syn-gas, depicted in (12.4) and (12.5), producing water and carbon dioxide respectively as by-products.

$$2CO + 4H_2 \rightarrow CH_3OCH_3 + H_2O \qquad \Delta H_{298K} = -49.0 \, kcal \, mol^{-1} \qquad (12.4)$$

$$3CO + 3H_2 \rightarrow CH_3OCH_3 + CO_2 \qquad \Delta H_{298K} = -58.7 \, kcal \, mol^{-1} \qquad (12.5)$$

$$2CO + 4H_2 \rightarrow 2CH_3OH \qquad \Delta H_{298K} = -43.5 \, kcal \, mol^{-1} \qquad (12.6)$$

$$2CH_3OH \rightarrow CH_3OCH_3 + H_2O \qquad \Delta H_{298K} = -5.5 \, kcal \, mol^{-1} \qquad (12.7)$$

$$CO + H_2O \rightarrow CO_2 + H_2 \qquad \Delta H_{298K} = -9.8 \, kcal \, mol^{-1} \qquad (12.8)$$

Route (12.4) is a combination of methanol synthesis (12.6) and methanol dehydration (12.7) to DME. Reaction (12.5) combines reaction (12.6) and (12.7) with the water gas shift reaction (12.8). Both routes are being utilized. Route (12.4) is used by Haldor Topsoe and others, whereas JFE follows route (12.5). The by-product of this route is CO_2, the separation of which from DME is much easier and less energy consuming than the separation of water from DME. Route (12.5) also allows a higher syn-gas conversion and has the advantage of using syn-gas with a H_2/CO ratio of 1. This ratio means that coal gasification or methane dry reforming can be used to produce the required syn-gas. The overall reaction combining methane dry reforming with DME synthesis through route (12.5) is basically the reaction of 3 moles of CH_4 with 1 mole of CO_2 with no hydrogen lost in water by-product:

$$3CH_4 + 3CO_2 \rightarrow 6CO + 6H_2$$
$$\underline{6CO + 6H_2 \rightarrow 2CH_3OCH_3 + 2CO_2}$$
$$3CH_4 + CO_2 \rightarrow 2CH_3OCH_3$$

In its DME plant, JFE uses an autothermal reforming unit combining dry reforming and partial oxidation of methane to produce syn-gas with a H_2/CO ratio of 1. The exothermic oxidation of methane generates the heat needed for the process but produces water as a by-product:

$$2CH_4 + O_2 + CO_2 \rightarrow 3CO + 3H_2 + H_2O$$

The discussed bi-reforming pathway can also effectively produce DME, either directly using the DME synthesis route (12.4) or through methanol. The water formed during DME synthesis can be recycled into the bi-reforming step, allowing all the hydrogen content of the used methane (or natural gas) to be utilized only in DME production [18]:

$$3CH_4 + 2H_2O + CO_2 \rightarrow 4CH_3OH \rightarrow 2CH_3OCH_3 + 2H_2O$$

The overall reaction for DME synthesis is accordingly:

$$3CH_4 + CO_2 \rightarrow 2CH_3OCH_3$$

If, instead of natural gas, coal is the utilized fossil fuel, CO_2 formed upon its combustion can be captured and converted into methanol using coalbed methane, frequently accompanying coal seams and readily available, for example, in the USA.

The discussed processes for the chemical recycling of carbon dioxide to methanol and DME depend on the availability of natural gas or other unconventional methane resources such as coalbed methane, methane hydrates, and so on. Ultimately, all of these resources are finite and non-renewable. Over time, they will become depleted or economically too prohibitive to exploit. In the long term, therefore, the large-scale, cost-effective production of hydrogen by electrolysis of water or other water cleavage processes is the key to the development of CO_2 to methanol production, and thus for the Methanol Economy.

The electrolysis of water to produce hydrogen is a well-developed and straight-forward process, and is achieved by applying an electric current between electrodes inserted into water with some electrolytes present. Hydrogen evolves at the cathode, and oxygen at the anode. The electricity needed for the process can be generated by any form of energy. At present, two-thirds of the electricity produced is still derived from fossil fuels, but in the future, to be sustainable and environmentally sound, the electric power required for large-scale electrolysis of water will be obtained from atomic energy (fission and even fusion, if it becomes technically feasible) or any renewable alternate energy source, primarily solar and wind, but also geothermal, hydro, waves, tides, and so on. Such production methods of hydrogen, avoiding the emission of CO_2 to the atmosphere, are discussed in Chapter 9.

12.5.4
Combining Chemical or Electrochemical Reduction and Hydrogenation of CO_2

Considering the chemical recycling of CO_2 to methanol the electrolysis of water is presently the only economically feasible alternative to fossil fuels for the production of hydrogen. However, as mentioned, the catalytic hydrogenation of CO_2 to methanol produces water as a by-product, using up $^1/_3$ of the needed electricity:

$$CO_2 + 3H_2 \rightleftharpoons CH_3OH + H_2O$$

Like methanol, DME can also be produced via the direct catalytic hydrogenation of CO_2. Similar to the discussed route from syn-gas to DME, CO_2 hydrogenation to DME uses a hybrid catalyst consisting of a combination of methanol synthesis and dehydration catalysts [306]. Water formed can, when needed, be recycled, particularly in arid areas or when the need for pure water would warrant it:

$$2CO_2 + 6H_2 \rightarrow 2CH_3OH + 2H_2O$$
$$\underline{2CH_3OH \rightarrow CH_3OCH_3 + H_2O}$$
$$2CO_2 + 6H_2 \rightarrow CH_3OCH_3 + 3H_2O$$

However, to utilize hydrogen more efficiently for CO_2 conversion into methanol and DME the initial chemical or electrochemical reduction of CO_2 to CO to minimize water formation is feasible. Carbon dioxide reduction to CO can be achieved by the reverse Boudouard reaction, that is, the thermal reaction of carbon dioxide with carbon (including coal):

$$CO_2 + C \xrightarrow{\Delta} 2CO \qquad \Delta H_{298K} = 40.8 \, \text{kcal mol}^{-1}$$

This endothermic reaction of coal gasification with CO_2 can be used at temperatures above $800\,^{\circ}C$. Its advantage over the steam reforming of coal, which is somewhat less endothermic ($31.3\,kcal\,mol^{-1}$), is that it allows recycling of CO_2. Coal gasification with CO_2 can be conducted using packed bed or fluidized bed reactors and molten salt media (such as Na_2CO_3 and K_2CO_3 mixtures) [425]. Two-step thermochemical coal gasification combined with metal oxide reduction has also been proposed and tested [426, 427]. Coal gasification with CO_2 has especially been investigated for the conversion of solar thermal heat into chemical fuels, which would allow solar energy to be stored and transported in the form of a convenient fuel such as methanol. The direct conversion of CO_2 into CO using a thermochemical cycle and solar energy is also being studied. Researchers at the Sandia National Laboratories recently developed a solar furnace, which heats a device containing cobalt-doped ferrite (Fe_3O_4) to around $1400-1500\,^{\circ}C$, driving off oxygen gas. At a lower temperature, the reduced material FeO is then exposed to CO_2, from which it absorbs oxygen, producing CO and ferrite, which can be recycled. This technology shows promise, but its viability on an industrial scale is still far away.

Another method to perform the reduction of CO_2 to CO that does not require high temperatures is electrochemical reduction in aqueous or organic solvent media:

$$CO_2 \xrightarrow{\;\;e^-\;\;} CO + \frac{1}{2}O_2$$

This approach has been studied using various metal electrodes in aqueous media [428, 429]. Similar reductions in some organic solvent media were also studied. Methanol in particular, used industrially as a physical absorber for CO_2 in the Rectisol process, has been extensively studied as a medium for the electrochemical reduction of CO_2 [430–432].

During the electrochemical reduction of CO_2 in water or methanol, hydrogen formation competes with CO_2 reduction, thereby reducing the faradic efficiency of the CO_2 reduction. Progress is being made to suppress hydrogen formation.

In our studies of electrochemical CO_2 recycling, instead of considering H_2 formation as a problem it was found advantageous to generate CO and H_2 concomitantly at the cathode in a H_2:CO ratio close to 2, producing a syn-gas mixture ("metgas"), which is then further transformed into methanol, allowing the energy to be used efficiently for CO_2 reduction [433, 434]. An additional advantage is the valuable pure oxygen produced at the anode. The electrochemical reduction reaction of CO_2, however, still has overpotential and efficiency problems, which must be overcome:

$$CO_2 + 2H_2O \xrightarrow{\;\;e^-\;\;} \begin{cases} \overset{\text{Syn-gas}}{[CO + 2H_2]} & \text{at the cathode} \rightarrow CH_3OH \\ 3/2O_2 & \text{at the anode} \end{cases}$$

Regardless, methanol and dimethyl ether can be produced selectively from CO_2 via electrochemically generated syn-gas (metgas) in the same way as it is done from natural gas or coal. The advantage is that no purification step is required and no impurities such as sulfur, which could deactivate the methanol production catalyst, are present. The reaction is preferably run under pressure to feed the metgas directly into the methanol synthesis reactor.

Carbon monoxide obtained by chemical (reverse Boudouard reaction) or electro-chemical reduction of CO_2 can also be reacted with H_2 generated from water to produce DME:

$$CO_2 + C \rightarrow 2CO \xrightarrow{4H_2} 2CH_3OH \rightarrow CH_3OCH_3 + H_2O$$
$$4H_2O \xrightarrow{e^-} 4H_2 + 2O_2$$

We have also found other ways of overcoming some of the discussed difficulties of converting CO_2 into methanol. When using electrochemical reduction of CO_2, besides methanol, inevitably formaldehyde and to some extent formic acid are also obtained. Formic acid and formaldehyde in a subsequent treatment step, not unlike in the previously discussed oxidative conversion of methane (*vide supra*), can be converted into methanol [385, 435, 436]

12.5.5
Separating Carbon Dioxide from Industrial and Natural Sources for Chemical Recycling

Carbon dioxide for chemical recycling to methanol or DME can come from various natural and industrial emissions, and eventually from the CO_2 content of the atmosphere. Presently, worldwide, more than 27 billion tonnes of CO_2 related to human activities are released into the atmosphere every year. Globally, CO_2 emissions from electricity generation, cement and fermentation plants, industries, the transportation sector, heating and cooling of buildings and others are contin-uously increasing despite the Kyoto agreement and other environmental efforts. They all contribute to the increase in CO_2 levels in the atmosphere, from 270 ppm before the beginning of the industrial era to about 380 ppm today. The resulting increased greenhouse effect clearly has a detrimental influence on the Earth's global climate and our ecological systems. For the coming decades, fossil fuels will, foreseeably, continue to provide the largest share of humanity's energy needs. The Kyoto agreement, although not yet approved by all countries, limits the level of CO_2 emissions. Cap and trade levels assigned to countries and industries, however, cannot solve the problem by themselves. To reduce CO_2 emissions, mitigation technologies must be developed and enforced. More energy-efficient technologies and conservation can help, but this will not be sufficient to stop the global increase of CO_2 emissions. To significantly reduce emissions, capture of CO_2 from industrial and natural sources is becoming essential. Carbon dioxide can be captured most readily from concentrated sources such as flue gases from fossil fuel-burning power plants, containing typically between 10 and 15% CO_2 by volume. Many industrial exhausts, including those from coke ovens, cement, iron, steel and aluminum factories and fermentation plants, also contain considerable concentra-tions of CO_2. The recovery of CO_2, though not yet employed on a large scale, is a well-developed procedure. CO_2 can be separated from gas streams by a range of known separation techniques. They are based on different physical and chemical processes, including absorption into a liquid solution, adsorption onto suitable solids, cryogenic separation and permeation through membranes [437]. Amine solution-based CO_2 absorption/desorption systems using monoethanolamine

(MEA) and diethanolamine (DEA) are some of the most widely employed for the separation of CO_2 from gas mixtures. High energy requirements for the regeneration step, limited loadings of amines due to corrosion problems and amine degradation are, however, major drawbacks warranting the development of more efficient, renewable CO_2 sorbents. In our own work, we have developed a new highly effective CO_2 absorption–desorption system consisting of polyethylenimine supported on nanostructured silica, which can even capture to some extent atmospheric CO_2 [438]. Metal-organic frameworks (MOFs) with CO_2 storage capacities up to $1.47\,g\,CO_2$ per g of MOF at a pressure of 30 atm have also been discovered [439]. Scale-up of these technologies, as well as further improvements, are necessary to reduce the cost of CO_2 capture, which presently is limited to highly concentrated industrial or natural sources.

Capturing CO_2 from such sources for recycling to methanol and its derived products also necessitates its purification from commonly accompanying pollutants (especially H_2S and SO_x). This purification is particularly significant in developing "clean coal" technologies. These pollutants, besides their environmental effect, also frequently tend to poison the catalyst systems used in the chemical CO_2 recycling processes discussed and are key to the Methanol Economy. Their removal is therefore also needed to allow technical CO_2 recycling. Methods for the removal of these contaminants have been described in detail in the literature [437]. In the Selexol process, introduced by Union Carbide/UOP over 30 years ago, most of the contaminants can be removed from the gas stream along with CO_2 in a single step. This technology, based on a physical solvent made of a dimethyl ether of polyethylene glycol, has been selected for the first clean coal IGCC power station in the United Kingdom, scheduled to start operation in 2013 [440].

Because they are large and concentrated sources of CO_2, most proposed methods for CO_2 capture, recovery and sequestration are currently focused on large fossil fuel-burning power plants and heavy industries. For such sources, on-site capture of CO_2 offers the most sensible and cost-effective approach. It can significantly mitigate our overall CO_2 emissions before excess atmospheric CO_2 chemical recycling from the air will be realized. On the other hand, more than half of the CO_2 emissions are the result of small dispersed sources such as home and office heating and cooling, and most importantly the transportation sector. The collection at source of CO_2 from millions, even billions, of small fossil fuel-burning units will be difficult if at all possible. For example, while capturing CO_2 from vehicles onboard may be technically feasible, it would be economically prohibitive. Moreover, CO_2, once captured, would have to be transported to a sequestration site, requiring the construction of a massive and expensive infrastructure. Capturing CO_2 onboard airplanes is even less feasible because of the added weight involved. In homes and offices, producing highly dispersed and limited amounts of CO_2, the capture and transportation of CO_2 would also require an extensive and costly infrastructure. While these dispersed CO_2 emissions can probably not be addressed at present, they represent a significant part of the global CO_2 emissions and their importance cannot be ignored in the long term. New feasible approaches for the mitigation of the CO_2 problem, including novel essential technologies, are therefore needed.

12.5.6
Separation of Carbon Dioxide from the Atmosphere

To deal with small and dispersed CO_2 emitters and to avoid the need to develop and construct an enormous CO_2-collecting infrastructure, CO_2 should be eventually captured from the atmosphere – an approach that has already been proposed by some in the past [441–444]. The atmosphere could thus serve as a means of transporting CO_2 emissions to the site of its capture. This would make the CO_2 collection independent of CO_2 sources, and CO_2 could be captured from any source – small or large, static or mobile. With the concentration of CO_2 in air being at equilibrium all around the world, CO_2 extraction facilities could be located anywhere, but to allow for any subsequent chemical recycling to methanol and derived products they should ideally be placed close to hydrogen production sites and major population centers. As mixing and equilibration of air is relatively rapid, local depletion in CO_2 is not likely to pose any problem. If this were not the case, emissions from power plants would cause much higher local concentration of CO_2 near the plants, which is not the case.

Despite the low concentration of CO_2 of presently only 0.038% in the atmosphere, Nature routinely recycles CO_2 by photosynthesis in plants, trees and algae to produce carbohydrates, cellulose and lipids, and eventually new plant life, while simultaneously releasing oxygen and thereby sustaining life on earth. Following Nature's example, humankind will be able to capture excess CO_2 from air and recycle it to generate fuels, synthetic hydrocarbons and their products. Carbon dioxide can already be captured from the atmosphere, although at a substantial expense, using basic absorbents such as calcium hydroxide [$Ca(OH)_2$], potassium hydroxide (KOH) or sodium hydroxide (NaOH), which react with CO_2 to form calcium carbonate ($CaCO_3$), potassium carbonate (K_2CO_3) and sodium carbonate (Na_2CO_3), respectively [446, 447]. Owing to its low CO_2 content, very large volumes of air must be contacted with the sorbent material, and this should be achieved with minimum energy input, preferably using natural air convection. After capture, CO_2 would be recovered from the sorbent by desorption, through heating, vacuum or electrochemically. Calcium carbonate, for example, as well known in the cement industry, can be thermally calcinated to release carbon dioxide. Carbon dioxide absorption is an exothermic reaction, which liberates heat, and is readily achieved by simply contacting CO_2 with an adequate base. The energy-demanding step is the endothermic desorption, requiring energy to regenerate the base and to recover CO_2. Calcium carbonate or sodium carbonate require high energy input for recovery and therefore are probably not the most suitable, practical candidates for CO_2 capture from air. Other bases might be more appropriate for this application. Research in this area, though still in its relatively early phases of development, should determine suitable absorbents and technologies to remove CO_2 from air, with the lowest possible energy input. For example, when using KOH as an absorbent, it has been shown that the electrolysis of K_2CO_3 in water could efficiently produce not only CO_2 but also H_2 with relatively modest energy input [448]. Using a similar electrochemical recovery process, potassium carbonate (K_2CO_3), which reacts with CO_2 in water to form potassium bicarbonate ($KHCO_3$), has also been proposed as an absorbent for CO_2

[449]. With further developments and improvements, CO_2 capture from the atmosphere, which has already been described as technically feasible, will also become economically viable [446]. The cost of capturing CO_2 from the atmosphere has been estimated at between $100 and $200 per tonne of CO_2, while further improvements could drive costs even lower [450–452]. In submarines and space flights, the essential removal of CO_2 to keep the air breathable is carried out using regenerable polymeric or liquid amine scrubbers. In our own work, as previously mentioned, a nanostructured silica-supported polyethylenimine absorbent was found to absorb CO_2 from air, although further work is needed to increase the efficiency of CO_2 capture at these low concentrations [438].

Among the various advantages of CO_2 extraction from air is the fact that CO_2 capture, in being independent of CO_2 sources, allows more CO_2 to be captured than is actually emitted from human activities. This technology could allow humankind to not only to stabilize CO_2 levels (making us carbon neutral), but also eventually to lower them, making our carbon emissions negative. By capturing more carbon than is emitted it would be even possible to reduce the atmospheric level of carbon dioxide. This capture would actually permit earth to return to a lower concentration of CO_2 in the air without having to wait for natural processes to absorb excess anthropogenic CO_2.

Our air also contains other essentials for mankind's sustainable future in considerably higher concentration than the low (0.038%) CO_2 content: (i) pure water vapor, in concentrations from 1 to 6%, depending on the moisture content of the air in varied location, essential to life and an inexhaustible and renewable source of hydrogen that can be generated using any available energy; (ii) some 79% nitrogen, for the synthesis of ammonia and derived synthetic nitrogen-containing compounds, especially fertilizers, as well as together with CO_2 and water for producing synthetic proteins; (iii) 20% oxygen, also essential to life and for combustion processes. Utilizing all these atmospheric resources can assure a sustainable future for most of humankind's needs.

The removal of even a fraction of the CO_2 from industrial and natural emissions and capture of CO_2 from the atmosphere would result in huge amounts of available CO_2. As proposed presently, the captured CO_2 could be stored/sequestered in depleted gas- and oil-fields, deep aquifers, underground cavities or at the bottom of the seas. This approach, however, does not provide a permanent solution, nor does it assist in mankind's future needs for fuels and hydrocarbons and their products. The recycling of CO_2 via its chemical reduction with hydrogen to produce methanol (i.e., the "Methanol Economy") is, therefore, an attractive alternative. As fossil fuels become scarcer, the capture and recycling of even atmospheric CO_2 would become and remain feasible for the production of methanol, DME and derived synthetic hydrocarbons and associated products. The hydrogen required would be obtained by the electrolysis of water (an unlimited renewable resource), while also releasing oxygen. The electrical power needed will be provided by any suitable renewable and/or atomic energy source. Upon their combustion and use, methanol and derived synthetic hydrocarbon products would be again transformed into CO_2 and water, thereby closing the methanol cycle. This would constitute mankind's artificial

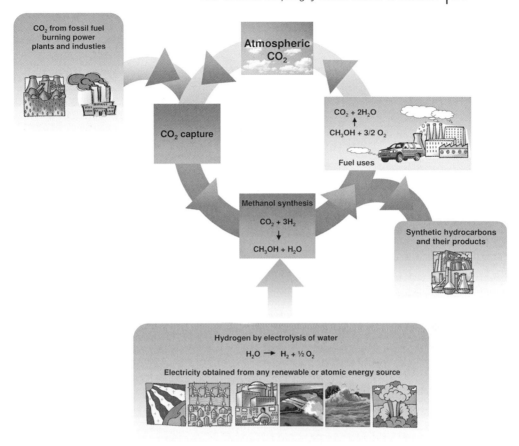

Figure 12.14 CO_2 recycling for methanol and synthetic hydrocarbons production.

version of Nature's CO_2 recycling via photosynthesis. By using this approach, there would be no need for any drastic change in the nature of our energy storage and transportation systems, as would be required by switching to the practically very difficult hydrogen economy, while also providing synthetic hydrocarbons (Figure 12.14). Furthermore, as CO_2 is available to everybody on earth, it would liberate us from the reliance on diminishing and non-renewable fossil fuels and all the geopolitical instability associated with them.

Figure 12.15 shows a possible outline for the transition to a sustainable future for carbon fuels and hydrocarbon products. The production of methanol and DME from biomass feedstocks and from recycling of CO_2 contained in flue gases of various industries could be the first steps towards this goal. However, as fossil fuel become less abundant or their use is regulated by stricter emission standards, related CO_2 emissions will eventually diminish. On the other hand, as discussed, the amount of biomass that can be generated in a sustainable way is large but nevertheless limited. These limitations all point to methanol, DME and derived products being

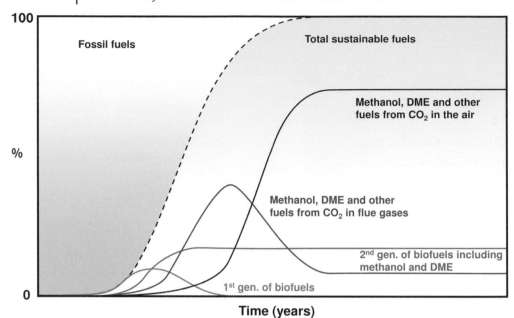

Figure 12.15 Possible transition to a sustainable fuel future including methanol and DME as key components. Based on Reference [257].

increasingly produced from CO_2 captured from the air. Depending on various factors, including policies, geographical locations, state of development and available resources, the timeline and layout for the transition to sustainable fuels and products will clearly vary from country to country.

13
Methanol-Based Chemicals, Synthetic Hydrocarbons and Materials

13.1
Methanol-Based Chemical Products and Materials

Methanol, besides becoming an important transportation fuel together with its derived product DME, is one of the most important starting materials for the petrochemical and chemical industry. Most of the 40 million tonnes of methanol produced yearly are used for the production of a large variety of diverse chemical products and materials, including such basic chemicals as formaldehyde, acetic acid and methyl *tert*-butyl ether (MTBE), as well as various polymers, paints, adhesives, construction materials and others (Figure 13.1).

In processes for the production of basic chemicals, raw material feedstocks constitute typically up to 60–70% of the manufacturing costs. The cost of feedstock therefore plays a significant economic role. In the past, for example, acetic acid was predominantly produced from ethylene using the Wacker process, but during the early 1970s Monsanto introduced a process, which by carbonylation of methanol using a Wilkinson's rhodium-phosphine catalyst and iodide (in the form of HI, CH_3I or I_2), produces acetic acid with a conversion and a selectivity close to 100%. Because of the high efficiency of this process and the lower cost of methanol compared to ethylene, most of the new acetic acid plants built worldwide since then are based on this technology [453]. Taking advantage of its lower cost, methanol is also considered as a potential feedstock for other processes currently utilizing ethylene. Rhodium-based catalysts promote the reductive carbonylation of methanol to acetaldehyde, with selectivities close to 90%. With the addition of ruthenium as a co-catalyst, the further reduction of acetaldehyde to ethanol is possible, providing a new catalytic route for the direct conversion of methanol into ethanol. The possibility of producing ethylene glycol via methanol oxidative coupling instead of the usual process using ethylene as a feedstock is also being pursued. Significant advances have also been made towards the synthesis of ethylene glycol from dimethyl ether, obtained by methanol dehydration. In addition to acetic acid, acetaldehyde, ethanol and ethylene glycol, other large-volume chemicals produced currently from ethylene or propylene, such as styrene and ethylbenzene, may also be manufactured from methanol in the future.

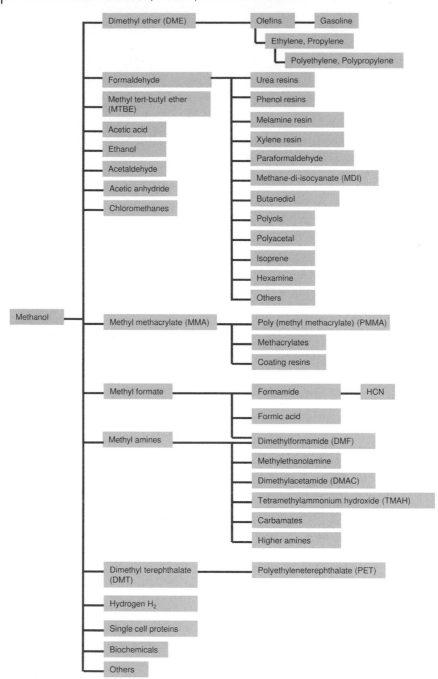

Figure 13.1 Methanol-derived chemical products and materials.

13.2
Methyl *tert*-butyl Ether and DME

Methyl *tert*-butyl ether is mainly produced by the reaction of methanol with isobutylene over a mildly acidic catalyst such as polymeric acidic resins at around 100 °C:

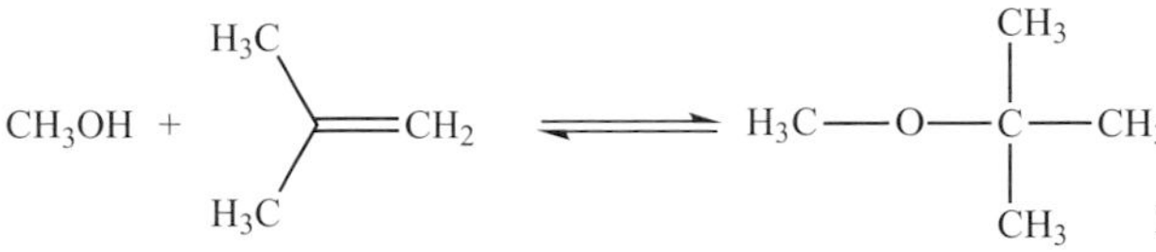

Methyl *tert*-butyl ether became a large scale methanol-derived product of the petrochemical industry in the 1980s when it gained increasing significance as an additive for gasoline due to its high octane rating of 116 and the necessity to find alternatives to the toxic and harmful tetraethyl-lead, which was used as an octane booster and antiknock compound [454, 455]. The use of MTBE in gasoline also reduced the need for alternative high octane aromatic components such as benzene and toluene, which are toxic and known carcinogens. At the same time the addition of an oxygenated compound such as MTBE in reformulated gasoline helped the fuel burn cleaner and more completely, reducing emissions of CO, hydrocarbons and ozone. Ethers and MTBE in particular were preferred over other oxygenated compounds due to their excellent vaporization properties. MTBE, however, has a relatively high solubility in water and is not easily degraded in nature. In the mid-1990s, MTBE contamination began to be detected in groundwater across the US. This accumulation of MTBE in groundwater was found to be mainly due to leakages from underground gasoline tanks in filling stations, due to lacking or inadequate regulation. The taste and foul odor of MTBE can be detected in water by humans even at very low concentration. Its toxicity is, however, much lower than other components of gasoline such as benzene, which might also be present. Nevertheless, contamination of groundwater prompted a ban on MTBE and other ethers as fuel additives in California, New York and many other states in the US. In Europe, MTBE has not been phased out. However, due to directives to increase the use of biofuels for transportation, European countries, starting in 2003, have moved from MTBE to "bio"-ethanol and derived ethyl *tert*-butyl ether (ETBE) as oxygenated additives for gasoline. ETBE is produced by the reaction of ethanol with isobutylene much like MTBE from methanol. In Europe, most MTBE plants were therefore retrofitted and are currently producing ETBE. In the rest of the world, however, especially in Asia, the use of MTBE and other ethers is still growing.

Dimethyl ether, as discussed in Chapter 11, is the simplest ether derived from methanol, whose use as transportation and heating fuel substitute is expanding rapidly, notably in China, Korea and Japan. Contrary to MTBE, which necessitates isobutylene for its production, DME is made solely from methanol. Furthermore, in contrast with MTBE, it is not only a fuel additive for gasoline or diesel fuel but rather an excellent fuel for compression ignition (CI) engines on its own. Dimethyl ether is currently produced mainly by the dehydration of methanol over mildly acidic catalysts, although its direct production from syn-gas is also being developed (Chapter 12):

$$2CH_3OH \rightarrow CH_3OCH_3 + H_2O$$

Dimethyl ether is also an important intermediate in the methanol to olefin (MTO) and methanol to gasoline (MTG) processes described hereafter.

13.3
Methanol Conversion into Light Olefins and Synthetic Hydrocarbons

While methanol can directly replace light olefins (ethylene and propylene) in some applications, these will remain indispensable building blocks for synthetic hydrocarbons and other products. Ethylene and propylene are by far the two largest volume chemicals produced by the petrochemical industry. In 2007, about 115 million tonnes of ethylene and 73 million tonnes propylene were consumed worldwide. They are important starting materials in the production of plastics, fibers and chemical intermediates such as ethylene oxide, ethylene dichloride, propylene oxide, acrylonitrile and others. The demand for light olefins C_2 to C_5, however, is primarily driven by polyolefin production. Today, almost 60% of ethylene and propylene is consumed in the manufacture of polyethylene (low-density polyethylene, LDPE, high-density polyethylene, HDPE, etc.) and polypropylene. Most light olefins are currently produced by the petrochemical industry as a by-product of steam cracking and fluid catalytic cracking (FCC) of naphtha and other gas liquids. The demand for propylene is growing at a faster rate than that for ethylene; about 6% per year compared to 4% per year for ethylene. The reason for the higher growth in propylene demand is the increasing popularity of polypropylene, which is being substituted for many other materials and more expensive polymers, especially for automobile parts. Because of the high cost of transporting light olefins by sea, and the proximity of downstream markets, they are mainly produced in North America and Western Europe. With lower feedstock costs, a large part of the new production capacity for light olefins, however, will be installed in the Middle East. Through steam cracking, light hydrocarbons (especially ethane), produced together with petroleum oil or from natural gas in that region, yield predominantly ethylene. Given the higher growth rate in propylene demand, this could lead to supply problems. Steam cracking and fluid catalytic cracking will not be able to cover the expected demand for propylene in the coming decades. Thus, the balance will have to be supplied by other sources, including propane dehydrogenation, metathesis, olefin cracking and significantly new methanol-to-olefin (MTO) technologies. MTO can also provide a part of the demand for ethylene. Because methanol is presently produced mainly from natural gas and in some countries from coal (through syn-gas) it can decrease the dependence of ethylene on petroleum feedstocks. Considering the very large market for ethylene and propylene, these technologies will also substantially increase methanol demand, calling for the construction of numerous mega-methanol plants, each able to produce between 1 and 3.5 million tonnes of methanol per year. With the development of new technologies to produce methanol – most significantly via the chemical hydrogenative recycling of CO_2 – the MTO processes already in commercial development will gain increased significance, allowing the production of light olefins from non-fossil sources and their subsequent conversion into synthetic hydrocarbons and their various products.

More than a century ago, LeBel and Greene first reported the formation of gaseous saturated hydrocarbons (and some hexamethylbenzene) by dripping methanol onto "hot" zinc chloride. Under higher pressure, significant amounts of light hydrocarbons were formed when methanol or dimethyl ether was reacted over zinc chloride at about 400 °C. Recently, it was reported that methanol reacts with zinc iodide at 200–240 °C to give a mixture of C_4–C_{13} hydrocarbons containing almost 50% 2,2,3-trimethylbutane (triptane, an excellent high-octane jet fuel) [456–459]. Other catalysts, including phosphorus pentoxide, polyphosphoric acid and, later, tantalum pentafluoride and other superacid systems, have also been reported for the production of hydrocarbons from methanol. Most of the described catalysts, however, deactivate rapidly. It was only during the 1970s that researchers at the Mobil Oil Company discovered that an acidic zeolite called ZSM-5 was able to catalyze the practical conversion of methanol into both olefins (MTO process) and hydrocarbons in the gasoline range (MTG process) [460–462].

13.4
Methanol to Olefin (MTO) Processes

The methanol-to-olefin technology, or MTO, was developed as a two-step process, in which natural gas or coal is first converted via syn-gas into methanol, followed by its transformation into light olefins. The initial driving force for the development of this technology was to utilize natural gas sources far from major consumer centers, such as New Zealand.

The conversion of methanol into olefins proceeds through the steps:

$$2CH_3OH \underset{+H_2O}{\overset{-H_2O}{\rightleftharpoons}} CH_3OCH_3 \overset{-H_2O}{\longrightarrow} \overset{\text{ethylene \& propylene}}{H_2C = CH_2 \ \& \ H_2C = CH - CH_3}$$

The initial step is the dehydration of methanol to dimethyl ether (DME), which then reacts further to form ethylene and propylene. In the process, smaller amounts of butenes, higher olefins, alkanes and some aromatics are also produced. Besides the synthetic aluminosilicate zeolite (ZSM-5) catalysts, numerous other catalysts were also subsequently studied. UOP developed silicoaluminophosphate (SAPO) molecular sieves such as SAPO-34 and SAPO-17, which have demonstrated high activity and selectivity for the MTO process. Both types of zeolites have defined three-dimensional crystalline structures. They are microporous solids permeated with channels and cages of very specific size. The many different zeolite catalysts differ by their chemical composition and the size and structure of their channels and cages, which have molecular dimensions ranging from 3 to 13 Å. The catalytic sites responsible for the catalyst's activity are in the pores and channels of the zeolite catalysts. Access to these sites will therefore be limited to reagents small enough to enter the zeolites pores and channels. At the same time, the size of the reaction products is also controlled by steric constraints imposed by the catalyst's structure. As a result, zeolites can be highly shape-selective catalysts. ZSM-5 has pore openings of some 5.5 Å. With a pore size of only 3.8 Å, SAPO-34, used in the MTO process, allows effective control of the size of the olefins that emerge from the catalyst. Larger olefins diffuse out at a lower rate, making smaller olefins such as ethylene and propylene the predominant products.

Independent from zeolites, bi-functional supported acid–base catalysts, such as tungsten oxide on alumina (WO_3/Al_2O_3) were found by Olah and coworkers during the 1980s to also be active for the conversion of methanol and other heterosubstituted methanes into ethylene and propylene at temperatures between 250 and 350 °C [463, 464]. These heterogeneous bi-functional catalysts catalyze the reaction despite the fact that they lack the well-defined three-dimensional structure of shape-selective zeolite catalysts.

Based on SAPO-34, an MTO process has been developed jointly by UOP and Norsk Hydro [461–463]. This process converts methanol with more than 80% selectivity into ethylene and propylene. In addition, about 10% is converted into butylenes, which are also valuable starting materials for various products. Depending on the operating conditions, the propylene-to-ethylene weight ratio can be modified from 0.77 to 1.33. This adjustment allows considerable flexibility and adaptation to changing market conditions. The technology has been extensively tested in a demonstration plant in Norway, and more than ten years of development have now been completed. Currently, the UOP/Hydro MTO process is commercialized in Nigeria and units are being considered in other locations. A methanol plant, built close to the capital city of Lagos, with a capacity of 2.5 million tons per year, will be the largest in the world. The methanol produced from natural gas via syn-gas will then be converted into ethylene and propylene to produce 400 000 tonnes per year of polyethylene and 400 000 tonnes per year of polypropylene.

Lurgi has also developed an MTO process [468], which, unlike the UOP/Hydro technology, is designed to yield mostly propylene. It is thus named as a methanol-to-propylene (MTP) process. In the first step, methanol is dehydrated over a slightly acidic catalyst to produce DME. The DME is then reacted over a ZSM-5-based catalyst under moderate pressure (1.3–1.6 atm) and temperatures between 420 and 490 °C to form light olefins. The achievable overall yield of propylene is above 70%. The process has been demonstrated at Statoil's Tjeldbergodden methanol plant in Norway, and is now ready for commercialization. Propylene obtained from this unit, with a purity of 99.7%, has also been successfully polymerized to polypropylene, showing the feasibility of producing polypropylene directly from methanol obtained from natural gas. Using the methanol synthesis and the MTO processes developed by Lurgi, the Caribbean country of Trinidad and Tobago is planning to produce 450 000 tonnes per year of polypropylene from methanol using its large natural gas resources. Lurgi has also licensed its technology to two Chinese companies, which are both building plants with a capacity of about 450 000 tonnes per year produced from methanol based on coal-derived syn-gas [30]. Other MTO plants based on syn-gas obtained from coal are also being constructed in China. One of these plants, built by the Shenhua Company, includes a 1.8 million tonne methanol unit and 600 000 tonne MTO unit, as well as facilities to produce polyethylene and polypropylene. The DMTO (dimethyl ether/methanol to olefins) technology used in this plant to produce ethylene and propylene was recently developed by the Dalian Institute of Chemistry and Physics. In the US, Eastman, which has been operating a coal-based methanol plant in Kingsport, Tennessee for more than 25 years, is planning the construction of several new methanol-to-chemical plants, including MTO plants.

Mobil (now ExxonMobil), which pioneered the MTO technology using the company's ZSM-5 catalyst, also demonstrated its technology on a 100 barrel-per-day scale in Wesseling, Germany. In addition, Mobil also developed the related olefins to gasoline and distillate (MOGD) process. In this process, which was originally developed as a refinery process, the olefins from the MTO unit are oligomerized over a ZSM-5 catalyst to yield, with selectivity greater than 95%, hydrocarbons in the gasoline and/or distillate range. Depending on the reaction conditions, the ratio between gasoline and distillate can be varied considerably, allowing a significant flexibility in production. When operated at relatively low temperatures and high pressures (200–300 °C, 20–105 bar), called the distillate mode, the products are higher molecular weight olefins that, after being subjected to hydrogenation, produce fuels, including diesel and premium quality jet fuels. Changing the operating conditions to higher temperatures and lower pressures affords products of lower molecular weight but with higher aromatic content (i.e., high-octane gasoline). Besides being a catalyst for the MTO and MOGD processes, ZSM-5 was also found to catalyze the direct conversion of methanol into hydro-carbons in the gasoline range.

13.5
Methanol to Gasoline (MTG) Processes

The MTG process was conceived and developed in response to the energy crisis of the 1970s. It was the first major new route to synthetic hydrocarbons since the introduction of the Fischer–Tropsch process before World War II, and provided an alternative pathway for the production of high-octane gasoline from coal or natural gas. Discovered accidentally by a research team at Mobil, the MTG process was actually developed before the MTO process. The MTO process can be considered as a modified MTG process designed to produce mainly olefins instead of gasoline. For the MTG conversion, medium-pore zeolites with considerable acidity are the most suitable catalysts, with ZSM-5 recognized as the most selective and stable. Because of the defined structure and geometry of their pores, channels and cavities they are shape-selective catalysts, able to control product selectivity, depending on their molecular size and shape. Over this catalyst, methanol is first dehydrated to an equilibrium mixture of DME, methanol and water. This mixture is then converted into light olefins, primarily ethylene and propylene. Once these small olefins are formed, they can undergo further transformations into higher olefins, C_3–C_6 alkanes and C_6–C_{10} aromatics:

$$2CH_3OH \underset{+H_2O}{\overset{-H_2O}{\rightleftharpoons}} CH_3OCH_3 \xrightarrow{-H_2O} \text{light olefins} \rightarrow \text{higher olefins} \quad \begin{array}{l} \text{alkanes} \\ \\ \text{aromatics} \end{array}$$

The mechanistic pathway of the MTG process could be more complex, going through the intermediate formation of polymethylbenzenes and their conversion involving a hydrocarbon pool. Regardless of the mechanism, these processes effectively produce the desired hydrocarbons.

Owing to the shape selectivity of ZSM-5, heavier hydrocarbons containing more than ten carbon atoms are practically not produced in the MTG process. This selectivity is fortuitous, as C_{10} is also the usual limit for conventional gasoline. At the same time, the process also produces aromatics (toluene, xylenes, trimethylbenzene, etc.), providing a route to aromatic hydrocarbons. Depending on the catalyst and conditions used, the product distribution can be modified if desired. Using zeolites with larger pores, channels and cavities, such as ZSM-12 or mordenite, leads to products with higher molecular weight.

Again, as studied by Olah and coworkers in the 1980s, non-zeolitic bi-functional acid–base catalysts such as tungsten oxide (WO_3) supported on alumina were found to be active above 350 °C for the conversion of methanol into a hydrocarbon mixture including alkanes in the gasoline range, alkenes and aromatic compounds.

In 1979, the New Zealand government selected the MTG process developed by Mobil [462] for the conversion of its natural gas from the large off-shore Maui field into gasoline. The New Zealand plant began operations in 1986, producing about 600 000 tonnes of gasoline per year, supplying one-third of New Zealand's gasoline needs. Methanol was produced using the ICI low-pressure methanol process in two production units, each capable of producing 2200 tonnes of methanol per day. Crude methanol from these units was fed directly to the MTG section, where it was first converted over an alumina catalyst into an equilibrium mixture of methanol, DME and water. This mixture was then transferred to the gasoline synthesis reactor where it was reacted over ZSM-5 at 350–400 °C and 20 atm [461]. The crude gasoline was then treated to remove any of the heavier components. Without further distillation or refining necessary, the high-octane gasoline obtained can be blended directly with the general gasoline pool.

As mentioned earlier, the MTG process was developed in response to the energy crisis of the 1970s, which saw the price of oil and its products rise dramatically. However, as the oil price fell again during the 1980s, dipping briefly under $10 in 1986, commercial interest for MTG dropped accordingly. Gasoline production at the New Zealand plant ceased because it was now cheaper to use low-cost gasoline derived from petroleum than to produce it from natural gas via methanol. Methanol production itself is, however, still in operation, providing methanol at a competitive cost. Increasing oil prices experienced during the past few years will probably revive interest in MTG.

In West Virginia, Consol Energy, a large coal producer, is planning the construction of a coal-to-liquid (CTL) plant that will produce about 720 000 tonnes of methanol per year. A part of the methanol will be used to produce about 2.5 million barrels of gasoline [271]. A comparable plant with a capacity of 5.5–7.5 million barrels a year, also using the ExxonMobil MTG process, is planned in Wyoming. In this plant, the excess CO_2 produced during the process will be captured, compressed and used for enhanced oil recovery [469]. Ongoing development in the Far East, particularly in China, of large scale methanol and DME production from coal, but also in the Middle East and elsewhere based on natural gas, indicates future potential for the development of MTO and MTG production capacities.

13.6
Methanol-Based Proteins

Methanol, interestingly, can also serve as the source for single-cell protein production. Single-cell proteins (SCPs) refer to proteins produced by various microorganisms degrading hydrocarbons while gaining energy [232, 237]. The protein content depends on the type of microorganism: bacteria, yeast, mold, and so on. The use of microorganisms in human alimentation has been practiced since ancient times, in the form of yeasts used in brewing and baking, and bacteria in cultured dairy products such as cheese, yogurt and sour cream. In modern times, the possibility of using proteins produced by microorganisms for animal and human alimentation first emerged in Germany during World War I. The main development of SCP, however, began during the 1950s when the petroleum industry realized that, by the degradation of hydrocarbons, some microorganisms could produce high-quality proteins that were suitable for animal feed or as a nutritional source for human food. Because of the possibility of contamination and build-up of carcinogenic compounds in animals fed with SCP produced from oil products, however, methanol was chosen as a substitute. In contrast with petroleum-derived feeds, methanol is non-carcinogenic, forms a homogeneous solution with the aqueous nitrogen-containing nutrient salt solutions used in the process, and can be readily separated from the protein products after their formation. Several companies, including Shell, Mitsubishi, Hoechst, Phillips Petroleum and ICI, have studied the bacterial fermentation process. For some time, ICI operated a commercial plant in Billingham, England, producing 70 000 tonnes of SCP from 100 000 tonnes of methanol per year. The bacteria obtained had a very high protein content, with the dried cells containing up to 80% protein (much higher than other types of food such as fish and soybean). As the structure of the bacteria was highly complex, a wide range of amino acids – including in particular aspartic and glutamic acids, alanine, leucine and lysine – was obtained. The overall quality of SCP produced from methanol by the ICI process was very high, and the product sold as an animal feed under the name Pruteen.

Besides serving as a medium for microorganism culture, methanol was also found to substantially increase the growth rate in various plants [232]. In plants with C_3 metabolism – for which the first product of photosynthesis is a three-carbon sugar – significantly higher photosynthetic productivity has been observed, especially in regions with high sunlight intensity such as the south-western United States. C_3

plants include sunflower, watermelon, tomato, strawberry, lettuce and eggplant. Methanol sprayed onto the plants is rapidly absorbed by the foliage and metabolized to CO_2, sugars, amino acids and other structural components. It is thus used as a concentrated source of carbon in place of CO_2 (1 mL of liquid methanol contains about as much carbon as 2 000 000 mL of air).

Methanol was also found to be an economic and effective means of inhibiting the process of photorespiration; that is, the plants uptake of oxygen, which is competing with CO_2 uptake for photosynthesis. Oxygen assimilation results in the breakdown of sugars, reversing the photosynthetic process. When exposed to stressful conditions such as high light intensity and high temperatures, the stomata (tiny pores used by the plant to absorb atmospheric CO_2) close, reducing the uptake of CO_2, and resulting in increased photorespiration. This can stop plant growth for several hours during the hottest period of day. The control of photorespiration is, therefore, key to enhancing the photosynthetic yield of plants. Further studies on the effects of methanol on the complex growth mechanism of plants are needed before large-scale applications can be envisioned. Methanol, however, has a good potential to effect significant improvements in crop productivity.

With an increasing world population, agricultural production might encounter difficulties in providing sufficient protein for food and animal feed. The "Methanol Economy" could, therefore, also supplement essential protein needs through SCP.

13.7
Outlook

Inevitably, with decreasing oil and gas reserves, synthetic hydrocarbons will play a major future role. Since methanol can be produced not only from still-available carbonaceous sources of fossil fuels (e.g., coal, natural gas), natural bio-sources but also from chemical recycling of natural and industrial CO_2 emissions and eventually from the air itself, methanol-based chemicals and products – and particularly synthetic hydrocarbons available through MTG and MTO processes – will assume increasing importance in replacing decreasing oil- and natural gas-based resources.

14
Conclusions and Outlook

14.1
Where We Stand Now

Much has been said of the extent of our available oil and gas resources (Chapters 4 and 5). Although our coal reserves may last for another two or three centuries, the mining of coal, except in areas suited to surface strip mining, necessitates difficult and dangerous labor and involves hazards and environmental difficulties. Moreover, our readily accessible oil and gas reserves will not last much longer than the end of the twenty-first century – even taking into consideration new discoveries, improved technologies and unconventional sources.

Besides finite accessible petroleum oil, natural gas and coal resources, we have additional unconventional hydrocarbon sources such as heavy oil deposits in Venezuela, oil shale in various geological formations, including the US Rocky Mountains, and vast tar sand deposits in Alberta, Canada. The hydrates of methane, as are found under the Siberian tundra and along the continental shelves of the oceans, also represent very significant resources for natural gas for the future. They too will all eventually be exploited, although the difficulties and costs involved are immense.

In addition to the size of the reserves, one must also consider the demand from an expanding world population, which currently nears seven billion and will most likely reach eight to ten billion during the twenty-first century. The consequences of increasing consumption of our reserves, also due to improved standards of living around the world, as well as increased demands in fast-developing countries such as China and India, are clear. Potential oil reserves estimated at some two trillion barrels or 270 billion tonnes must be considered when taking into account these factors. The best present estimates of our relatively easily accessible oil reserves imply that they would last for no more than 70 years at the current rate of consumption. Natural gas reserves are somewhat larger, and may last for another 80–100 years. New discoveries, savings, improved technologies and enhanced recovery methods such as directional drilling can extend these estimates, while increased oil consumption due to population growth and increasing standards of living will place more pressure on our reserves. In any case, humankind must begin to prepare itself for the future, and find new sources and solutions.

Beyond Oil and Gas: The Methanol Economy, Second updated and enlarged edition
George A. Olah, Alain Goeppert, and G. K. Surya Prakash
Copyright © 2009 WILEY-VCH Verlag GmbH & Co. KGaA, Weinheim
ISBN: 978-3-527-32422-4

All fossil fuels are mixtures of hydrocarbons, which contain varying ratios of carbon and hydrogen. Upon their combustion, carbon is converted into CO_2 and hydrogen into water. Consequently, when burned, these fuels are irreversibly used up. The increase in CO_2 content of the atmosphere that results from human activities and the excessive combustion of fossil fuel is considered to be a major man-made cause of global warming superimposed on Nature's own cyclical changes.

To satisfy humankind's ever-increasing energy needs and to provide them in an environmentally adaptable way, the use of all feasible alternative energy sources will be necessary. Hydro and geothermal energy are already well used where Nature makes them available, but relatively few major new suitable locations are expected to be found even in developing countries. The energy of the sun, wind, waves, tides of the seas and others all have great potential and are increasingly being exploited, though their large-scale use as substitutes for fossil fuels will have only a gradual impact on our energy picture in the foreseeable future.

Perhaps one of most remarkable technological achievements of the twentieth century has been humankind's ability to harness the energy of the atom. Regretfully, as this was first achieved for the building of the atom bomb, public opinion has in subsequent years increasingly turned against atomic energy, even when considering its peaceful uses. During the past few decades, relatively few new atomic power plants have been constructed, and none in the United States. There is even strong sentiment in many countries to close them down altogether, although other countries such as France depend on them for some 80% of their electricity needs. Much progress has been made to limit the use of atomic energy only to peaceful uses and to improve safety aspects, including radioactive waste storage and disposal. Our society can and will solve these problems. The decline of the atomic energy industry in most industrialized countries is most regrettable and shortsighted. Whether or not one likes atomic energy, it is for the foreseeable future, beside solar and wind energy, one of the most feasible and massive energy sources available to humankind. Of course it should be made even safer and more effective, solving the problem of reuse and storage of radioactive wastes as well as developing new improved reactor designs and eventually even the use of atomic fusion. At the same time, atomic energy is not polluting our atmosphere or contributing to global warming. There have been recent hopeful signs that the generally negative perception of atomic energy is slowly changing. Conservation and use of alternate energy sources are most desirable, but none of these on their own can satisfy our enormous appetite for energy.

Despite their non-renewable nature and diminishing resources, fossil fuels will maintain their leading role as long as they are readily available. A vast infrastructure exists for their transport and distribution. As transportation fuels for our cars, trucks and airplanes, oil and gas are used in the form of their convenient products: gasoline, diesel fuel or as compressed natural gas. Natural gas and heating oil are essential for heating our homes and offices, and in providing energy for industry. Oil and gas are also the raw materials for chemical products and materials essential for our everyday life. However, a large part of the fossil fuels are still utilized to generate electricity. Coal is still dominating the generation of electricity in many parts of the world, and will continue to do so for a long time. Continuing progress is being made to make coal "clean," but this focuses generally only on removing toxic and harmful pollutants. Carbon dioxide formed upon the combustion of coal remains a major greenhouse gas, greatly contributing to global warming caused by human activity. From whatever source electricity is generated, its storage on a large scale remains also unresolved. Batteries, for example, are inefficient and bulky, and it is therefore necessary to find, besides new energy sources, efficient ways for energy storage and distribution. We also need to develop new and efficient ways to produce synthetic hydrocarbons and their varied products from non-fossil fuel sources. It is ironic that fully knowing that we will need to produce synthetic hydrocarbons at a significant cost and major technological efforts we continue to burn much of the still-existing natural fossil fuel resources to provide energy.

One environmentally advantageous approach that has been proposed and much discussed recently is the use of hydrogen as a clean fuel in the context of the so-called "Hydrogen Economy" (Chapter 9). Free hydrogen is, however, not a natural energy source on earth as it is incompatible with the high oxygen content of our atmosphere. While hydrogen is indeed "clean burning," forming only water, its generation from its bound forms (hydrocarbons, water) is a highly energy-consuming process that is at present far from clean, as hydrogen is mainly produced by the reforming of natural gas, oil or coal (i.e., fossil fuels) to syn-gas, a mixture of CO and H_2. Hydrogen, however, can also be produced by the electrolysis of water, which can be generated using any energy source. Our oceans represent an inexhaustible source of water (although it must first be desalinized) that can be split by electricity, or by other means, to produce hydrogen and oxygen. Highly volatile hydrogen is, however, not a convenient fuel to handle. Its storage, transportation and distribution are difficult, costly and even dangerous. The essential infrastructure for a Hydrogen Economy is also yet to be developed, at extreme cost. Realistically a wide and safe use of hydrogen as a transportation and household fuel is unlikely. As mentioned before, realizing these difficulties, the DOE announced in May 2009 its intention to drastically cut funding for research into hydrogen fuel cell for automotive applications [225]. It is therefore most desirable to find and develop more acceptable and feasible ways for solving these problems.

14.2
The "Methanol Economy", a Solution for the Future

A feasible alternative to the Hydrogen Economy and other proposals is what we have called the "Methanol Economy." The concept and relevant chemistry have been discussed extensively in this book and in our publications and patents (see Figure 14.1 below). Methanol is a convenient and the simplest oxygenated liquid hydrocarbon that at present is prepared from fossil fuel-based syn-gas. As discussed, however, new methods are being developed for its production by the direct oxidative conversion of still-existing large natural gas (methane) sources and the chemical hydrogenative recycling of exhausts of fossil fuel-burning power plants and other industrial plants as well as natural sources rich in CO_2. Eventually, it will be even feasible to separate and

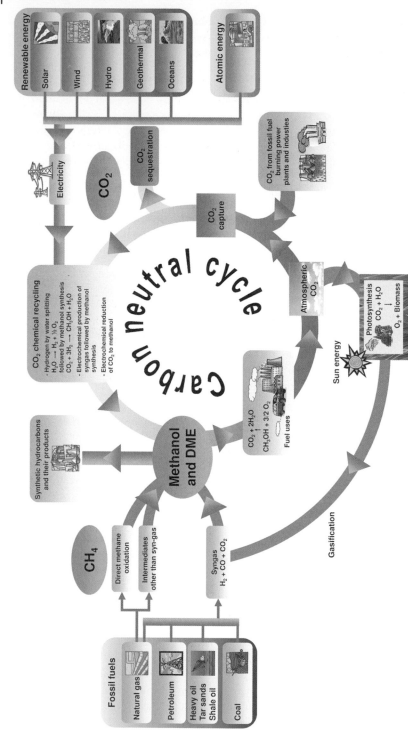

Figure 14.1 The Methanol Economy.

chemically recycle atmospheric CO_2 to methanol. The required hydrogen will be obtained from water (an inexhaustible resource) using any energy source. In this way, extremely volatile hydrogen gas will be conveniently and safely stored by converting it with CO_2 into liquid methanol.

Methanol represents not only a convenient and safe means of storing energy, but both it and readily derived dimethyl ether (DME) are also excellent fuels for transportation and other applications such as substitute household gas. Methanol and DME can be blended with gasoline/diesel fuel and used in internal combustion engines, as well as in electricity generators. Methanol is particularly efficient when used in the direct methanol fuel cell (DMFC) (Chapter 11), allowing methanol to be oxidized directly with air while producing electricity. In addition to its many uses for diverse chemical products and materials and even as a feedstock for protein production, methanol can also be readily converted into ethylene and/or propylene (the MTO process), which then can be used to produce synthetic hydrocarbons and their products that presently are obtained from oil and gas:

$$CH_3OH \rightarrow CH_2 = CH_2 \text{ and/or} \rightarrow \text{hydrocarbons}$$
$$CH_3CH = CH_2$$

Today, methanol is prepared exclusively from fossil fuel, mostly natural gas and coal-based syn-gas. As long as natural gas remains available it would be reasonable to convert it directly into methanol and DME, without first going through the syn-gas stage. This developing technology would not only greatly simplify its production but also extend the availability of natural gas. Coal being our most abundant fossil fuel, its use via syn-gas will prevail in major coal producing areas. Solutions must, however, be found to mitigate the serious environmental effects of generated CO_2.

Methanol, importantly, can be obtained from CO_2 by catalytic recycling using hydrogen, or by electrochemical reduction in water. The emissions of fossil fuel-burning power plants and varied chemical plants contain high concentrations of CO_2 that can be readily captured and recycled to methanol. As mentioned, water can provide the hydrogen for converting CO_2 into methanol using any energy source (alternatives include photochemical and even bacterial conversions). At the same time, the recycling of atmospheric CO_2 via its reductive hydrogenation to methanol offers an inexhaustible future carbon source for humankind's continued need for fuels, synthetic hydrocarbons and their products.

The CO_2 content of the atmosphere is low (presently 0.038%). New and efficient ways for the technical separation of CO_2 from the air, supplementing Nature's photosynthetic cycle, are therefore needed. Selective absorption and other separation methods are being developed to allow the capture and chemical recycling of atmospheric CO_2 from the air on a practical scale. Methanol produced efficiently from atmospheric CO_2 and hydrogen generated from water can eventually replace oil and gas as a convenient way to store energy, as a suitable fuel, as a chemical raw material for synthetic hydrocarbons and their varied products, and even for proteins production. From a detrimental greenhouse gas causing global warming, captured and chemically recycled carbon dioxide will be transformed into a valuable, renewable and inexhaustible carbon source of the future, allowing for the environmentally neutral use of carbon fuels and derived hydrocarbon products. The "Methanol Economy" can liberate humankind from its reliance on diminishing and non-renewable fossil fuels while mitigating the threat of human activity enhanced global warming.

Nature's photosynthesis uses the sun's energy with chlorophyll of the plants as a catalyst to recycle CO_2 and water into new plant life. The subsequent natural formation of fossil fuels from plant life is, though, a very slow process requiring many millions of years – although industrial biofuel production is also possible. Hence, in a way the proposed technological "Methanol Economy" can provide a viable alternative to Nature's own photosynthetic CO_2 recycling.

Ultimately, as discussed, most of our energy on earth comes from the sun. It is foreseen that the sun will last for at least another 4.5 billion years. In the future, if humankind survives it will surely devise more efficient ways of harnessing and using this limitless energy. As we cannot even begin to imagine the advances that could be made by future generations, our present discussions must be limited to what might be achieved in the more foreseeable future, based on present and developing knowledge and technologies.

We conclude with an optimistic view for the future. Humankind is an ingenious species that always seems to find ways for overcoming adversities and challenges. As history teaches us, however, our response to significant major problems and challenges usually comes only when a crisis is already upon us. Many believe that the problem of our diminishing oil and gas, as well as our still more abundant coal reserves, is not yet at the crisis stage, and therefore we do not need to unduly worry about it. Past pessimistic predictions concerning rapidly diminishing fossil fuel resources always turned out to be "false alarms." This is substantially true but only within the short time spans to which they were applied. With regard to the longer range, the outlook is different. We must face the fact that our Nature-given non-renewable fossil fuel resources are finite and diminishing, while worldwide population and consumption are both growing. We need to find new technological solutions if we wish to continue our lives at a comparable or even higher standard of living. We need to start developing these new solutions now, while we still have the time and resources to do it in an orderly and timely fashion to avoid major crises.

Specifically, as pointed out, one way of extending our petroleum oil and natural gas reserves is their better and more economical use, conservation measures and the introduction of improved, more efficient technologies, particularly in the transportation area where oil-based fuels continue to be primarily used. Fuel savings, together with more efficient vehicles, such as using electric or hybrid propulsion systems combining internal combustion engines with on-board generated electricity driven electric motors, can reduce gasoline and diesel fuel use and extend their availability. Fuel cells based not only on hydrogen but also on the direct use of methanol (DMFC) can provide cars with great fuel efficiency. The wider use of hydrogen for energy storage and as a fuel in the context of the so-called "Hydrogen Economy" – except perhaps for larger static installations – is, however, considered less feasible, as volatile hydrogen gas is *inter alia* extremely difficult to handle and use. This would necessitate not only the development of an entirely new and extremely expensive infrastructure but also the recognition and control of serious safety hazards. Instead, hydrogen can be converted with ubiquitous carbon dioxide into methanol and its products in the context of the proposed and discussed "Methanol Economy" depicted schematically in Figure 14.1. It represents a feasible new approach that extends beyond the era of still relatively easily available and affordable oil and gas. We hope that our book will further raise interest in its development. We do not suggest that this is the only approach to be followed, or is even necessarily the most feasible in all aspects and conditions. Rather, humankind will need to rely on all possible solutions available to satisfy its energy needs. Particular attention must be given to make our fuels and derived products regenerable and environmentally benign. In these aspects, we believe that the "Methanol Economy" is a feasible new approach and warrants extensive further study, development and implementation.

Humankind is facing both short and long range problems related to carbon based fuels and feedstocks to which we must find solutions. In the short range, one must provide sufficient clean energy and material sources to supplement existing fossil fuel reserves and simultaneously mitigate the harmful effects of excessive carbon dioxide release resulting in global warming. In the longer range our fossil fuels will last for only a relatively short period of time, at most few centuries, a fleeting moment on the geological timescale of earth. Therefore, the major emphasis is already beginning to shift to provide clean and sustainable energy sources for future generations, as well as effective recycling of carbon, through CO_2, supplementing natures own biological cycles to supply all the carbon based materials, fuels etc. We believe that the "Methanol Economy" concept is capable of offering feasible solutions to both challenges. We can essentially deal with these problems based on our presently available science and derived technology. Future generations will clearly vastly extend our knowledge and technology base, primarily by capturing our Sun's inexhaustible energy in a more effective way as well as by achieving controlled nuclear fusion. Looking back on our present efforts, we hope that future generations will credit us for laying firm foundations on which to build more advanced solutions.

References

1 Schobert, H.H. (2002) *Energy and Society, an Introduction*, Taylor and Francis, New York.

2 BP (2007) BP Statistical Review of World Energy, available online at www.bp.com/statisticalreview.

3 United Nations, (2007) World Population Prospects. The 2006 Revision, ST/ESA/SE.R.A/261/ES, United Nations Department of Economic and Social Affairs, Population Division, New York.

4 Energy Information Administration (2007) International Energy Outlook 2007, Energy Information Administration, U.S. Department of Energy, Washington.

5 Bartlett, R. (2008) Peak Oil Special Order Speech Transcript, Congressional Record, U.S. House of Representatives, February 28 H1172.

6 U.S. Census Bureau (2007) *Statistical Abstract of the United States: 2008*. Section 19, Energy, Utilities, 127 edn, U.S. Census Bureau, Washington, DC.

7 Energy Information Administration (2007) International Energy Annual 2005, Energy Information Administration, U.S. Department of, Energy, Washington.

8 Olah, G.A. and Molnár, Á. (2003) *Hydrocarbon Chemistry*, 2nd edn, John Wiley & Sons, Inc., Hoboken, New Jersey.

9 Steynberg, A.P. and Dry, M.E. (eds) (2004) *Stud. Surf. Sci. Catal.*, **152**.

10 (2008) Proceedings of the High Level Conference on World Food Security: The Challenges of Climate Change and Bioenergy. Soaring Food Prices: Facts, Perspectives, Impacts and Actions Required (HLC/08/INF/1), Rome, 3–5 June. Conference organized by the Food and Agricultural Organization of the United Nations.

11 Olah, G.A. Methanol Economy® (trademark) No 78/692,647.

12 Olah, G.A., Goeppert, A. and Prakash, G.K.S. (2009) Chemical recycling of carbon dioxide to methanol and dimethyl ether: from greenhouse gas to renewable, environmentally carbon neutral fuels and synthetic hydrocarbons. *J. Org. Chem.*, **74**, 487.

13 Romm, J.J. (2004) *The Hype about Hydrogen. Fact and Fiction in the Race to Save the Climate*, Island Press, Washington, DC.

14 Rifkin, J. (2002) *The Hydrogen Economy*, Tarcher/Putnam, New York.

15 Arrhenius, S. (1896) On the influence of carbonic acid in the air upon the temperature of the ground. *Philos. Mag.*, **41**, 237.

16 Surampudi, S., Narayanan, S.R., Vamos, E. *et al.* (1994) Advances in direct oxidation methanol fuel cells. *J. Power Sources*, **47**, 377.

17 Surampudi, S., Narayanan, S.R., Vamos, E. *et al.* (1997) U.S. Patent 5,599,638; (2001) U.S. Patent 6,248,460; (2004) U.S. Patent 6,740,434; (2004) U.S. Patent 6,821,659.

18 Olah, G.A. and Prakash, G.K.S. (2008) Conversion of carbon dioxide to methanol

Beyond Oil and Gas: The Methanol Economy, Second updated and enlarged edition
George A. Olah, Alain Goeppert, and G. K. Surya Prakash
Copyright © 2009 WILEY-VCH Verlag GmbH & Co. KGaA, Weinheim
ISBN: 978-3-527-32422-4

and/or dimethyl ether using bi-reforming of methane or natural gas, U.S. Patent Application, 20080319093.

19 Freese, B. (2003) *Coal, A Human History*, Perseus Publishing, Cambridge.

20 Stearns, P.N. (2007) *The Industrial Revolution in World History*, 3 edn, Westview Press, Boulder, Colorado.

21 International Energy Agency (2004) *World Energy Outlook 2004*, International Energy Agency, Paris.

22 Survey of Energy Resources (2007), World Energy Council (WEC), http://www.worldenergy.org/wec-geis/.

23 Smil, V. (2003) *Energy at the Crossroads, Global Perspectives and Uncertainties*, MIT Press, Cambridge.

24 International Energy Agency (2001) *World Energy Outlook 2001: Insights*, International Energy Agency, Paris.

25 U.S. Geological Survey (2000) Health Impact of Coal Combustion, Fact Sheet USGS FS-094-00, U.S. Geological Survey.

26 Greenwood, N.N. and Earnshaw, A. (1984) *Chemistry of the Elements*, Pergamon Press, p. 297.

27 Katzer, J. (Ed.) (2007) *The Future of Coal – Option for a Carbon-Constrained World*, Massachusetts Institute of Technology.

28 International Energy Agency (1997) *Energy Technologies for the 21st Century*, International Energy Agency, Paris.

29 Johnson, J. (2004) Getting to "clean coal", *Chem. Eng. News*, **82**, 20.

30 Tullo, A.H. and Tremblay, J.-F. (2008) Coal: the new black, *Chem. Eng. News*, (March 17), **86**, p. 15.

31 Campbell, C.J. (1997) *The Coming Oil Crisis*, Multi-science Publishing, Brentwood, England.

32 Gold, T. (1999) *The Deep Hot Biosphere*, Copernicus Press, New York.

33 Black, B. (2000) *Petrolia, the Landscape of America's First Oil Boom*, The Johns Hopkins University Press, Baltimore.

34 American Petroleum Institute, All About Petroleum, brochure from American Petroleum Institute, available at http://api-ec.api.org.

35 Economides, M. and Oligney, R. (2000) *The Color of Oil*, Round Oak Publishing Co., Katy, Texas.

36 Yergin, D. (1991) *The Prize: The Epic Quest for Oil, Money & Power*, Simon & Schuster.

37 Brantly, J.E. (1971) *History of Oil Well Drilling*, Gulf Publishing Company, Houston, Texas.

38 Johnson, J. (2005) LNG weighs anchor. *Chem. Eng. News*, **83**, 19.

39 Hightower, M., Gritzo, L., Anay, L.-H., Covan, J., Tieszen, S. et al. (2004) *Guidance on Risk Analysis and Safety Implications of Large Liquefied Natural Gas (LNG) Spill over Water*, Sandia National Laboratories.

40 Seddon, D. (2006) *Gas Usage & Value: The Technology and Economics of Natural Gas Use in the Process Industries*, PennWell, Tulsa, Oklahoma.

41 Campbell, C.J. and Laherrère, J.H. (1998) The end of cheap oil, *Sci. Am.*, (March) 78.

42 International Energy Agency (2008) *Key World Energy Statistics 2008*, International Energy Agency (IEA), Paris.

43 World Coal Institute, http://www.worldcoal.org/.

44 Energy Information Administration (2008) *Annual Energy Outlook 2008 with Projections to 2030*, Energy Information Administration, U.S. Department of Energy, Washington, DC, available at http://www.eia.doe.gov/oiaf/aeo.

45 Bradsher, K. and Barboza, D. (2006) Pollution from Chinese coal casts a global shadow, The New York Times, June 11.

46 Energy Information Administration (2005) *Annual Energy Outlook 2005 With Projections to 2025*, Energy Information Administration, U.S. Department of, Energy, Washington, DC.

47 U.S. Environmental Protection Agency, (2005) Mountaintop Mining/Valley Fills in Appalachia. Final Programmatic Environmental Impact Statement, EPA 9-03-R-05002.

48 International Energy Agency (2007) *World Energy Outlook 2007: China and India Insights*, International Energy Agency, Paris.

49 International Energy Agency (2001) *World Energy Outlook 2001: Insights*, International Energy Agency, Paris.

50 International Energy Agency (2008) *Coal Information 2008*, International Energy Agency, Paris.

51 US Geological Survey (2000) Health Impact of Coal Combustion, Fact Sheet USGS FS-094-00, U.S. Geological Survey.

52 U.S. Bureau of Transportation Statistics, http://www.bts.gov/.

53 Simmons, M.R. (2005) *Twilight in the Desert: The Coming Saudi Oil Shock and the World Economy*, John Wiley & Sons, Hoboken, New Jersey.

54 ExxonMobil Corp., (2007) 2007 Summary Annual Report.

55 Hunt, C.G. (2004) Nuclear Power an Attractive Option for Tar Sands: Alberta Chamber Report, Nuclear Canada, Vol 5, February 10, p. 2.

56 European Academies Science Advisory Council (2007) A Study on the EU Oil Shale Industry - Viewed in the Light of the Estonian Experience, Report by the European Academies Science Advisory Council for the Committee on Industry and Energy of the European Parliament.

57 Snyder, R.E. (2004) Oil shale back in the picture, worldOil Magazine online, August, vol. 225.

58 Hess, G. (2006) Oil shale research is moving forward,*Chem. Eng. News*, (April 24), **84**, 29.

59 US Department of Energy (2007) Secure Fuel from Domestic Resources - The Continuing Evolution of America's Oil Shale and Tar Sands Industries, U.S. Department of Energy, Office of Petroleum Reserves and Office of Naval Petroleum and Oil Shale Reserves.

60 Houghton, J.T., Ding, Y., Griggs, D.J. and Noguer, M. (eds) (2001) *IPCC Third Assessment Report: Climate Change 2001: The Scientific Basis*, Cambridge University Press, Cambridge, UK.

61 Forster, P., Ramaswamy, V., Artaxo, P., Berntsen, T.Betts, R. *et al.* (2007) Changes in Atmospheric Constituents and in Radiative Forcing, in *Climate Change 2007: The Physical Science Basis. Contribution of Working Group I to the Fourth Assessment Report of the Intergovernmental Panel on Climate Change* (eds S. Salomon, D. Qin, M. Manning, Z. Chen, M. Marquis, K.B. Averyt, M. Tignor and H.L. Miller), Cambridge University Press, Cambridge, UK.

62 International Energy Agency (2008) *Natural Gas Information 2008*, International Energy Agency, Paris.

63 Sasol, http://www.sasol.com/.

64 Gold, R. (2005) In Qatar, oil firms make huge bet on alternative fuel. *The Wall Street Journal*, (February 15).

65 ExxonMobil, http://www.exxonmobil.com.

66 International Energy Agency (2007) *World Energy Outlook 2007: China and India Insights*, International Energy Agency, Paris.

67 U.S. Geological Survey (2000) U.S. Geological Survey World Petroleum Assessment 2000, USGS, Denver, Colorado, available at: http://pubs.usgs.gov/dds/dds-060/.

68 Hubbert, M.K. (1956) Nuclear energy and the fossil fuels. American Petroleum Institute drilling and production practice, Proceedings of the Spring Meeting, San Antonio, March 7–9.

69 Energy Information Administration (EIA), http://www.eia.doe.gov/.

70 Deffeyes, K.S. (2001) *Hubbert's Peak, the Impending World Oil Shortage*, Princeton University Press, Princeton.

71 Hirsch, R.L., Bezdek, R. and Wendling, R. (2005) Peaking of World Oil Production: Impact, Mitigation & Risk Management, prepared for the U.S. DOE's National Energy Technology Laboratory (NETL) by Science Applications International Corporation (SAIC).

72 Bentley, R.W. (2002) Oil & gas depletion: an overview. *Energy Policy*, **30**, 189.

73 International Energy Agency (2001) *World Energy Outlook 2001: Insights*, International Energy Agency, Paris.

74 Donnely, J.K. and Pendergast, D.R. (1999) Nuclear energy in industry: application to oil production. Climate Change and Energy Options Symposium, Canadian Nuclear Society, Ottawa, Canada, November 17–19.

75 Alberta Chamber of Commerce (2004) Oil Sands Technology Roadmap. Unlocking the Potential, Alberta Chamber of Resources, Edmonton, Alberta, Available at http://www.acr-alberta.com/.

76 Odell, P.R. and Rosing, K.E. (1984) *The Future of Oil. World Oil Resources and Use*, Kogan Page, London.

77 Smolowe, J. (1986) Cameroon the Lake of Death. *Time Magazine*, September 8.

78 Greene, D.L., Hopson, J.L. and Li, J. (2003) Running Out of and Into Oil: Analyzing Global Oil Depletion and Transition Through 2050, prepared by Oak Ridge National Laboratory for the U.S. DOE.

79 Laherrère, J. (2004) Natural gas future supply. International Institute for Applied Systems Analysis (IIASA) International Energy Workshop, June 22–24, Paris, available at: http://www.hubbertpeak.com/laherrere/.

80 Rogner, H.H. (1997) An assessment of world hydrocarbon resources. *Annu. Rev. Energy Environ.*, **22**, 217.

81 Odell, P.R. (1999) *Fossil Fuel Resources in the 21st Century*, Financial Times Energy, London.

82 Odell, P.R. (2004) *Why Carbon Fuels will Dominate the 21st Century's Global Energy Economy*, Multi-Science Publishing, Brentwood, UK.

83 Tomasko, M.G., Archinal, B., Becker, T., Bezard, B., Bushroe, M. *et al.* (2005) Rain, winds and haze during the Huygens probe's descent to Titan's surface. *Nature*, **438**, 765.

84 Raulin, F. (2008) Planetary science: organic lakes on Titan. *Nature*, **454**, 587.

85 Salomon, S., Qin, D., Manning, M., Chen, Z., Marquis, M., Averyt, K.B., Tignor, M. and Miller, H.L. (eds) (2007) *Climate Change 2007: The Physical Science Basis. Contribution of Working Group I to the Fourth Assessment Report of the Intergovernmental Panel on Climate Change*, Cambridge University Press, Cambridge, UK.

86 Rapp, D. (2008) *Assessing Climate Change. Temperatures, Solar Radiation, and Heat Balance*, Praxis Publishing, Chichester, UK.

87 Kerr, R.A. (2006) Yes, it's been getting warmer in here since the CO_2 began to rise. *Science*, **312**, 1854.

88 Solanki, S.K., Usoskin, I.G., Kromer, B., Scluessler, M. and Beer, J. (2004) Unusual activity of the Sun during recent decades compared to the previous 11,000 years. *Nature*, **431**, 1084.

89 Essenhigh, R.H. (2001) Does CO_2 really drive global warming? *Chem. Innovation*, **31**, 44.

90 IPCC Climate Change 2007, Synthesis Report, IPCC Fourth Assessment Report.

91 Kharecha, P.a. and Hansen, J.E. (2008) Implications of "peak oil" for atmospheric CO_2 and climate. *Global Biogeochem. Cycles*, **22**, GB3012.

92 Idso, S.B. (1997) Biological Consequences of Increased Concentrations of Atmospheric CO_2, in *Global Warming: The Science and the Politics* (ed. L. Jones), The Fraser Institute, Vancouver.

93 Thayer, A.M. (2009) Chemicals to Help Coal Come Clean, *Chem. Eng. News*, July 13, **87**, 19.

94 Socolow, R.H. (2005) Can we bury global warming? *Sci. Am.*, July, 49–55.

95 Johnson, J. (2004) Putting a lid on carbon dioxide. *Chem. Eng. News*, **82**, 36.

96 Stern, N. (2007) *The economics of Climate Change: The Stern Review*, Cambridge University Press.

97 Stern, N. (2009) *A Blueprint for a Safer Planet: How to Manage Climate Change and Create a New Era of Progress and Prosperity*, The Bodley Head, London.

98 United Nations Climate Change Conference (2009), Copenhagen, Dec. 7–18, http://en.cop15.dk.

99 United Nations Framework Convention on Climate Change, http://unfccc.int.

100 International Energy Agency (2001) *World Energy Outlook 2001: Insights*, International Energy Agency, Paris.

101 World Commission on Dams, http://www.dams.org/.

102 Wonder of the World Databank, http://www.pbs.org.

103 Bertani, R. (2005) World geothermal power generation 2001–2005. *Geothermics*, **34** (6), 651.

104 European Deep Geothermal Energy Programme, http://www.soultz.net/.

105 Bertani, R. (2002) Geothermal power generation plant CO_2 emission survey, IGA News, July-September, vol. 49.

106 Bloomfield, K., Moore, J. and Neilson, R.M.J. (2003) Geothermal Energy Reduces Greenhouse Gases, GRC Bulletin, April/March, p. 77.

107 MIT (2006) *The Future of Geothermal Energy*, Massachusetts Institute of Technology.

108 Global Wind Energy Council (GWEC), http://www.gwec.net/.

109 Zervos, A. and Kjaer, C. (2008) *Pure Energy. Wind Energy Scenarios up to 2030*, European Wind Energy Association.

110 (2008) *Global Wind Energy Outlook 2008*, Global Wind Energy Council.

111 Chandler, H. (ed.), (2003) *Wind Energy: The Facts. An Analysis of Wind Energy in the EU-25*, European Wind Energy Association (EWEA), available at: http://www.ewea.org/.

112 US Department of Energy (2004) *Solar Energy Technologies Program: Multi-Year Technical Plan 2003–2007 and Beyond*, Energy Efficiency and Renewable Energy (EERE) U.S DOE.

113 Tao, M. (2008) Inorganic photovoltaic solar cells: silicon and beyond. *Electrochem. Soc. Interface*, **17** (4), 30.

114 IEA Photovoltaic Power Systems Programme, http://www.iea-pvps.org/.

115 First Solar, Inc., http://www.firstsolar.com.

116 Dickerson, M. (2009) Solar farm cuts gap with fossil fuel, Los Angeles Times, January 5.

117 Michael, G.W. (2007) 2007 Minerals Yearbook - Selenium and Tellurium, U.S. Geological Survey.

118 Brabec, C., Dyakonov, V. and Scherf, U. (2008) *Organic Photovoltaics: Materials, Device Physics, and Manufacturing Technologies*, Wiley-VCH GmbH, Weinheim, Germany.

119 IEA Photovoltaic Power Systems Programme, (2008) Trends in photovoltaic applications. Survey report of selected IEA countries between 1992 and 2007. Report IEA-PVPS T1-T17: 2008.

120 Sayigh, A. (2003) Spotlight on PV Energy: as Commercialisation Grows, Solar needs Attention, in *Renewable Energy 2003*, An official publication of the World Renewable Energy Network, UNESCO.

121 International Energy Agency (2007) *Renewables Information 2007*, International Energy Agency (IEA), Paris.

122 (1997) *Renewable Energy Technology Characterizations*, Electric Power Research Institute (EPRI) and U.S. DOE.

123 Johnson, J. (2008) U.S. solar energy heats up. *Chem. Eng. News*, **86**, 40.

124 Zweibel, K., Mason, J. and Fthenakis, V. (2008) A solar grand plan. *Sci. Am.*, January, 64.

125 Port, O. (2005) Power from the sunbaked desert: solar generator may be the hot source of plentiful electricity, BusinessWeek, September 12, p. 76.

126 Stirling Energy Systems, http://www.stirlingenergy.com/.

127 Einav, A. (2004) Solar energy research and development achievements in Israel and their practical significance. *J. Solar Energy Eng.*, **126**, 921.

128 Ton, D.T., Hanley, C.J., Peek, G.H. and Boyes, J.D. (2008) Solar Energy Grid Integration Systems - Energy Storage (SEGIS-ES), SAND2008-4247, Sandia National Laboratories, Albuquerque, New Mexico and Livermore, California.

129 Electric Power Research Institute, http://www.epri.com.

130 EPRI-DOE (2003) *EPRI-DOE Handbook of Energy Storage for Transmission and*

Distribution Applications, 1001834, EPRI, Palo Alto, California and U.S. Department of Energy, Washington, DC.

131 Succar, S., Greenblatt, J.B., Denkenberger, D. and Williams, R.H. (2006) An integrated optimization of large-scale wind with variable rating coupled to compressed air energy storage. Conference Proceedings of Windpower 2006, Pittsburgh, Pennsylvania, June 4–7th.

132 EPRI (2004) *Wind Power Integration Technology Assessment and Case Studies, 1004806*, EPRI, Palo Alto, California.

133 EPRI (2004) *EPRI-DOE Handbook Supplement of Energy Storage for Grid Connected Wind Generation Applications, 1008703*, EPRI, Palo Alto, California and U.S. Department of Energy, Washington, DC.

134 Rosillo-Calle, F., Bajay, S.V. and Rothman, H. (eds) (2000) *Industrial Uses of Biomass Energy. The Example of Brazil*, Taylor & Francis, London.

135 Buarque de Hollanda, J. and Dougals Poole, A. *Sugarcane as an Energy Source in Brazil*, Instituto Nacional de Eficiencia Energetica, http://www.inee.org.br.

136 São Paulo Sugarcane Agroindustry Union (UNICA), http://www.unica.com.br/.

137 Dickerson, M. (2005) Homegrown fuel supply helps Brazil breathe easy, Los Angeles Times, June 15.

138 Koplow, D. (2006) *Biofuels - At What Cost? Government Support for Ethanol and Biodiesel in the United States*, Prepared by Earthtrack, Inc. for The Global Subsidies Initiatives (GSI) of the International Institute for Sustainable Development, Geneva, Switzerland.

139 Shapouri, H., Duffield, J.A. and Wang, M. (2002) *The Energy Balance of Corn Ethanol: An Update*, U.S. Department of Agriculture, Washington, DC.

140 Hess, G. (2005) Ethanol wins big in energy policy. *Chem. Eng. News*, **83**, 28.

141 Patzek, T.W. (2004) Thermodynamics of the corn-ethanol biofuel cycle. *Crit. Rev. Plant Sci.*, **23**, 519.

142 Pimentel, D. (2003) Ethanol fuels: energy balance, economics, and environmental impacts are negative. *Nat. Resources Res.*, **12**, 127.

143 Pimentel, D. and Patzek, T.W. (2005) Ethanol production using corn, switchgrass, and wood; biodiesel production using soybean and sunflower. *Nat. Resources Res.*, **14**, 65.

144 Pimentel, D., Patzek, T. and Cecil, G. (2007) Ethanol production: energy, economic, and environmental losses. *Rev. Environ. Contam. Toxicol.*, **189**, 25.

145 EurObserv'ER (2007) Biofuels barometer, systemes solaires. *J. Energies Renouvelables*, **179**, 63.

146 Bensaïd, B. (2005) *Panorama 2005: Road Transport Fuels in Europe: the Explosion of Demand for Diesel Fuel*, Institut Francais du Petrole (IFP).

147 Padgett, T. (2009) The next big biofuels? Jatropha seeds produce clean-burning diesel (without driving up your grocery bill), Time Magazine, February 9, p. 50.

148 Achten, W.M.J., Verchot, L., Franken, Y.J., Mathijs, E., Singh, V.P. *et al.* (2008) Jatropha bio-diesel production and use. *Biomass Bioenergy*, **32** (12), 1063.

149 Searchinger, T., Heimlich, R., Houghton, R.A., Fengxia, D., Elobeid, A. *et al.* (2008) Use of U.S. croplands for biofuels increases greenhouse gases through emissions from land use change. *Science*, **319**, 1238.

150 Fargione, J., Hill, J., Tilman, D., Polasky, S. and Hawthorne, P. (2008) Land Clearing and biofuel carbon debt. *Science*, **319**, 1235.

151 Gallagher, E. (2008) *The Gallagher Review of the Indirect Effects of Biofuels Production*, Renewable Fuels Agency, St. Leonards-on-Sea, UK.

152 European Parliament (2008) Draft Report on the Proposal for a Directive of the European Parliament and of the Council on the Promotion of the Use of Energy From Renewable Sources,PR\722155EN, PE 405. 949 v 01-00. European Parliament. Committee on Industry, Research and, Energy.

153 Lee, H., Clark, W.C. and Devereaux, C. (2008) Biofuels and sustainable development: report of an executive session on the grand challenges of a sustainability transition. San Servolo Island, Venice, Italy, May 19–20.

154 OECD (2008) *Biofuel Support Policies. An Economic Assessment*, OECD.

155 King, C.W. and Webber, M.E. (2008) Water intensity in transportation. *Environ. Sci. Technol.*, **42**, 7866.

156 Zah, R., Böni, H., Gauch, M., Hischier, R., Lehmann, M. *et al.* (2008) *Ökobilanz von Energieprodukten: Ökologische Bewertung von Biotreibstoffen. Schlussbericht*, Empa, Abteilung Technologie und Gesellschaft im Auftrag, des Bundesamtes für Energie, des Bundesamtes für Umwelt und des Bundesamtes für Landwirtschaft, Switzerland.

157 Scharlemann, J.P.W. and Laurance, W.F. (2008) How green are biofuels? *Science*, **319**, 43.

158 Doornbosch, R. and Steenblik, R. (2007) *Biofuels: Is the Cure Worse than the Disease, SG/SD/RT(2007)3*, OECD, Paris.

159 International Energy Agency (2004) *Energy Technologies for a Sustainable Future: Transport*, International Energy Agency (IEA), Paris.

160 Voith, M. (2009) Up from the slime. *Chem. Eng. News*, (January 26), **87**, 22.

161 Pontes, T. and António, F. (2001) Ocean energies: resources and utilisation. 18th World Energy Conference, Buenos Aires, Argentina, October 21–25.

162 International Energy Agency, Ocean Energy Systems (IEA-OES), http://www.iea-oceans.org/.

163 Johnson, J. (2004) Power from moving water. *Chem. Eng. News*, (October 4), **82**, 23.

164 Ordóñez, I. (2008) Everybody into the ocean, *The Wall Street Journal*, (October 6).

165 Fraenkel, P. (October 2008) SeaGen - the world's first commercial-scale tidal current turbine, Newsletter International Energy Agency, Ocean Energy, Systems.

166 Clément, A., McCullen, P., Falcão, A., Fiorentino, A., Gardner, F. *et al.* (2002) Wave energy in Europe: current status and perspectives. *Renewable Sustainable Energy Rev.*, **6**, 405.

167 Pelamis Wave Power, http://www.pelamiswave.com.

168 Jha, A. (2008) Making waves: UK firm harnesses power of the sea. . . in Portugal, The Guardian, September 25.

169 Cowan, G.A. (1976) A natural fission reactor. *Sci. Am.*, July, 36.

170 (2002) *A Technology Roadmap for Generation IV Nuclear Energy Systems*, U.S. DOE Nuclear Energy Research Advisory Committee, and the Generation IV International Forum.

171 (2007) GEN IV International Forum 2007 Annual Report, OECD Nuclear Energy Agency, Generation IV, International Forum, available from: http://www.gen-4.org.

172 Hecht, J.(2004) US Plans 'Take-Away' Nuclear Power Plants: Can a Sealed, Mobile Nuclear Power Plant Prevent Proliferation by Rogue States?, *New Scientist*, September 4, p. 17.

173 Hyperion Power Generation, http://www.hyperionpowergeneration.com/.

174 Claude, B. (1999) *Superphénix, le Nucléaire à la Française*, l'Harmattan, Paris.

175 Argonne National Laboratory, http://www.anl.gov/.

176 (2004) *Uranium 2003: Resources, Production and Demand*, OECD Nuclear Energy Agency and the International, Atomic Energy Agency.

177 Lidsky, L.M. and Miller, M.M. (2002) Nuclear power and energy security: a revised strategy for Japan. *Sci. Global Security*, **10**, 127.

178 United Nations (2005) Chernobyl's Legacy: Health, Environmental and Socio-Economic Impacts and Recommendations to the Governments of Belarus, the Russian Federation and Ukraine.

179 U.S. Geological Survey (1997) Radioactive Elements in Coal and Fly Ash:

Abundance, Forms, and Environmental Significance, USGS Fact Sheet FS-163-97, U.S. Geological Survey.

180 Tadmor, J. (1986) Radioactivity from coal-fired power plants: a review. *J. Environ. Radioactivity*, **4**, 177.

181 Gabbard, A. (1993) Coal Combustion: Nuclear Resource or Danger?, vol. 26, ORNL Review, available at: http://www.ornl.gov/info/ornlreview/rev26-34/text/colmain.html.

182 U.S. Office of Civilian Radioactive Waste Management, http://www.ocrwm.doe.gov.

183 Yucca Mountain Standards, EPA, http://www.epa.gov/radiation/yucca/.

184 Nuclear Energy Institute, http://www.nei.org/.

185 Morris, R.C. (2000) *The Environmental Case for Nuclear Power. Economical, Medical and Political Considerations*, Paragon House, St. Paul, Minnesota.

186 Dewan, S. (2009) Hundreds of coal ash dumps lack regulation, The New York Times, January 6.

187 Schmitt, H.H. (2006) *Return to the Moon: Exploration, Enterprise and Energy in the Human Settlement of Space*, Praxis Publishing Ltd.

188 http://www.hydrogen.gov.

189 (2006) *The Hydrogen Economy. A Non-Technical Review*, United Nations Environment Programme (UNEP).

190 Hoffmann, P. (2002) *Tomorrow's Energy. Hydrogen, Fuel Cells, and the Prospects for a Cleaner Planet*, The MIT Press, Cambridge.

191 Pohl, H.W. (ed.) (1995) *Hydrogen and Other Alternative Fuels for Air and Ground Transportation*, John Wiley & Sons, Ltd., Chichester, England.

192 (2004) *The Hydrogen Economy: Opportunities, Costs, Barriers and R&D Needs*, National Research Council and National Academy, Engineering, The National Academic Press, Washington, DC.

193 (2008) Hydrogen From Coal Program. Research, Development, and Demonstration Plan for the Period 2008 Through 2016, External Draft, U.S. Department of Energy.

194 (1999) Vision 21 Program Plan. Clean Energy Plants for the 21st Century, Federal Energy Technology Center, Office of Fossil Energy, - U.S. Department of Energy.

195 U.S. Department of Energy, http://www.energy.gov/.

196 U.S. Department of Energy, (2004) FutureGen. Integrated Hydrogen, Electric Power Production and Carbon Sequestration Research Initiative. Report to the Congress, available at http://www.energy.gov.

197 Toward a hydrogen economy, editorial and special issue. *Science*, **305**, (2004) 957.

198 Milne, T.A., Elam, C.C. and Evans, R.J. (2002) Hydrogen from biomass: state of the art and research challenges, IEA/H2/TR-02/001, International Energy Agency (IEA).

199 NREL (2007) Photobiological Production of Hydrogen, NREL/FS-560-42285, National Renewable Energy Laboratory, Golden, Colorado.

200 Ghirardi, M.L., Maness, C.P. and Seibert, M. (2008) Photobiological Methods of Renewable Hydrogen Production, in *Solar Hydrogen Generation* (eds K. Rajeshwar, R. McConnell and S. Licht), Springer, New York, p. 229.

201 Simbeck, D.R. and Chang, E. (2002) Hydrogen Supply: Cost Estimate for Hydrogen Pathways - Scoping Analysis, NREL/SR-540-32525. National Renewable Energy Laboratory, Golden, Colorado.

202 Ivy, J. (April 2004) Summary of Electrolytic Hydrogen Production, Milestone Completion Report, NREL/MP-560-35948, NREL, Golden, Colorado.

203 Borgschulte, A., Züttel, A. and Wittstadt, U. (2008) Hydrogen Production, in *Hydrogen as a Future Energy Carrier* (eds A. Züttel, A. Borgschulte and L. Schlapbach), Wiley-VCH GmbH, Weinheim, p. 149.

204 Grimes, C.A., Varghese, O.K. and Ranjan, S. (2008) *Light, Water, Hydrogen. The Solar Generation of Hydrogen by Water Photoelectrolysis*, Springer, New York.

205 Licht, S. (2008) Thermochemical and Thermo/photo Hybrid Solar Water Splitting, in *Solar Hydrogen Generation* (eds K. Rajeshwar, R. McConnell and S. Licht), Springer, New York, p. 87.

206 Shenoy, A. (1995) Modular helium reactor for non-electric applications, IAEA-TECDOC–923. Presented at the Advisory Group Meeting on Non-Electric Applications of Nuclear Energy, Jakarta, Indonesia, November 21–23.

207 Bossel, U., Eliasson, B. and Taylor, G. (2003) The Future of the Hydrogen Economy: Bright or Bleak? Available from http://www.efcf.com/reports/.

208 Roswell, J.L.C. and Yaghi, O.M. (2005) Strategies for hydrogen storage in metal-organic framework. *Angew. Chem. Int. Ed.*, **44**, 4670.

209 Ritter, S. (2007) Hydrogen storage gets a boost. *Chem. Eng. News*, (January 1), **85**, 11.

210 The Online Fuel Cell Information Resource, http://www.fuelcells.org/.

211 Schwarzenegger, A. (2004) Transcript of Governor Arnold Schwarzenegger's hydrogen highways network announcement. Hydrogen Highways Network Announcement, UC Davis, Davis, California, April 20.

212 Altmann, M., Gaus, S., Landinger, H., Stiller, C. and Wurster, R. (2001) Wasserstofferzeugung in Offshore Windparks "Killer-Kriterien", Grobe Auslegung und Kostenabschaetzung, Studie im Auftrag von GEO Gesellschaft fuer Energie und Oekologie mbH, L-B-Systemtechnik GmbH, Ottobrunn, Germany.

213 His, S. (2004) Panorama 2004: Hydrogen: An Energy Vector for the Future?, Institut Francais du Petrole (IFP), available at http://www.ifp.fr/IFP/en/aa.htm.

214 AREVA (2004) Bientôt l'ère hydrogène, Alternatives Magazine, vol. 7, p. 8, available at http://www.areva.com/.

215 Zimmerman, M. (2008) Toyota suspends plan to build the Prius in the U.S., Los Angeles Times.

216 U.S. Department of Energy, Energy Efficiency and Renewable Energy (EERE). http://www.eere.energy.gov.

217 National Energy Technology Laboratory (2007) 2007 Office of Fossil Energy. Fuel Cell Program Annual Report, DOE/NETL-2007/1288, U.S. Department of Energy, Office of Fossil Energy, National Energy Technology Laboratory.

218 The Unitized Regenerative Fuel Cell, Lawrence Livermore National Laboratory, http://www.llnl.gov/str/Mitlit.html.

219 Burke, K.A. (2003) Unitized Regenerative Fuel Cell System Development, NASA/TM-2003-212739, NASA, Glenn Research Center, Cleveland, Ohio, prepared for the First International Energy Conversion Engineering Conference, Portsmouth, Virginia, August 17-21.

220 Mitlitsky, F., Myers, B., Weisberg, A.H., (1996), Lightweight Pressure Vessels and Unitized Regenerative Fuel Cells, UCRL-JC-125220, Lawrence Livermore National Laboratory, presented at 1996 Fuel Cell Seminar, Orlando, Florida, November 17-20.

221 Fairley, P. (2003) Recharging the power grid. *Technol. Rev.*, 50.

222 VRB-ESS: the great leveller. *Modern Power Systems*, (2005), June, 55.

223 Williams, B. and Hennesy, T. (2005) Electric oasis. *IEE Power Eng.*, February–March, 28.

224 Wilks, N. (2004) Whatever the weather. Advances in battery technology could hold the key to successful development of alternative sources of energy. *Professional Eng.*, 33.

225 Service, R.F. (2009), Hydrogen Cars: Fad of the Future?, *Science*, **324**, 1257.

226 Olah, G.A. (2005) Beyond oil and gas: the methanol economy. *Angew. Chem. Int. Ed.*, **44**, 2636.

227 Olah, G.A. (2003) The methanol economy. *Chem. Eng. News*, (September 22), **81**, 5.

228 Olah, G.A. (1998) Oil and Hydrocarbons in the 21st Century, in *Chemical Research -*

2000 and Beyond: Challenges and Vision (ed. P. Barkan), American Chemical Society, Washington DC, and Oxford, University Press, Oxford.

229 Olah, G.A. (2004) After oil and gas: methanol economy. *Catal. Lett.*, **93**, 1.

230 Boyle, R. (1661) *The Sceptical Chymist*, F. Cadwell for F. Crooke, London.

231 Stiles, A.B. (1977) Methanol, past, present, and speculation on the future. *AIChE J.*, **23**, 362.

232 Cheng, W-.H. and Kung, H.H. (eds) (1994) *Methanol Production and Use*, Marcel Dekker, New York.

233 Fiedler, E., Grossmann, G., Kersebohm, D.B., Weiss, G. and Witte, C. (2003) Methanol, in *Ullmann's Encyclopedia of Industrial Chemistry*, 6th edn, vol. 21, Wiley-VCH GmbH, Weinheim, p. 611.

234 Fischer, F. and Tropsch, H. (1923) Synthesis of higher members of the aliphatic series from carbon monoxide. *Berichte*, **56**, 2428.

235 Fischer, F. and Tropsch, H. (1926) Direct synthesis of petroleum hydrocarbons at ordinary pressure. *Berichte*, **59**, 830.

236 Edmonds, W.J. (1932) Synthetic methanol process, Commercial Solvents Corporation U.S. Patent 1,875,714.

237 Weissermel, K. and Arpe, H.-J. (2003) *Industrial Organic Chemistry*, 4th edn, Wiley-VCH GmbH, Weinheim.

238 Heward, A. (2006) Upgraded MERLIN Spies Cloud of Alcohol Spanning 288 Billion Miles, Royal Astronomical Society PN 06/14 (NAM7).

239 Watanabe, N., Nagaoka, A., Shiraki, T. and Kouchi, A. (2004) Hydrogenation of CO on pure solid CO and CO-H_2O mixed ice. *Astrophys. J.*, November 20, 638.

240 Peeters, Z., Rodgers, S.D., Charnley, S.B., Schriver-Mazzuoli, L., Schriver, A. *et al.* (2006) Astrochemistry of dimethyl ether. *Astron. Astrophys.*, **445**, 197.

241 (2005) Product Focus,*Chemical Week*, June 22, p. 33.

242 Methanol in our Lives, available from: http://www.methanex.com. Brochure by

methanol producer, Methanex, illustrating the presence of methanol in many products and materials of our daily lives.

243 On the Road with Methanol: The Present and Future Benefits of Methanol Fuel, Prepared for the Methanol Institute, available at http://www.methanol.org (1994).

244 Bernton, H., Kovarik, W. and Sklar, S. (1982) *The Forbidden Fuel. Power Alcohol in the Twentieth Century*, Boyd Griffin, New York.

245 Reed, T.B. and Lerner, R.M. (1973) Methanol: A versatile fuel for immediate use. *Science*, **182**, 1299.

246 Rosillo-Calle, F., Bajay, S.V. and Rothman, H. (eds) (2000) Industrial Uses of Biomass: The Example of Brazil, Taylor & Francis, London.

247 Beyond the Internal Combustion Engine: The Promise of Methanol Fuel Cell Vehicles, Brochure published by the American Methanol Institute, available from: http://www.methanol. org/.

248 Perry, J.H. and Perry, C.P. (1990) *Methanol, Bridge to a Renewable Energy Future*, University Press of America, Lanham, Maryland.

249 Moffat, A.S. (1991) Methanol-powered. *Science*, **251**, 514.

250 Energy Information Administration (1998) Alternative to Traditional Transportation Fuels 1998, DOE/EIA-0585(98), Washington, DC.

251 Alternative Fuels for Vehicles Fleet Demonstration Program Volume 3, Technical Reports, New York State Energy Research and Development Authority, (1997).

252 Green, C.J., King, L., Mueller, S. and Cockshutt, N.A. (1990) Dimethyl Ether as a Methanol Ignition Improver – Substitution Requirements and Exhaust Emission Impact, SAE paper 902155, October.

253 Kozole, K.H. and Wallace, J.S. (1988) The Use of Dimethyl Ether as a Starting Aid for Methanol-Fuelled SI Engines At Low

Temperatures, SAE Paper 881677, October.

254 Lotus Researches Cars Running on CO_2 – Exige 270E Tri-Fuel is the Next Stage of Lotus Engineering's Long-Term Sustainable, Synthetic Alcohol Research, News Release Lotus Engineering, January, (2008).

255 Turner, J.W.G., Pearson, R.J., Holland, B. and Peck, R. (2007) Alcohol-based fuels in high performance engines. SAE paper 2007-01-0056, presented at the 2007 Fuels and Emissions Conference, Cape Town, South Africa, January 23rd–25th, 2007.

256 Group Lotus, Norwich, UK, http://www.grouplotus.com/.

257 Pearson, R.J., Turner, J.W.G. and Peck, A.J. (2009) *Gasoline-Ethanol-Methanol Tri-Fuel Vehicle Development and Its Role in Expediting Sustainable Organic Fuels for Transport*, Lotus Engineering, Norwich, UK.

258 U.S. Environmental Protection agency (1995), *Waiver Requests Under Section 211 (f) of the Clean Air Act, Revised August 22, 1995*, U.S. Environmental Protection Agency (EPA), Washington, DC.

259 Olah, G. A., Prakash, G. K. S. (2008), *Environmentally Friendly Ternary Transportation Flex-Fuel of Gasoline, Methanol and Biodiesel*, U.S. Patent Application 12/345697, December 30th.

260 Häpp, H.J. and Truong, H.-S. (1983) The Effect of Methanol/Diesel Fuel Emulsions on the Mixture Formation in Direct-Injection Diesel Engines: a Theory on Spontaneous Evaporation, SAE Technical Paper Series No. 830376.

261 Alternative Fuel: Transit Buses. Final Results from the National Renewable Energy Laboratory Vehicle Evaluation Program, Produced for the U.S. DOE, (1996).

262 Waterland, L.R., Venkatesh, S. and Unnasch, S. (2003) *Safety and Performance Assessment of Ethanol/Diesel Blends (E-Diesel)*, NREL/SR-540-34817, NREL, Golden, Colorado.

263 Merritt, P.M., Ulmet, V., McCormick, R.L., Mitchell, W.E. and Baumgard, K.J. (2005) Regulated and unregulated exhaust emissions comparison for three tier II non-road diesel engines operating on ethanol-diesel blends, NREL/CP-540-38493. NREL, SAE paper 2005-01-2193, presented at the 2005 SAE Brasil Fuels & Lubricants Meeting, Rio de Janeiro, Brasil, May, 2005.

264 Ogawa, T., Inoue, N., Shikada, T. and Ohno, Y. (2003) Direct dimethyl ether synthesis. *J. Nat. Gas Chem.*, **12**, 219.

265 Hirano, M., Imai, T., Yasutake, T. and Kuroda, K. (2004) Dimethyl ether synthesis from carbon dioxide by catalytic hydrogenation (Part 2). Hybrid catalyst consisting of methanol synthesis and methanol dehydration catalysts. *J. Jpn. Petrol. Inst.*, **47**, 11.

266 Xia, L. (2008) China DME market outlook. Proceedings of the 3rd International DME Conference & 5th Asian DME Conference, Shanghai, China September 21st–24th, 2008.

267 Hansen, J.B. and Mikkelsen, S.-E. (Halder Topsøe AIS) (2001) DME as a Transportation Fuel, Project Carried out for the Danish Road Safety & Transport Agency and the Danish Environmental Protection Agency. http://www.dieselnet.com/links/fuel_dme.html.

268 Volvo Bus Corporation Company Presentation, (2004).

269 Paas, M. (1997) Safety Assessment of DME Fuel, prepared for the Transportation Development Centre Safety and Security Transport Canada, TP 12998E. http://www.tc.gc.ca.

270 Arcoumanis, C., Bae, C., Crookes, R. and Kinoshita, E. (2008) The potential of di-methyl ether (DME) as an alternative fuel for compression-ignition engines: a review. *Fuel*, **87**, 1014.

271 Coal to Methanol Plant Set for West Virginia (2008), *Chem. Eng. News*, **86**, 22.

272 Basu, A., Fleisch, T.H., McCarthy, C.I. and Udovich, C.A. (1997) Process and fuel for

spark ignition engines, U.S. Patent
5,632,786.

273 JFE Holdings, Inc, http://www.jfe-
holdings.co.jp/en/dme/.

274 Basu, A. and Wainwright, J.M. (2001)
DME as a power generation fuel:
performance in gas turbines. Presented
at the PETROTECH-2001 Conference,
New Dehli, India, January, 2001.

275 Ohno, Y. and Omiya, M. (2003) Coal
conversion into dimethyl ether as an
innovative clean fuel. Presented at the
12th International Conference on Coal
Science, November, 2003.

276 World Energy Council (WEC): http://
www.worldenergy.org/wec-geis/.

277 Pavone, A. (2003) Mega Methanol Plants,
Report No. 43D, Process Economics
Program, SRI Consulting, Menlo Park,
California.

278 Ryu, J.Y. and Gelbein, A.P. (2000)
Producing dimethyl carbonate from CO_2
and methanol. A green chemistry
alternative to phosgene as a chemical
intermediate. 4th Annual Green
Chemistry and Engineering Conference
Proceedings, Washington, DC, June
27–29, 2000, p. 33.

279 Zhu, R., Wang, X., Miao, H., Huang, Z.,
Gao, J. *et al.* (2009) Performance and
emission characteristics of diesel
engines fueled with diesel-
dimethoxymethane (DMM) blends.
Energy Fuels, **23**, 286.

280 Ball, J., Lapin, C.A., Buckingham, J.P.,
Frame, E.A., Yost, D.M. *et al.* (2001)
Dimethoxy Methane in Diesel Fuel: Part
1. The Effect of Fuels and Engine
Operating Modes on Emissions of Toxic
Air Pollutants and Gas/Solid Phase Pah,
SAE Paper 2001-01-36276. Presented at
SAE International Fall Fuel & Lubricants
Meeting & Exhibition, September 2001,
San Antonio, TX, USA.

281 Zhang, Q., Tan, Y., Yang, C., Han, Y.,
Shamoto, J. *et al.* (2007) Catalytic oxidation
of dimethyl ether to dimethoxymethane
over Cs modified $H_3PW_{12}O_{40}/SiO_2$
catalysts. *J. Nat. Gas Chem.*, **16** (3), 322.

282 (2004) Methanol Institute Comments to
U.S. DOE On-Board Fuel Processing
Review Panel, Methanol Institute,
available from: http://www.methanol.org.

283 See Fuel Cell Vehicles Chart (from
Auto Manufacturers), http://www.
fuelcells.org.

284 See Daimler, http://www.daimler.com.

285 See Ford Motor Company, http://www.
ford.com/.

286 For more information see the Georgetown
University web site on fuel cell buses,
http://fuelcellbus.georgetown.edu/.

287 Methanol to Hydrogen Fueling Stations,
Fact Sheet, Methanol Institute, (2003),
http://www.methanol.org.

288 Dolan, G., Vassar, M.A. (2004), Methanol
to Hydrogen Fueling Station, SAE
Hydrogen Economy TOPTEC,
Sacramento, CA, February 19.

289 Japan Hydrogen & Fuel Cell
Demonstration Project (JHFC), http://
www.jhfc.jp.

290 Semelsberger, T.A., Borup, R.L. and
Greene, H.L. (2006) Dimethyl ether
(DME) as an alternative fuel. *J. Power
Sources*, **156**, 497.

291 Apanel, G. and Johnson, E. (2004) Direct
methanol fuel cells: ready to go
commercial? *Fuel Cells Bull.*, **11**, 12.

292 Voss, D. (2001) A fuel cell in your phone.
Technol. Rev., (November), 68.

293 Surampudi, S., Narayanan, S.R., Vamos,
E., Frank, H., Halpert, G. *et al.* (1994)
Advances in direct oxidation methanol
fuel cells. *J. Power Sources*, **47**, 377.

294 Surampudi, S., Narayanan, S.R., Vamos,
E., Frank, H., Halpert, G. *et al.* (1997) U.S.
Patent 5,599,638; U.S. Patent 6,248,460
(2001); U.S. Patent 6,740,434 (2004); U.S.
Patent 6,821,659 (2004).

295 Prakash, G.K.S., Smart, M.C., Wang, Q.-J.,
Atti, A. and Pleynet, V. (2004) High
efficiency direct methanol fuel cell based
on poly(styrenesulfonic) acid (PSSA) – poly
(vinylidenefluoride) (PVDF) composite
membranes. *J. Fluorine Chem.*, **125**, 1217.

296 Prakash, G.K.S., Olah, G.A., Smart, M.C.,
Narayanan, S.R., Wang, Q.S. *et al.* (2002)

Polymer electrolyte membranes for use in fuel cells, U.S. Patent 6,444,343.

297 Dillon, R., Srinivasan, S., Arico, A.S. and Antonucci, V. (2004) International activities in DMFC R&D: status of technologies and potential applications. *J. Power Sources*, **127**, 112.

298 McGrath, K.M., Prakash, G.K.S., Olah, G.A. (2004), Direct Methanol Fuel Cells, *J. Ind. Eng. Chem.*, **10**, 1063.

299 Aricò, A.S., Srinivasan, S., Antonucci, V. (2001), DMFCs: From Fundamental Aspects to Technology Development, *Fuel Cell*, **1**, 133.

300 Narayanan, S.R., Valdez, T.I., Clara, F. (2000), Design and Development of Miniature Direct Methanol Fuel Cell Sources for Cellular Phone Application, Proceedings of the Fuel Cell Seminar, Portland, Oregon, p. 795.

301 Jung, D.H., Jo, Y.-K., Jung, J.-H., Cho, C.-H. *et al.* (2000) A 10W Class Liquid-Feed Direct Methanol Fuel Cell for Portable Application. Proceedings of the Fuel Cell Seminar, Portland, Oregon, p. 420.

302 Bostaph, J., Koripella, R., Fisher, A. Zindel, D., Hallmark, J., *et al.* (2001) Microfluid Fuel Delivery Systems for 100 mW DMFC, Direct Methanol Fuel Cell. Proceedings of the 199th Meeting of Direct Methanol Fuel Cells, Electro-chemical Society, Washington, DC March 25–29, 2001.

303 Kim, D., Cho, E.A., Hong, S.-A. and Oh, I.H. (2004) Recent progress in the passive direct methanol fuel cell at KIST. *J. Power Sources*, **130**, 172.

304 Dolan, G.A. (2002) In search of the perfect clean-fuel options. *Hydrocarb. Process*, (March), 1.

305 JuVOMe Presentation, Research Center Julich, http://www.fz-juelich.de/portal/angebote/pressemitteilungen/scooter.

306 Yamaha Motor Co.: http://www.yamaha-motor.co.jp/motorshow/html/0003.html.

307 Yamaha Motor Co.: http://www.yamaha-motor.co.jp.

308 Geiger, S. and Jollie, D. (2003) Report from the 2003 Fuel Cell Seminar, Miami,

Fuel Cell Today, 14 November, http://www.fuelcelltoday.com.

309 Cai, K.-D., Yin, G.-P., Zhang, J., Wang, Z.-B., Du, C.-Y. *et al.* (2008) Investigation of a novel MEA for direct dimethyl ether fuel cell. *Electrochem. Commun.*, **10**, 238.

310 Kéranguéven, G., Coutanceau, C., Sibert, E., Hahn, F., Léger, J.-M. *et al.* (2006) Mechanism of di(methyl)ether (DME) electrooxidation at platinum electrodes in acid medium. *J. Appl. Electrochem.*, **36**, 441.

311 Logan, B.E., Hamelers, B., Rozendal, R., Schroeder, U., Keller, J. *et al.* (2006) Microbial fuel cells: methodology and technology. *Environ. Sci. Technol.*, **40**, 5181.

312 Olah, G.A. and Prakash, G.K.S. (1999) Recycling of carbon dioxide into methyl alcohol and related oxygenates for hydrocarbons, U.S. Patent 5,928,806.

313 Sea fairer: maritime transport and CO_2 emissions, OECD Observer, 267, May 26, (2008).

314 (2008) 58th session of the Marine Environment Protection Committee (MEPC), International Maritime Organization, London, October 6th to 10th, 2008.

315 Endresen, Ø., Sørgård, E., Sundet, J.K., Dalsøren, S.B., Isaksen, I.S.A. *et al.* (2003) Emission from international sea transportation and environmental impact. *J. Geophys. Res.*, **108** (D17), 4560.

316 Árnason, B. and Sigfússon, T.I. (1999) Converting CO_2 emissions and hydrogen into methanol vehicle fuel. *JOM*, **51** (May), 46.

317 METHAPU: http://www.methapu.eu/.

318 Temchin, J. (2003) Analysis of Market Characteristics for Conversion of Liquid Fueled Turbines to Methanol, Prepared for The Methanol Foundation and Methanex by Electrotek Concepts.

319 GE Position Paper: Feasibility of Methanol as Gas Turbine Fuel, General Electric (2001). http://www.methanol.org.

320 Jones, G.R.J., Holm-Larsen, H., Romani, D. and Sills, R.A. (2001) DME for power generation fuel: supplying India's southern region. Presented at the PETROTECH-2001 Conf., New Dehli, India, January, 2001.

321 http://www.fuelcelltoday.com/.

322 Cocco, D. and Tola, V. (2008) SOFC-MGT hybrid power plants fuelled by methanol and DME. *J. Appl. Electrochem.*, **38**, p. 955.

323 Hansen, J.B. (2005) Oxygenates as fuels for SOFC auxiliary power units. 15th International Symposia on Alcohol Fuels, San Diego, California, September 26–28, 2005.

324 Stokes, H. (2004) Commercialization of a New Stove and Fuel System for Household Energy in Ethiopia Using Ethanol from Sugar Cane Residues and Methanol from Natural Gas, Presented to the Ethiopian Society of Chemical Engineers (ESChE) at the Forum on "Alcohol as an Alternative Energy Resource for Household Use", October 30.

325 Ebbeson, B., Stokes, H.C. and Stokes, C.A. (2000) Methanol – The Other Alcohol: A Bridge to a Sustainable Clean Liquid Fuel.

326 Prepared by EA engineering, Science and Technology, Inc. for the American Methanol Foundation (1999) Methanol Refueling Station Costs, available from http://www.methanol.org.

327 Ashley, S. (2005) On the road to fuel cell cars. *Sci. Am.*, (March), 62.

328 Methanol Market Distribution Infrastructure in the United States, Prepared by DeWitt & Company, Inc for the Methanol Institute, (2002), available from: http://www.methanol.org.

329 Methanol Institute (1993) Methanol Fact Sheets, American Methanol Institute, Washington, DC.

330 Methanex website: http://www.methanex.com/.

331 Yotaro, O. (2001) A new DME production technology and operation results. 4th Doha Conference on Natural Gas, Doha, Qatar March 11–15th, 2001.

332 Energy Information Administration (EIA): http://www.eia.doe.gov/.

333 Specht, M., Staiss, F., Bandi, A. and Weimer, T. (1998) Comparison of the renewable transportation fuels, liquid hydrogen and methanol, with gasoline – energetic and economic aspects. *Int. J. Hydrogen Energy*, **23** (5), 387.

334 Zeman, F.S. and Keith, D.W. (2008) Carbon neutral hydrocarbons. *Phil. Trans. R. Soc. A*, **366**, 3901.

335 Prepared by Statoil, Norway (2001) Methanol in Fuel Cell Vehicles: Human Toxicity and Risk Evaluation (Revised).

336 Methanol Health Risk Fact Sheet, Methanol Institute, available at http://www.methanol.org.

337 Methanol Institute Methanol Health Effects Fact Sheet, available at http://www.methanol.org.

338 U.S. Environmental Protection Agency (EPA) (1994) Methanol Fuels and Fire Safety, Fact Sheet OMS-8, EPA 400-F-92-010, U.S. Environmental Protection Agency (EPA), Office of Mobile Sources, Washington, DC.

339 Paas, M. (Consulting Ltd.) (1998) Safety Assessment of DME Fuel Addendum, Prepared for Transportation Development Centre Safety and Security Transport Canada TP 12998 E Addendum.

340 Prepared by DuPont for the U.S. EPA, Chemical Right to Know Program (2000) Robust Summary for Dimethyl Ether.

341 DuPont (1987) Toxicity Summary for Dimethyl ether (DME); Dymel a Propellant, Technical Information.

342 U.S. Environmental Protection Agency (EPA), Transportation and Air Quality (2002) Clean Alternative Fuels: Methanol, Fact Sheet EPA 420-F-00-040.

343 Pollutant Emissions from Georgetown University Methanol Powered Fuel Cell Buses: http://fuelcellbus.georgetown.edu/overview3.cfm.

344 Gray, C. and Webster, G. (2001) A Study of Dimethyl Ether (DME) as an Alternative Fuel for Diesel Applications, Prepared for CANMET Energy Technology Centre, Natural Resources Canada and Transportation Development Centre, Transport Canada TP 13788E by Advanced Engine Technology Ltd.

345 Kajitani, S. (2006) Prospects of fuel DME. Conference on the development and Promotion of Environmental Friendly Heavy Duty Vehicles Such as DME Trucks, Washington DC, March 17th, 2006.

346 Conference on the Development and Promotion of Environmentally Friendly Heavy Duty Vehicles such as DME Trucks, Co-hosted by the Japan International Transport Institute and National Traffic Safety and Environment Laboratory. Washington DC March 17, 2006.

347 WHO (1997) Methanol, Health and Safety Guide (HSG 105, 1997), International Programme on Chemical Safety (IPCS), http://www.inchem.org/.

348 Wastewater Treatment with Methanol Denitrification, Fact Sheet, Methanol Institute, available at http://www.methanol.org.

349 Malcolm Pirnie, Inc. for the Methanol Institute (1999) Evaluation of the Fate and Transport of Methanol in the Environment, available from http://www.methanol.org/.

350 Good, D.A., Francisco, J.S., Jain, A.K. and Wuebbles, D.J. (1998) Lifetimes and global warming potential for dimethyl ether and for fluorinated ethers: CH_3OCF_3 (E143a), CHF_2OCHF_2 (E134), CHF_2OCF_3 (E125). *J. Geophys. Res.*, **103**, 28181.

351 Brown, R. (2005) Methanol Market Quiet, Chemical Market Reporter, section 2, January 31, p. 5.

352 Source: Methanol Institute. http://www.methanol.org.

353 Pavone, A. (2003) Mega Methanol Plants, Report No. 43D, Process Economics Program, SRI Consulting, Menlo Park, California.

354 Brown, R. (2004) Methanol pricing steady as supply situation changes, Chemical Market Reporter, section 2, October 4, 19.

355 Plouchart, G. (2005) *Panorama 2005: Energy Consumption in the Transportation Sector,* Institut Francais du Petrole (IFP), available from http://www.ifp.fr/IFP/en/aa.htm.

356 Air Products Liquid Phase Conversion Company for the U.S. DOE National Energy Technology Laboratory (2003) Commercial-Scale Demonstration of the Liquid Phase Methanol (LPMEOH™) Process: Final Report.

357 Kochloefl, K. (1997) Steam Reforming, in *Handbook of Heterogeneous Catalysis*, vol. 4 (eds G. Ertl, H. Knözinger and J. Weitkamp), Wiley-VCH GmbH, Weinheim, p. 1819.

358 Choudhary, T.V. and Choudhary, V.R. (2008) Energy-efficient syngas production through catalytic oxy-methane reforming reactions. *Angew. Chem. Int. Ed.*, **47**, 1828.

359 Hansen, J.B. (1997) Methanol Synthesis, in *Handbook of Heterogeneous Catalysis*, vol. 4 (eds G. Ertl, H. Knözinger and J. Weitkamp), Wiley-VCH GmbH, Weinheim, p. 1856.

360 Bradford, M.C.J. and Vannice, M.A. (1999) CO_2 reforming of CH_4. *Catal. Rev. – Sci. Eng.*, **41** (1), 1.

361 Turek, T., Trimm, D.L. and Cant, N.W. (1994) The catalytic hydrogenolysis of esters to alcohols. *Cat. Rev. – Sci. Eng.*, **36**, 645.

362 Christiansen, J.A. (1919) Method of producing methyl alcohol from alkyl formate, U.S. Patent 1,302,011.

363 Marchionna, M., Lami, M. and Raspolli Galleti, A.M. (1997) Synthesizing methanol at lower temperature. *Chemtech*, (April), 27.

364 Crabtree, R.H. (1995) Aspects of methane chemistry. *Chem. Rev.*, **95**, 987.

365 Lunsford, J.H. (2000) Catalytic conversion of methane to more useful chemicals and

fuels: a challenge for the 21st century. *Catal. Today*, **63**, 165.

366 Otsuka, K. and Wang, Y. (2001) Direct conversion of methane into oxygenates. *Appl. Catal. A*, **222**, 145.

367 Olah, G.A. (1987) Electrophilic methane conversion. *Acc. Chem. Res.*, **20**, 422.

368 Weng, T. and Wolf, E.E. (1993) Partial oxidation of methane on Mo/Sn/P silica supported catalysts. *Appl. Catal. A*, **96** (2), 383.

369 Sugino, T., Kido, A., Azuma, N., Ueno, A. and Udagawa, Y. (2000) Partial oxidation of methane on silica-supported silicomolybdic acid catalysts in an excess amount of water vapor. *J. Catal.*, **190** (1), 118.

370 Olah, G.A. and Prakash, G.K.S. (2006) Selective oxidative conversion of methane to methanol, dimethyl ether and derived products, U.S. Patent Application, 20060235088.

371 Periana, R.A., Bhalla, G., Tenn, W.J.III, Young, K.J.H., Liu, X.Y. *et al.* (2004) Perspective on some challenges and approaches for developing the next generation of selective, low temperature, oxidation catalysts for alkane hydroxylation based on CH activation reaction. *J. Mol. Catal. A: Chem.*, **220**, 7.

372 Conley, B., Tenn, W.J.III, Young, K.J.H., Ganesh, S., Meier, S. *et al.* (2006) Methane Functionalization, in *Activation of Small Molecules* (ed. W.B. Tolman), Wiley-VCH GmbH, Weinheim, p. 235.

373 Periana, R.A., Taube, T.J., Evitt, E.R., Löffler, D.G., Wentrcek, P.R. *et al.* (1993) A mercury-catalyzed, high-yield system for the oxidation of methane to methanol. *Science*, **259**, 340.

374 DeVos, D.E. and Sels, B.F. (2005) Gold redox catalysis for selective oxidation of methane to methanol. *Angew. Chem. Int. Ed.*, **44**, 30.

375 Jones, C.J., Taube, D., Ziatdinov, V.R., Periana, R.A., Nielsen, R.J. *et al.* (2004) Selective oxidation of methane to methanol catalyzed, with C-H activation,

by homogeneous, cationic gold. *Angew. Chem. Int. Ed.*, **43**, 4626.

376 Olah, G.A., Gupta, B., Farina, M., Felberg, J.D., Ip, W.M. *et al.* (1985) Selective monohalogenation of methane over supported acid or platinum metal catalysts and hydrolysis of methyl halides over gamma-alumina-supported metal oxide/hydroxide catalysts. A feasible path for the oxidative conversion of methane into methyl alcohol/dimethyl ether. *J. Am. Chem. Soc.*, **107**, 7097.

377 Olah, G.A. (1985) Methyl halides and methyl alcohol from methane, U.S. Patent 4,523,040.

378 Olah, G.A. and Mo, Y.K. (1972) Electrophilic reaction at single bonds. XIII. Chlorination and chlorolysis of alkanes in SbF_5-Cl_2-SO_2ClF solution at low temperature. *J. Am. Chem. Soc.*, **94**, 6864.

379 Olah, G.A., Renner, R., Schilling, P. and Mo, Y.K. (1973) Electrophilic reactions at single bonds. XVII. SbF_5, $AlCl_3$, and $AgSbF_6$ catalyzed chlorination and chlorolysis of alkanes and cycloalkanes. *J. Am. Chem. Soc.*, **95**, 7686.

380 Pan, H.Y., Minet, R.G., Benson, S.W. and Tsotsis, T.T. (1994) Process for converting hydrogen chloride to chlorine. *Ind. Eng. Chem. Res.*, **33**, 2996.

381 Mortensen, M., Minet, R.G., Tsotsis, T.T. and Benson, S.W. (1999) The development of dual fluidized-bed reactor system for the conversion of hydrogen chloride to chlorine. *Chem. Eng. Sci.*, **54**, 2131.

382 Schweizer, A.E., Jones, M.E. and Hickman, D.A. (2002) Oxidative halogenation of C_1 hydrocarbons into halogenated C_1 hydrocarbons and integrated processes related thereto, U.S. Patent 6,452,058.

383 Lorkovic, I., Noy, M., Weiss, M., Sherman, J., McFarland, E. *et al.* (2004) C_1 coupling via bromine activation and tandem catalytic condensation and neutralization over CaO/zeolite composite. *Chem. Commun.*, 566.

384 Yilmaz, A., Yilmaz, G.A., Lorkovic, I.M., Stucky, G.D., Ford, P.C. *et al.* (2004)

Integrated process for synthesizing alcohols, ethers, aldehydes, and olefins from alkanes, U.S. Patent, 6,713,655.

385 Sherman, J.H., McFarland, E., Weiss, M.J., Lorkovic, I.M., Laverman, L.E. *et al.* (2007) Method and apparatus for synthesizing olefins, alcohols, ethers, and aldehydes, U.S. Patent, 7,161,050.

386 Zhou, X.P., Stucky, G.D. and Sherman, J. (2002) Integrated process for synthesizing alcohols, ethers, and olefins from alkanes, U.S. Patent, 6,465,696.

387 Periana, R.A., Mirinov, O., Taube, D.J. and Gamble, S. (2002) High yield conversion of methane to methyl bisulfate catalyzed by iodine cations. *Chem. Commun.*, 2376.

388 Hanson, R.S. and Hanson, T.E. (1996) Metanotrophic bacteria. *Microbiol. Rev.*, **60**, 439.

389 Wallar, B.J. and Lipscomb, J.D. (1996) Dioxygen activation by enzymes containing binuclear non-heme iron clusters. *Chem. Rev.*, **96**, 2625.

390 Baik, M.-H., Newcomb, M., Friesner, R.A. and Lippard, S.J. (2003) Mechanistic studies on the hydroxylation of methane by methane monooxygenase. *Chem. Rev.*, **103**, 2385.

391 Ayala, M. and Torres, E. (2004) Enzymatic activation of alkanes: constraints and prospective. *Appl. Catal. A*, **272**, 1.

392 Xu, F., Bell, S.G., Lednik, J., Insley, A. and Rao, Z. (2005) *et al.* The heme monooxygenase cytochrome P450$_{cam}$ can be engineered to oxidize ethane to ethanol. *Angew. Chem. Int. Ed.*, **44**, 4029.

393 Meinhold, P., Peters, M.W., Chen, M.M.Y., Takahashi, K. and Arnold, F.H. (2005) Direct conversion of ethane to ethanol by engineering cytochrome P450 BM3. *ChemBioChem*, **6**, 1765.

394 Süss-Fink, G., Stanislas, S., Shul'pin, G.B. and Nizova, G.V. (2000) Catalytic functionalization of methane. *Appl. Organomet. Chem.*, **14**, 623.

395 Milne, T.A., Evans, R.J. and Abatzoglou, N. (1998) Biomass Gasifier "Tars": Their Nature, Formation, and Conversion NREL/TP-570-25357, National Renewable Energy Laboratory.

396 Adinberg, R., Epstein, M. and Karni, J. (2004) Solar gasification of biomass: a molten salt pyrolysis study. *Trans. ASME*, **126**, 850.

397 Hamelinck, C.N. and Faaij, A.P.C. (2001) *Future Prospects for Production of Methanol and Hydrogen from Biomass*, University Utrecht, Copernicus Institute, The Netherlands.

398 Swaaij, W.P.M., Kersten, S.R.A. and Van denAarsen, F.G. (2004) Routes for methanol from biomass. International Business Conference on Sustainable Industrial Developments, Delfzijl, The Netherlands, April, 2004.

399 Henrich, E. (2002) Kraftstoff aus Stroh, NRW Fachtagung "Was Tanken wir Morgen?", Oberhausen, November 25–26.

400 Norbeck, J.M. and Johnson, K. (2000) *The Hynol Process: A Promising Pathway for Renewable Production of Methanol*, College of Engineering, Center for Environmental Research, and Technology, University of California, Riverside.

401 Ekbom, T., Lindblom, M., Berglin, N. and Ahlvik, P. (2003) Cost-Competitive, Efficient Bio-Methanol Production from Biomass via Black Liquor Gasification, Alterner Program of the European Union.

402 Brochure from Volvo (2004) Future Fuels for Commercial Vehicles, ref No 011-949-007.

403 Brochure from Volvo (2006) Powerful Ways to the Future, Brochure from Volvo, ref No 011-949-012.

404 Concawe, European Council for Automotive R&D and European Commission Joint Research Centre (2004) Well-to-Wheels Analysis of Future Automotive Fuels and Powertrains in the European Context, Tank-to-Wheels Report, Version 1b.

405 Renewable Fuels for Advanced Powertrains, European Project: http://www.renew-fuel.com.

406 Rudberg, J. (2008) The Future of 2nd Generation Biofuels in Europe. Toward a New Vision for the Pulp and Paper Mill, Answers from the RENEW Project to the Biofuel Debate, Brussels, September 15th.

407 Cassedy, E.S. (2000) *Prospects for Sustainable Energy. A Critical Assessment,* Cambridge University Press, Cambridge.

408 Intergovernmental Panel on Climate Change (IPCC), Climate Change 2001, Mitigation, IPCC Third Assessment Report.

409 Huggins, D.R. and Reganold, J.P. (2008) No-till: the quiet revolution. *Sci. Am.,* (July), 70.

410 Tilman, D., Hill, J. and Lehman, C. (2006) Carbon-negative biofuels from low-input high-diversity grassland biomass. *Science,* 314, 1598.

411 Nyns, E.-J. (2003) Methane, in *Ullmann's Encyclopedia of Industrial Chemistry,* vol. 21, Wiley-VCH GmbH, Weinheim, Germany, p. 599.

412 U.S. Environmental Protection Agency (EPA), Municipal Solid Waste and Landfill Methane Outreach Program, http://www.epa.gov.

413 Sheehan, J., Dunahay, T., Benemann, J. and Roessler, P. (1998) *A Look Back at the U.S. Department of Energy's Aquatic Species Program – Biodiesel from Algae,* National Renewable Energy Laboratory (NREL), NREL/TP-580-24190, Golden, Colorado.

414 Ostrovskii, V.E. (2002) Mechanism of methanol synthesis from hydrogen and carbon oxides at Cu-Zn containing catalysts in the context of some fundamental problems of heterogeneous catalysis. *Catal. Today,* 77, 141.

415 Rozovskii, A.Y. and Lin, G.I. (2003) Fundamentals of methanol synthesis and decomposition. *Top. Catal.,* 22 (3–4), 137.

416 Goehna, H. and Koenig, P. (1994) Producing methanol from CO_2. *Chemtech,* (June), 39.

417 Saito, M. (1998) R&D activities in Japan on methanol synthesis from CO_2 and H_2. *Catal. Surv. Jpn.,* 2, 175.

418 Shulenberger, A.M., Jonsson, F.R., Ingolfsson, O. and Tran, K.-C. (2007) Process for producing liquid fuel from carbon dioxide and water, U.S. Patent Appl., 2007/0244208 A1.

419 Tremblay, J.-F. (2008) CO_2 as feedstock. Mitsui will make methanol from the greenhouse gas. *Chem. Eng. News,* 86 (35), 13.

420 Xiaoding, X. and Moulijn, J.A. (1996) Mitigation of CO_2 by chemical conversion: plausible chemical reactions and promising products. *Energy & Fuels,* 10, 305.

421 Saito, M. and Murata, K. (2004) Development of high performance Cu/ZnO-based catalysts for methanol synthesis and water-gas shift reaction. *Catal. Surv. Asia,* 8 (4), 285.

422 Steinberg, M. (1999) Fossil fuel decarbonization technology for mitigating global warming. *Int. J. Hydrogen Energy,* 24, 771.

423 Song, C. and Pan, W. (2004) Tri-reforming of methane: a novel concept for catalytic production of industrially useful synthesis gas with desired H_2/CO ratios. *Catal. Today,* 98, 463.

424 Adachi, Y., Komoto, M., Watanabe, I., Ohno, Y. and Fujimoto, K. (2000) Effective utilization of remote coal through dimethyl ether synthesis. *Fuel,* 79, 229.

425 Matsunami, J., Yoshida, S., Oku, Y., Yokota, O., Tamaura, Y. *et al.* (2000) Coal gasification by CO_2 gas bubbling in molten salt for solar/fossil energy hybridization. *Solar Energy,* 68, 257.

426 Kodama, T., Funatoh, A., Shimizu, K. and Kitayama, Y. (2001) Kinetics of metal oxide-catalyzed CO_2 gasification of coal in a fluidized-bed reactor for solar thermochemical process. *Energy & Fuels,* 15, 1200.

427 Kodama, T., Aoki, A., Ohtake, H., Funatoh, A., Shimizu, T. *et al.* (2000) Thermochemical CO_2 gasification of coal using a reactive coal-In_2O_3 system. *Energy & Fuels,* 14, 202.

428 Jitaru, M., Lowy, D.A., Toma, M. and Oniciu, L. (1997) Electrochemical reduction of carbon dioxide on flat metallic cathodes. *J. Appl. Electrochem.*, **27**, 875.

429 Gattrell, M., Gupta, N. and Co, A. (2006) A review of the aqueous electrochemical reduction of CO_2 to hydrocarbons at copper. *J. Electroanal. Chem.*, **594**, 1.

430 Kaneco, S., Iiba, K., Suzuki, S.K., Ohta, K. and Mizuno, T. (1999) Electrochemical reduction of carbon dioxide to hydrocarbons with high faradaic efficiency in LiOH/methanol. *J. Phys. Chem. B*, **103**, 7456.

431 Kaneco, S., Iwao, R., Iiba, K., Itoh, S.I., Ohata, K. *et al.* (1999) Electrochemical reduction of carbon dioxide on an indium wire in a KOH/methanol-based electrolyte at ambient temperature and pressure. *Environ. Eng. Sci.*, **16**, 131.

432 Kaneco, S., Katsumata, H., Suzuki, T. and Ohta, K. (2006) Photoelectrochemical reduction of carbon dioxide at p-type gallium arsenide and p-type indium phosphide electrodes in methanol. *Chem. Eng. J.*, **116**, 227.

433 Olah, G.A. and Prakash, G.K.S. (2007) Electrolysis of carbon dioxide in aqueous media to carbon monoxide and hydrogen for production of methanol, U.S. Provisional Pat. Appl., 60/949,723.

434 Jensen, S.H., Larsen, P.H. and Mogensen, M. (2007) Hydrogen and synthetic fuel production from renewable energy sources. *Int. J. Hydrogen Energy*, **32**, 3253.

435 Augustynski, J., Sartoretti, C.J. and Kedzierzawski, P. (2003) Electrochemical Conversion of Carbon Dioxide, in *Carbon Dioxide Recovery and Utilization* (ed. M. Aresta), Kluwer Academic Publisher, Dordrecht, p. 279.

436 Bagotzky, V.S. and Osetrova, N.V. (1995) Electrochemical reduction of carbon dioxide. *Russ. J. Electrochem.*, **31**, 409.

437 Kohl, A. and Nielsen, R. (1997) *Gas Purification*, 5th edn, Gulf Publishing Company, Houston.

438 Olah, G.A., Goeppert, A., Meth, S. and Prakash, G.K.S. (2008) Nano-structure supported solid regenerative polyamine and polyamine polyol absorbents for the separation of carbon dioxide from gas mixtures including the air, International patent application, 2008021700.

439 Millward, A.R. and Yaghi, O.M. (2005) Metal-organic frameworks with exceptionally high capacity for storage of carbon dioxide at room temperature. *J. Am. Chem. Soc.*, **127**, 17998.

440 UOP: http://www.uop.com/. Solexol™ Process, Fact Sheet (2002).

441 Specht, M. and Bandi, A. (1995) Herstellung von Fluessigen Kraftstoffen aus Atmosphaerischem Kohlendioxid, in *Forschungsverbund Sonnenenergie, Köln, Themen, 1994–1995, Energiespeicherung*, p. 41.

442 Asinger, F. (1987) *Methanol, Chemie- und Energierohstoff. Die Mobilisation der Kohle*, Springer-Verlag, Heidelberg.

443 Pasel, J., Peters, R. and Specht, M. (2000) Methanol Herstellung und Einsatz Als Energietraeger fuer Brennstoffzellen, in *Forschungsverbund Sonnenenergie, Themen 1999–2000: Zukunftstechnologie Brennstoffzelle*, p. 46.

444 Specht, M. and Bandi, A."*The Methanol Cycle*" – *Sustainable Supply of Liquid Fuels*, Center for Solar Energy and Hydrogen Research (ZSW), Stuttgart, Germany.

445 Specht, M., Bandi, A. (1999) "Der Methanol Kreislauf" nachhaltige Bereitstellung flüssiger Kraftstoffe, in Forschungsverbund Sonnenenergie, Themen 1998 – 1999, Nachhaltigheit und Energie, p. 59.

446 Lackner, K.S., Ziock, H.-J. and Grimes, P. (1999) The case for carbon dioxide extraction from air. *SourceBook*, **57**, 6.

447 Zeman, F. (2007) Energy and material balance of CO_2 capture from ambient air. *Environ. Sci. Technol.*, **41**, 7558.

448 Schuler, S.S. and Constantinescu, M. (1995) Coupled CO_2 recovery from the atmosphere and water electrolysis:

feasibility of a new process for hydrogen storage. *Int. J. Hydrogen. Energy*, **20**, 653.

449 Martin, J.F. and Kubic, W.L. (2007) *Green Freedom™ – A Concept for Producing Carbon-Neutral Synthetic Fuels and Chemicals LA-UR-07-7897*, Los Alamos National Laboratory.

450 Keith, D.W., Ha-Duong, M. and Stolaroff, J.K. (2006) Climate strategy with CO_2 capture from the air. *Clim. Change*, **74**, 17.

451 Zeman, F.S. and Keith, D.W. (2008) Carbon neutral hydrocarbons. *Phil. Trans. R. Soc. A*, **366**, 3901.

452 Zarembo, A. (2008) It's a tidy answer to global warming. Los Angeles Times.

453 Rabo, J.A. (1993) Catalysis: past, present and future. Proceedings of the 10th International Congress on Catalysis, Budapest, Hungary, July 19–24 (Elsevier Science Publishers).

454 Peters, U., Nierlich, F., Schulte-Körne, E., Sakuth, M., Deeb, R. *et al.* (2007) Methyl tert-Butyl Ether and Ethyl tert-Butyl Ether, in *Handbook of Fuels* (ed. B. Elvers), Wiley-VCH GmbH, p. 253.

455 Hamid, H. and Aliin, M.A. (eds) (2004) *Handbook of MTBE and Other Gasoline Oxygenates*, Marcel Dekker, New York.

456 Olah, G.A. and Molnár, Á. (2003) *Hydrocarbon Chemistry*, 2nd edn, John Wiley & Sons Inc., Hoboken, New Jersey.

457 Bercaw, J.E., Diaconescu, P.L., Grubbs, R.H., Hazari, N., Kay, R.D. *et al.* (2007) Conversion of methanol to 2,2,3-trimethylbutane (triptane) over indium(III) iodide. *Inorg. Chem.*, **46** (26), 11373.

458 Bercaw, J.E., Grubbs, R.H., Hazari, N., Labinger, J.A. and Li, X. (2007) Enhanced selectivity in the conversion of methanol to 2,2,3-trimethylbutane (triptane) over zinc iodide by added phosphorous or hypophosphorous acid. *Chem. Commun.*, 2974.

459 Walspurger, S., Prakash, G.K.S. and Olah, G.A. (2008) Zinc catalyzed conversion of methanol-methyl iodide to hydrocarbons with increased formation of triptane. *Appl. Catal. A*, **336**, 48.

460 Chang, C.D. (1994) Methanol to Gasoline and Olefins, in *Methanol Production and Use* (eds W-.H. Cheng and H.H. Kung), Marcel Dekker, New York, p. 133.

461 Chang, C.D. (1997) Methanol to Hydrocarbons, in *Handbook of Heterogeneous Catalysis*, vol. 4 (eds G. Ertl, H. Knözinger and J. Weitkamp), Wiley-VCH GmbH, Weinheim, p. 1894.

462 Stocker, M. and Weitkamp, J. (eds) (1999) *Microporous Mesoporous Mater.*, **29**, (1–2). Special issue covering methanol to hydrocarbons technologies and processes.

463 Olah, G.A., Doggweiler, H., Felberg, J.D., Frohlich, S., Grdina, M.J. *et al.* (1984) Onium ylide chemistry. 1. bifunctional acid-base-catalyzed conversion of heterosubstituted methanes into ethylene and derived hydrocarbons. The onium ylide mechanism of the C_1 to C_2 conversion. *J. Am. Chem. Soc.*, **106**, 2143.

464 Olah, G.A. (1983) Bifunctional acid-base catalyzed conversion of heterosubtituted methanes into olefins, U.S. Patent 4,373,109.

465 Chen, J.Q., Bozzano, A., Glover, B., Fuglerud, T. and Kvisle, S. (2005) Recent advancements in ethylene and propylene production using the UOP/Hydro MTO process. *Catal. Today*, **106**, 103.

466 Andersen, J., Bakas, S., Kvisle, H., Reier, N. *et al.* (2003) MTO: meeting the needs for ethylene and propylene production. presented at ERTC Petrochemical Conference, Paris, France, March 3–5, 2003.

467 UOP: http://www.uop.com/. UOP/ HYDRO MTO Process Methanol to Olefins Conversion, (2007), Fact Sheet.

468 Lurgi: http://www.lurgi.de.

469 DKRW Advanced Fuels Secures Exxon Mobil MTG Technology. Industrial Siting Permit Granted (2007). Press Release DKRW Advanced Fuels, Houston, Texas. Available at http://www.dkrwadvancedFuels.com.

For Further Reading and Information

General Information on Energy

Schobert, H.H. (2002) *Energy and Society, an Introduction*, Taylor and Francis, New York.

Smil, V. (2003) Energy at the Crossroads, *Global Perspectives and Uncertainties*, MIT Press, Cambridge.

Smil, V. (1994) *Energy in World History*, Westview Press, Boulder, Colorado.

Bent, R., Orr, L. and Baker, R. (2002) *Energy. Science, Policy, and the Pursuit of Sustainability*, Island Press, Washington, DC.

International Energy Agency (1997) *Energy Technologies for the 21st Century*, International Energy Agency, Paris.

International Energy Agency, *Key World Energy Statistics 2004*, International Energy Agency, Paris, 2004.

International Energy Agency, *World Energy Outlook 2001: Insights*, International Energy Agency, Paris, 2001.

International Energy Agency, *World Energy Outlook 2004*, International Energy Agency, Paris, 2004.

Annual Energy Outlook 2009, Energy Information Agency, Washington, DC, 2009, available at http://www. eia.doe.gov/oiaf/aeo.

World Energy Council (WEC): http://www. worldenergy.org/wec-geis/.

International Energy Agency (IEA): http:// www.iea.org/.

U.S. Energy Information Administration (EIA): http://www.eia.doe.gov/.

BP Statistical Review of World Energy: http:// www.bp.com/statisticalreview.

U.S. Department of Energy (DOE): http:// www.energy.gov/.

Coal

To learn more about the history of coal:
Freese, B. (2003) *Coal, a Human History*, Perseus Publishing, Cambridge.

General Information about Coal, Including Statistics on Production, Uses and Impact on the Environment

International Energy Agency, *Coal Information 2008*, International Energy Agency, Paris, 2008.

World Coal Institute: http://www.worldcoal. org/.

National Mining Association (U.S.): http:// www.nma.org/.

Clean Coal Technology

NETL clean coal technology: http://www.netl. doe.gov/cctc/.

IEA clean coal centre: http://www.iea-coal.org. uk/site/ieaccc/home.

Massachusetts Institute of Technology (2007) The Future of Coal, an Interdisciplinary MIT study, Massachusetts Institute of Technology. http://web.mit.edu/coal/.

Beyond Oil and Gas: The Methanol Economy, Second updated and enlarged edition
George A. Olah, Alain Goeppert, and G. K. Surya Prakash
Copyright © 2009 WILEY-VCH Verlag GmbH & Co. KGaA, Weinheim
ISBN: 978-3-527-32422-4

Petroleum Oil and Natural Gas

To learn more about the history of oil:
Black, B. (2000) *Petrolia, the Landscape of America's First Oil Boom*, The Johns Hopkins University Press, Baltimore.
Yergin, D. (1991) *The Prize: The Epic Quest for Oil, Money and Power*, Simon & Schuster.

To learn more about discovery and exploitation of oil and natural gas:
Conaway, C.F. (1999) *The Petroleum Industry. A Nontechnical Guide*, Pennwell, Tulsa, Oklahoma.
Hyne, N.J. (1995) *Nontechnical Guide to Petroleum Geology, Exploration, Drilling and Production*, Pennwell, Tulsa, Oklahoma.
Campbell, C.J. (1988) *The Coming Oil Crisis*, Multi-science Publishing, Brentwood, England.
Deffeyes, K.S. (2001) *Hubbert's Peak, the Impending World Oil Shortage*, Princeton University Press, Princeton.
Leffler, W.L., Pattarozzi, R. and Sterling, G. (2003) *Deepwater Petroleum Exploitation & Production: A Nontechnical Guide*, PennWell, Tulsa, Oklahoma.
Wiley Critical Content: Petroleum Technology, Wiley Interscience, Hoboken, 2007.
Seddon, D. (2006) *Gas Usage & Value*, PennWell, Tulsa, Oklahoma.
All About Petroleum, available at http://api-ec.api.org. Brochure from the American Petroleum Institute.
American Petroleum Institute (API): http://www.api.org.

Oil and Natural Gas Statistics on Production, Reserves and Uses

International Energy Agency , *Oil Information 2008*, International Energy Agency, Paris, 2008.
International Energy Agency , *Natural Gas Information 2008*, International Energy Agency, Paris, 2008.

Liquefied Natural Gas (LNG)

The Center for LNG: http://www.lngfacts.org/.

Department of Energy, information on LNG: http://www.fossil.energy.gov/programs/oilgas/storage/lng/feature/index.html.
Jensen, J.T. (2004) *The Development of a Global LNG Market. Is it Likely? If So When?* Oxford Institute for Energy Studies, Alden Press.

Unconventional Oil and Gas Resources

Tar Sands

Alberta Chamber of Resources (2004) Oil Sands Technology Roadmap. Unlocking the Potential, Alberta Chamber of Resources, Edmonton, Alberta, available at http://www.acr-alberta.com/.
For information about bitumen and extra-heavy oil: World Energy Council: http://www.worldenergy.org/publications.

Oil Shale

Is oil shale America's answer to peak-oil challenge. *Oil & Gas J.*, 2004.
Dyni, J.R. (2006) Geology and resources of some world oil-shale deposits. U.S. Geological Survey, Scientific Investigation Report 2005-5294.
U.S. Department of Energy (2007) Secure Fuels from Domestic Resources. The continuing Evolution of America's Oil Shale and Tar Sands Industries.
AOC Petroleum Support Services. Llc (2004) Strategic Significance of America's Oil Shale Resource, Naval Petroleum and Oil Shale Reserves & U.S. Department of Energy, Washington, DC.
European Academies Science Advisory Council (2007) A Study on the EU Oil Shale Industry – Viewed in the Light of Estonian Experience.
World energy Council, information about oil shale: http://www.worldenergy.org/publications.
Loucks, R.A. (2002) *Shale Oil: Tapping the Treasure*, Xlibris.

Coal Bed Methane, Tight Gas Sands and Shale Gas

Canadian Society for Unconventional Gas: http://www.csug.ca/.

Fischer, P.A. (2004) Unconventional gas resources fill the gap in future supplies, *worldOil Mag.*, August, vol. 225, available at http://worldoil.com.

Perry, K.F., Cleary, M.P. and Curtis, J.B. (1998) New Technology for Tight Gas Sands, 17th World Energy Congress, Houston, September 13–18, available at http://www.worldenergy.org.

Garbutt, D. (2004) Unconventional Gas, Schlumberger white paper, available at http://www.oilfield.slb.com.

Methane Hydrate

To learn more about methane hydrate:
The National Energy Technology Laboratory (NETL, U.S.): http://www.netl.doe.gov/scngo/NaturalGas/hydrates/.

U.S. Department of Energy, methane hydrate program: http://www.fe.doe.gov/programs/oilgas/hydrates/index.html.

Boswell, R. (2007) Resource potential of methane hydrate coming into focus. *J. Petrol. Sci. Technol.*, **56**, 9.

Diminishing Oil and Natural Gas Resources, Production Peak and Shortage

Appenzeller, T. (2004) The end of cheap oil. *Natl. Geographic Mag.*, June, 80.

Campbell, C.J. (1997) *The Coming Oil Crisis*, Multi-science Publishing, Brentwood, England.

Deffeyes, K.S. (2001) *Hubbert's Peak, the Impending World Oil Shortage*, Princeton University Press, Princeton.

Deffeyes, K.S. (2005) *Beyond Oil, the View from Hubbert's Peak*, Hill and Wang, New York.

Heinberg, R. (2003) *The Party's Over. Oil, War and the Fate of Industrial Societies*, New Society Publishers, Gabriola Island, Canada.

Odell, P.R. and Rosing, K.E. (1983) *The Future of Oil. World Oil Resources and Use*, Kogan Page, London.

Roberts, P. (2004) *The End of Oil*, Houghton Mifflin Company, New York.

Campbell, C.J. and Laherrère, J.H. (1998) The end of cheap oil. *Sci. Am.*, 78.

Goodstein, D. (2004) *Out of Oil: The End of the Age of Oil*, W. W. Norton & Company, New York.

Ruppert, M.C. (2004) *Crossing the Rubicon: The Decline of the American Empire at the End of the Age of Oil*, New Society Publishers.

Bentley, R.W. (2002) Oil & gas depletion: an overview. *Energy Policy*, **30**, 189.

Odell, P.R. and Rosing, K.E. (1983) *The Future of Oil. World Oil Resources and Use*, Kogan Page, London.

Odell, P.R. (2004) *Why Carbon Fuels Will Dominate the 21st Century's Global Energy Economy*, Multi-Science Publishing, Brentwood, England.

Huber, P. and Mills, M.P. (2005) *The Bottomless Well. The Twilight of Fuel, The Virtue of Waste, and Why we Will Never Run Out of Energy*, Basic Books, Cambridge.

Simmons, M.R. (2005) *Twilight in the Desert – The Coming Saudi Oil Shock and the World Economy*, John Wiley & Sons.

Maugeri, L. (2004) Oil: never cry wolf. Why the petroleum age is far from over. *Science*, **304**, 1115.

On the Internet

See Wikipedia under "Hubbert peak theory" and numerous references therein: http://en.wikipedia.org/.

Association for the study of peak oil & gas: http://www.peakoil.net/.

Other sites for information about oil peak: http://www.hubbertpeak.com/, http://www.peakoil.com/ and http://www.oilscenarios.info/.

Hydrocarbons and their Products

Olah, G.A. and Molnár, Á. (2003) *Hydrocarbon Chemistry*, 2nd edn, John Wiley & Sons, Inc., Hoboken, New Jersey.
Weissermel, K. and Arpe, H.-J. (2003) *Industrial Organic Chemistry*, 8th edn, Wiley-VCH GmbH, Weinheim, Germany.
Wiley Critical Content: Petroleum Technology, Wiley Interscience, Hoboken, 2007.

To learn more about petrochemistry:
http://www.petrochemistry.net/.

To learn more about plastics:
American Plastic Council: http://www.plastics.org/.
Plastic Europe, association of plastic manufacturers: http://www.plasticseurope.org.

Climate Change

Intergovernmental Panel on Climate Change (IPCC) (2007) Fourth Assessment Report: Climate Change 2007, available at http://www.ipcc.ch/.
Intergovernmental Panel on Climate Change (IPCC), (2001) Third Assessment Report: Climate Change 2001, available at http://www.ipcc.ch/.
Johansen, B.E. (2002) *The Global Warming Desk Reference*, Greenwood Press, Westport, Connecticut.
Kelly, R.C. (2002) *The Carbon Conundrum. Global Warming and Energy Policy in the Third Millennium*, CountryWatch, Houston.
Leggett, J. (2001) *The Carbon War. Global Warming and the End of the Oil Era*, Routledge, New York.
Crichton, M. (2004) *State of Fear*, HarperCollins Publisher, New York.
International Energy Agency, *Beyond Kyoto. Energy Dynamics and Climate Stabilisation*, International Energy Agency (IEA), Paris, 2002, available at http://www.iea.org/.
Lomborg, B. (2007) *Cool It: The Skeptical Environmentalist's Guide to Global Warming*, Knopf Edition.
Rapp, D. (2008) *Assessing Climate Change – Temperatures, Solar Radiation, and Heat Balance*, Spinger, Praxis.

Intergovernmental Panel on Climate Change (IPCC): http://www.ipcc.ch/.
U.S. Environmental Protection Agency (EPA) global warming site: http://epa.gov/climatechange/.

CO_2 Capture and Storage

Socolow, R.H. (2005) Can we bury global warming? *Sci. Am.*, 49.
Herzog, H., Eliasson, B. and Kaarstad, O. (2000) Capturing greenhouse gases. *Sci. Am.*, 72.
International Energy Agency (2004) *Prospects for CO_2 Capture and Storage*, International Energy Agency (IEA), Paris.
Anderson, S. and Newell, R. (2004) Prospects for carbon capture and storage technologies. *Annu. Rev. Environ. Resources*, **29**, 109.
Johnson, J. (2004) Putting a lid on carbon dioxide. *Chem. & Eng. News*, **82**, 51.
U.S. Department of Energy (2007) Carbon Sequestration Technology Roadmap and Program Plan.
International Energy Agency (2008) *CO2 Capture and Storage – A Key Carbon Abatement Option*, International Energy Agency (IEA), Paris.
Steeneveldt, R., Berger, B. and Torp, T.A. (2006) CO_2 capture and storage. Closing the knowing–doing gap. *Chem. Eng. Res. Des.*, **84** (A9), 739.
IEA Greenhouse Gas Research & Development Program: http://www.ieagreen.org.uk/.
Department of Energy (U.S.), Office of Fossil Energy, Carbon Sequestration R&D: http://www.fe.doe.gov/programs/sequestration/.
Princeton University Carbon Mitigation initiative: http://www.princeton.edu/~cmi/.
CO_2 Capture Project: http://www.co2captureproject.org/.

Renewable Energies

General Information about Renewable Energies

Cassedy, E.S. (2000) *Prospects for Sustainable Energy. A Critical Assessment*, Cambridge University Press, Cambridge.

Sørensen, B. (2000) *Renewable Energy. Its Physics, Engineering, Use, Environmental Impacts, Economy and Planning Aspects*, 2nd edn, Academic Press, London.

International Energy Agency (IEA) (2008) *Renewables Information 2008*, International Energy Agency (IEA), Paris.

International Energy Agency (IEA) (2003) *Renewables for Power Generation: Status and Prospects*, International Energy Agency (IEA), Paris.

International Energy Agency (IEA) (2004) Renewable Energy Outlook, *World Energy Outlook 2004*, International Energy Agency (IEA), Paris, p. 225.

International Energy Agency (IEA) (2001) Global Renewable Energy Supply Outlook, *World Energy Outlook 2001*, International Energy Agency (IEA), Paris.

Electric Power Research Institute (EPRI) and U.S. Department of Energy (DOE) (1997) Renewable Energy Technology Characterization.

Berinstein, P. (2001) *Alternative Energy. Facts, Statistics, and Issues*, Oryx Press, Westport, Connecticut.

Patel, M.R. (1999) *Wind and Solar Power Systems*, CRC Press, Boca Raton, Florida.

European Union website for renewable energies: http://ec.europa.eu/energy/renewables/ and http://ec.europa.eu/energy/atlas/html/renewables.html.

U.S. Department of Energy (DOE), Energy Efficiency and Renewable Energy (EERE) website: http://www.eere.energy.gov/.

Renewable Energy Journal, available for free from http://www.energies-renouvelables.org/.

Hydropower

International Energy Agency (IEA), Hydropower technologies: http://www.ieahydro.org/.

World Commission on Dams: http://www.dams.org/.

International Commission on Large Dams: http://www.icold-cigb.net/.

Geothermal

International Energy Agency (IEA), Geothermal energy: http://www.iea-gia.org/.

International Geothermal Association: http://iga.igg.cnr.it/index.php.

World Bank, Geothermal energy information: http://www.worldbank.org/html/fpd/energy/geothermal/.

Geo-Heat Center, Oregon Institute of Technology: http://geoheat.oit.edu/.

Wind

Archer, C.L. and Jacobson, M.Z. (2005) Evaluation of global wind power. *J. Geophys. Res.*, **110**, D12110.

International Energy Agency (IEA), Wind Energy Systems: http://www.ieawind.org/.

U.S. Department of Energy, Energy efficiency and Renewable Energy (EERE), wind technology program: http://www.eere.energy.gov/windandhydro/.

Global Wind Energy Council: http://www.gwec.net/.

European Wind Energy Association: http://www.ewea.org/.

American Wind Energy Association: http://www.awea.org/.

Solar Energy

Photovoltaic Technology Research Advisory Council (PV-TRAC) (2005), A Vision for Photovoltaic Technology, EUR 21242, European Commission, available from: http://ec.europa.eu/research/energy/pdf/vision-report-final.pdf.

Komp, R.J. (1995) *Practical Photovoltaic. Electricity from Solar Cells*, Aatec Publications, Ann Arbor.

U.S. Department of Energy, Energy Efficiency and Renewable Energy (EERE), solar energy topics: http://www.eere.energy.gov/solar.

Office of Science, U.S. Department of Energy (2005) Basic Research Needs for Solar Energy Utilization. Report of the Basic Energy Sciences Workshop on Solar Energy Utilization, available at http://www.sc.doe.gov/bes/reports/files/SEU_rpt.pdf.

Energy Efficiency and Renewable Energy (EERE) U.S. DOE (2004) Solar Energy Technologies Program: Multi-Year Technical Plan 2003–2007 and Beyond,

International Energy Agency (IEA), photovoltaic power program: http://www.iea-pvps.org/.

Solar thermal for electricity production: International Energy Agency (IEA), concentrated solar power for electricity generation: http://www.solarpaces.org/.

Solar for heat production: International Energy Agency (IEA), solar heating & cooling program: http://www.iea-shc.org/.

Biomass

Rosillo-Calle, F., Bajay, S.V. and Rothman, H. (eds) (2000) *Industrial Uses of Biomass Energy. The Example of Brazil*, Taylor & Francis, London.

Rothman, H., Greenshields, R. and Rosillo Callé, F. (1983) *Energy from Alcohol, the Brazilian Experience*, University Press of Kentucky, Lexington, Kentucky.

Schobert, H.H. (2002) Renewable Energy from Biomass, in *Energy and Society, an Introduction*, Taylor and Francis, New York.

U.S. Department of Agriculture (USDA) and U.S. Department of Energy (DoE) (2005) Biomass as a Feedstock for a Bioenergy and Bioproducts Industry: The Technical Feasibility of a Billion-Ton Annual Supply.

Rosillo Calle, F. (ed.) (2007) *The Biomass Assessment Handbook. Bioenergy for a Sustainable Environment*, Earthscan, London.

Soetaert, W. and Vandamme, E.J. (2009) *Biofuels*, John Wiley & Sons.

National Science Foundation (2008) Huber G.W. (Ed), Breaking the Chemical and Engineering Barriers to Lignocellulosis Biofuels: Next Generation Hydrocarbon Biorefineries, Report based on the June 25–26, 2007, Workshop, Washington, D.C., University of Massachusetts, Amherst.

International Energy Agency (IEA) bioenergy website: http://www.ieabioenergy.com/.

European Union website for biomass energy: http://ec.europa.eu/energy/renewables and also: http://ec.europa.eu/energy/atlas/html/renewables.html.

U.S. Department of Energy, National Renewable Energy Laboratory (NREL) Biomass Research: http://www.nrel.gov/biomass/ and Energy Efficiency and Renewable Energy (EERE), Biomass Program: http://www.eere.energy.gov/biomass/.

Ocean Energy

General information about ocean energy from: International Energy Agency (IEA), ocean energy systems: http://www.iea-oceans.org/.

Tidal and Current Power

European Union website for tidal energy: http://ec.europa.eu./energy/atlas/html/tidal.html.

Johnson, J. (2004) Power from moving water. *Chem. & Eng. News*, 23.

Wave Energy

The Electric Power Research Institute (EPRI) has studied the potential of offshore devices to produce electricity from waves. Reports available from on the web include (a) Previsic, M., Bedard, R. and Hagerman, G. (2004) E21 EPRI Assessment. Offshore Wave Energy Conversion Devices, available from http://my.epri.com. (b) Bedard, R., Hagerman, G., Previsic, M. et al. (2005) Offshore Wave Power Feasibility Demonstration Project. Project Definition Study. Final Summary Report, available from http://my.epri.com.

European Wave Energy Network: http://www.wave-energy.net/.

European Union website for wave energy: http://ec.europa.eu/energy/atlas/html/vawe.html.

Ocean Power Delivery Ltd, energy production with the Pelamis device: http://www.oceanpd.com/.

Ocean Thermal Energy

U.S. National Renewable Energy Laboratory (NREL), Ocean Thermal Energy Conversion (OTEC) web site: http://www.nrel.gov/otec/.
Sea Solar Power International, OTEC technology: http://www.seasolarpower.com/.

Nuclear Energy

Nuclear Energy Agency (NEA) (2003) *Nuclear Energy Today*, OECD Publication, Paris. Available from: http://www.nea.fr/html/pub/nuclearenergytoday/welcome.html.
Morris, R.C. (2000) *The Environmental Case for Nuclear Power. Economical. Medical and Political Considerations*, Paragon House, St. Paul, Minnesota.
Ramsey, C.B. and Modarres, M. (1998) *Commercial Nuclear Power. Assuring Safety for the Future*, John Wiley & Sons, Inc., New York.
University of Chicago. The Economic Future of Nuclear Power, A study conducted at the University of Chicago. Tolley, G.S., Jones, D.W. (eds.). Committee on Alternatives and Strategies for Future Hydrogen Production and Use (2004), Available from: http://www.ne.doe.gov/np2010/reports/NuclIndustryStudy-Summary.pdf. This study demonstrates that future nuclear power plants in the United States can be competitive with either natural gas or coal.
MIT (Cambridge) (2003) *The Future of Nuclear Power. An Interdisciplinary MIT Study* available from: http://web.mit.edu/nuclearpower/.
U.S. DOE Nuclear Energy Research Advisory Committee and the Generation IV International Forum (2002) A Technology Roadmap for Generation IV Nuclear Systems, available from: http://gif.inel.gov/roadmap/.

OECD Nuclear Energy Agency and the International Atomic Energy Agency (2004) Uranium 2003: Resources, Production and Demand.
World Nuclear Association: http://www.world-nuclear.org/.
OECD Nuclear Energy Agency (NEA): http://www.nea.fr/welcome.html.
Nuclear Energy Institute (NEI): http://www.nei.org/.
International Atomic Energy Agency: http://www.iaea.org/.
AREVA, http://www.areva.com/. worldwide leader in nuclear energy.
Commisariat à l'Energie Atomique (CEA): http://www.cea.fr/.
Generation IV Nuclear Energy Systems: http://gif.inel.gov/.

Nuclear Fusion

International Thermonuclear Experimental Reactor (ITER) to be constructed in Cadarache, France: http://www.iter.org/.
International Energy Agency (IEA), fusion section: http://www.iea.org/textbase/techno/technologies/index_fusion.asp.
Commissariat à l'énergie atomique (CEA) (2004) The Sun on the Earth *Clefs CEA No. 49*, available online from: www.cea.fr/var/cea/storage/static/gb/library/clefs49/contents.htm.

Hydrogen

Romm, J.J. (2004) *The Hype about Hydrogen. Fact and Fiction in the Race to Save the Climate*, Island Press, Washington, DC.
Hoffmann, P. (2002) *Tomorrow's Energy. Hydrogen, Fuel Cells, and the Prospects for a Cleaner Planet*, The MIT Press, Cambridge.
Rifkin, J. (2002) *The Hydrogen Economy*, Tarcher/Putnam, New York.
Gupta, R.M. (ed.) (2009) *Hydrogen Fuel – Production, Transport and Storage*, CRC Press.
Sperling, D. and Cannon, J. (2004) *The Hydrogen Energy Transition: Moving Toward*

the *Post Petroleum Age in Transportation*, Elsevier Academic Press.

The Hydrogen Economy: Opportunities, Costs, Barriers and R&D Needs, National Research Council and National Academy Engineering, The National Academic Press, Washington, DC, 2004.

UNEP The Hydrogen Economy – A Non-Technical Review, United Nation Environmental Programme, 2006.

Pohl, H.W. (ed.) (1995) *Hydrogen and Other Alternative Fuels for Air and Ground Transportation*, John Wiley & Sons, Ltd., Chichester, England.

Bossel, U., Eliasson, B. and Taylor, G. (2003) The Future of the Hydrogen Economy: Bright or Bleak? Available from http://www.efcf.com/reports/.

Toward a hydrogen economy, editorial and special issue: *Science*, 2004, **305**, 957.

Wald, M.L. (2004) Questions about a hydrogen economy. *Sci. Am.*, 66.

Oak Ridge National Laboratory , (2008) Analysis of the Transition to Hydrogen Fuel Cell Vehicles & the Potential Hydrogen Energy Infrastructure Requirements, ORNL/TM-2008/30.

U.S. Department of Energy Hydrogen Program: http://www.hydrogen.energy.gov/.

U.S. Federal government's central source of information on R&D activities related to hydrogen and fuel cells: http://www.hydrogen.gov/.

International Energy Agency (IEA) Hydrogen Program: http://www.ieahia.org/.

National Hydrogen Association (U.S.): http://www.hydrogenassociation.org/.

Hydrogen and fuel cell information system: www.netinform.net/h2/.

European Hydrogen Association: http://www.h2euro.org/.

Fuel Cell

EG&G Technical Services, Inc *Fuel Cell Handbook*, 7th edn, U.S. DOE, National Energy Technology Laboratory (NREL) , 2004.

Koppel, T. (1999) *Powering the Future. The Ballard Fuel Cell and the Race to Change the World*, John Wiley & Sons Canada Ltd, Toronto.

Online fuel cell information resources: http://www.fuelcells.org/.

Fuel Cell Europe: http://www.fuelcelleurope.org.

Fuel Cell Today: http://www.fuelcelltoday.com/.

U.S. Department of Energy, Energy Efficiency and Renewable Energy (EERE), fuel cells: http://www.eere.energy.gov/hydrogenandfuelcells/.

Methanol and the Methanol Economy

Cheng, W-.H. and Kung, H.H. (eds) (1994) *Methanol Production and Use*, Marcel Dekker, New York.

Asinger, F. (1987) *Methanol, Chemie- und Energierohstoff. Die Mobilisation der Kohle*, Springer-Verlag, Heidelberg.

Perry, J.H. and Perry, C.P. (1990) *Methanol, Bridge to a Renewable Energy Future*, University Press of America, Lanham, Maryland.

Bernton, H., Kovarik, W. and Sklar, S. (1982) *The Forbidden Fuel. Power Alcohol in the Twentieth Century*, Boyd Griffin, New York.

Dovring, F. (1988) *Farming for Fuel*, Praeger, New York.

Gray, C.L. Jr. and Alson, J.A. (1985) *Moving America to Methanol*, The University of Michigan Press, Ann Arbor.

Kohl, W.L. (1990) *Methanol as an Alternative Fuel Choice: An Assessment*, The Johns Hopkins University, Washington, DC.

Supp, E. (1990) *How to Produce Methanol from Coal*, Springer-Verlag, Berlin.

Pavone, A. (2003) Mega methanol plants. Report No. 43D, Process Economics Program, SRI Consulting, Menlo Park, California.

Lee, S. (1990) *Methanol Synthesis Technology*, CRC Press, Boca Raton, Florida.

Fiedler, E., Grossmann, G., Kersebohm, D.B. et al. (2003) Methanol, in *Ullmann's Encyclopedia of Industrial Chemistry*,

6th edn, vol. **21**, Wiley-VCH GmbH, Weinheim, Germany, p. 611.

Hansen, J.B. (1997) Methanol Synthesis, in *Handbook of Heterogeneous Catalysis*, vol. 4 (eds G. Ertl, H. Knözinger and J. Weitkamp), Wiley-VCH GmbH, Weinheim, Germany, p. 1856.

Weissermel, K. and Arpe, H.-J. (2003) *Industrial Organic Chemistry*, 4th edn, Wiley-VCH GmbH, Weinheim, Germany, p. 30.

Olah, G.A. and Molnár, Á. (2003) *Hydrocarbon Chemistry*, 2nd edn, John Wiley & Sons, Inc., Hoboken, New Jersey.

Olah, G.A. (1987) Electrophilic methane conversion. *Acc. Chem. Res.*, **20**, 422.

Olah, G.A. (1998) Oil and Hydrocarbons in the 21st Century, in *Chemical Research – 2000 and Beyond: Challenges and Vision* (ed. P. Barkan), American Chemical Society, Washington DC, and Oxford University Press, Oxford.

Olah, G.A., Goeppert, A. and Prakash, G.K.S. (2009) Chemical recycling of carbon dioxide to methanol and dimethyl ether: from greenhouse gas to renewable, environmentally carbon neutral fuels and synthetic hydrocarbons. *J. Org. Chem.*, **74**, 487.

Olah, G.A. (2003) The methanol economy. *Chem. & Eng. News*, September 22, **81**, 5.

Olah, G.A. (2005) Beyond oil and gas: the methanol economy. *Angew. Chem. Int. Ed.*, **44**, 2636.

Olah, G.A. and Prakash, G.K.S. (1999) Recycling of carbon dioxide into methyl alcohol and related oxygenates for hydrocarbons, U.S. Patent 5,928,806.

Reed, T.B. and Lerner, R.M. (1973) Methanol: A versatile fuel for immediate use. *Science*, **182**, 1299.

Stiles, A.B. (1977) Methanol, past, present, and speculation on the future. *AIChE J.*, **23**, 362.

Beyond the Internal Combustion Engine: The Promise of Methanol Fuel Cell Vehicles, the American Methanol Institute, available from: http://www.methanol.org/. Brochure published by the American Methanol Institute.

Methanol in our Lives, Methanex, available from: http://www.methanex.com. Brochure by the world's leading methanol producer, Methanex, illustrating the presence of methanol in many products and materials of our daily lives.

Malcolm Pirnie, Inc . for the Methanol Institute, (1999) Evaluation of the Fate and Transport of Methanol in the Environment. Available from http://www.methanol.org/.

American Methanol Institute: http://www.methanol.org/.

Methanex: http://www.methanex.com. World's leading methanol producer website.

Methanol to Hydrocarbons

Stocker, M. and Weitkamp, J. (eds) (1999) *Microporous Mesoporous Mater.*, **29**, (1–2). A special issue covering methanol to hydrocarbons technologies and processes.

Chang, C.D. (1997) Methanol to Hydrocarbons, in *Handbook of Heterogeneous Catalysis*, vol. 4 (eds G. Ertl, H. Knözinger and J. Weitkamp), Wiley-VCH GmbH, Weinheim, Germany, p. 1894.

Chang, C.D. (1994) Methanol to Gasoline and Olefins, in *Methanol Production and Use* (eds W.-H. Cheng and H.H. Kung), Marcel Dekker, New York, p. 133.

Olah, G.A., Goeppert, A. and Prakash, G.K.S. (2009) Chemical recycling of carbon dioxide to methanol and dimethyl ether: from greenhouse gas to renewable, environmentally carbon neutral fuels and synthetic hydrocarbons. *J. Org. Chem.*, **74**, 487.

Direct Methanol Fuel Cells (DMFC)

Apanel, G. and Johnson, E. (2004) *Direct methanol fuel cells: ready to go commercial? Fuel Cells Bull.*, 12.

McGrath, K.M., Prakash, G.K.S. and Olah, G.A. (2004) Direct methanol fuel cells. *J. Ind. Eng. Chem.*, **10**, 1063.

Surampudi, S., Narayanan, S.R., Vamos, E. *et al.* (1994) Advances in direct oxidation methanol fuel cells. *J. Power Sources*, **47**, 377.

Narayan, S.R. and Valdez, T.I. (2008) High-energy portable fuel cell power sources. *Interfaces, Winter*, 40.

Dimethyl Ether (DME)

Japan DME Forum (2007) *DME Handbook (English version)* (ed. Japan DME Forum).

International DME Association: http://www.aboutdme.org.

For current information on DME: http://www.greencarcongress.com/dme/.

Haldor Topsoe DME information: http://www.topsoe.com/.

Other Reading

Lomborg, B. (2001) *The Skeptical Environmentalist. Measuring the Real State of the World*, Cambridge University Press, Cambridge.

Professor Olah's Nobel lecture: Olah, G.A. (1995) Carbocations and their role in chemistry. *Angew. Chem. Int. Ed.*, **34**, 1393.

Transportation

International Energy Agency (IEA) (1999) *Automotive Fuels for the Future: The Search for Alternatives*, International Energy Agency, Paris.

International Energy Agency (IEA) (2004) *Energy Technologies for a Sustainable Future: Transport*, International Energy Agency, Paris.

Concawe, European Council for Automotive R&D and European Commission Joint Research Centre (2003) Well-to-Wheels Analysis of Future Automotive Fuels and Powertrains in the European Context, Tank-to-Wheels Report, Version 1.

International Transport Forum, OECD, Transport Research Centre . (2008). Oil Dependence: Is Transport Running Out of Affordable Fuel? Roundtable 139, International Transport Forum, OECD, Transport Research Centre.

Index

Beyond Oil and Gas: The Methanol Economy, Second updated and enlarged edition
George A. Olah, Alain Goeppert, and G. K. Surya Prakash
Copyright © 2009 WILEY-VCH Verlag GmbH & Co. KGaA, Weinheim
ISBN: 978-3-527-32422-4